Carbon Nanotube Electronics

Series on Integrated Circuits and Systems

Series Editor: Anantha Chandrakasan
Massachusetts Institute of Technology
Cambridge, Massachusetts

Carbon Nanotube Electronics
Ali Javey and Jing Kong (Eds.)
ISBN 978-0-387-36833-7

Wafer Level 3-D ICs Process Technology
Chuan Seng Tan, Ronald J. Gutmann, and L. Rafael Reif (Eds.)
ISBN 978-0-387-76532-7

Adaptive Techniques for Dynamic Processor Optimization: Theory and Practice
Alice Wang and Samuel Naffziger (Eds.)
ISBN 978-0-387-76471-9

mm-Wave Silicon Technology: 60 GHz and Beyond
Ali M. Niknejad and Hossein Hashemi (Eds.)
ISBN 978-0-387-76558-7

Ultra Wideband: Circuits, Transceivers, and Systems
Ranjit Gharpurey and Peter Kinget (Eds.)
ISBN 978-0-387-37238-9

Creating Assertion-Based IP
Harry D. Foster and Adam C. Krolnik
ISBN 978-0-387-36641-8

Design for Manufacturability and Statistical Design: A Constructive Approach
Michael Orshansky, Sani R. Nassif, and Duane Boning
ISBN 978-0-387-30928-6

Low Power Methodology Manual: For System-on-Chip Design
Michael Keating, David Flynn, Rob Aitken, Alan Gibbons, and Kaijian Shi
ISBN 978-0-387-71818-7

Modern Circuit Placement: Best Practices and Results
Gi-Joon Nam and Jason Cong
ISBN 978-0-387-36837-5

CMOS Biotechnology
Hakho Lee, Donhee Ham and Robert M. Westervelt
ISBN 978-0-387-36836-8

SAT-Based Scalable Formal Verification Solutions
Malay Ganai and Aarti Gupta
ISBN 978-0-387-69166-4, 2007

Continued after index

Ali Javey · Jing Kong
Editors

Carbon Nanotube Electronics

Springer

Editors
Ali Javey
University of California
Berkeley, CA
USA
ajavey@fas.harvard.edu

Jing Kong
Massachusetts Institute of Technology
Cambridge, MA
USA
jingkong@mit.edu

ISBN 978-0-387-36833-7 e-ISBN 978-0-387-69285-2
DOI 10.1007/978-0-387-69285-2

Library of Congress Control Number: 2008932042

Printed on acid-free paper

springer.com

Preface

Innovation of new materials with novel properties presents the driving force for technology development. Materials properties are not only governed by the atomic composition and the chemical bonding, but also by the dimensions of the material. Interesting properties arise when a material system approaches the molecular scales. At such small nm-scale dimensions, materials inherit some of the remarkable properties of molecules, resulting in unique physical and chemical characteristics. This miniaturization phenomenon was first envisioned by the Nobel Laureate Richard Feynman when he said, "there is plenty of room at the bottom." Since then, novel synthetic approaches have yielded a spectrum of materials with reduced dimensions. One particularly interesting example of miniaturized structures is the carbon nanotubes. Nanotubes are chemically derived synthetic nanomaterials with atomically smooth and well-defined surfaces, and 1-D structures. In recent years, significant progress has been made in their synthesis, purification and assembly; understanding the fundamental properties; developing novel electronic device designs; and utilization for a wide range of technological applications. The purpose of this book is to summarize some of the explosive research progress that has been made in the field of nanotube electronics in the recent years.

Tubular carbon nanostructures were first observed as early as 1952 by Radushkevich and Lukyanovich. However, it was not until nearly four decades later when Sumio Iijima reported the observation of carbon nanotubes in the journal of *Nature* that created a world-wide interest and excitement, and resulted in the development of the nanotube field that is in existence today. The first observed nanotubes were synthesized by the arc-discharge method, but since then a number of other approaches such as chemical vapor deposition and laser ablation have been demonstrated for enabling higher purity and yield of nanotubes. In this book, we present an overview of the nanotubes growth methods, mainly focusing on the direct synthesis of nanotubes on substrates for electronics integration.

Carbon nanotubes are perhaps closest analog to an ideal 1-D system with diameters as small as ~0.4 nm and lengths as long as a few cm. Their unique C-C bonding and 1-D structures results novel properties, including remarkable electron transport properties and band structures. An overview of such characteristics is given in this book as they have direct implications for integration of nanotubes for electronic applications.

The high electron and hole mobility ($\mu \sim 10,000$ cm^2/Vs) of semiconductor nanotubes; their compatibility with high-κ gate dielectrics for coaxially-gated devices; the enhanced electrostatics and reduced short channel effects due to their molecular-scale diameters; and ability to readily form metal ohmic contacts make these miniaturized structures an ideal material for high performance, nanoscale transistors. To date, significant progress has been made, both experimentally and theoretically, in detailed understanding of nanotube 1-D transistors while exploiting their ultimate performance limits. Some of the advancements in the field of nanotube transistors for both nanoelectronics and macroelectronics are presented in this book. The expected and experimentally observed DC and RF characteristics at both the device and circuit level are described, and various metrics for benchmarking their performance limits as compared to the state-of-the-art Si technology are summarized. Furthermore, the high conductance of metallic nanotubes and their immunity from electromigration makes nanotubes highly promising for nanoscale interconnects of future integrated circuits. In this book, the design considerations and performance metrics of nanotube interconnects are discussed while comparing them to the conventional copper wires of similar dimensions.

Another unique property of carbon nanotubes is their large surface-area-to-volume ratio with every atom being exposed to the surface. As a result, carbon nanotubes are highly sensitive to the environment, and unpassivated (i.e., the surface is exposed to the environment) nanotube devices are shown to enable highly sensitive detection of a wide range of analytes. The sensing mechanism, specificity, sensitivity, and device design for nanotube sensors are described in this book. Finally, the challenges facing the large-scale integration and manufacturing of nanotube devices are summarized while presenting an outlook for the field.

This book would have not been possible without the kind contributions of the authors for each chapter. We are indebted for all the efforts that they invested in bringing this book together by both writing their own chapters and reviewing other chapters. Furthermore, we would like to acknowledge our former Ph.D. advisor, Professor Hongjie Dai, whose encouragements and supports were the key in the development of this book.

Berkeley, California
Cambridge, Massachusetts

Ali Javey
Jing Kong

Contents

1 **Band Structure and Electron Transport Physics of One-Dimensional SWNTs** ... 1
Ji-Yong Park

2 **Direct Synthesis and Integration of SWNT Devices** ... 43
Mario Hofmann, Sreekar Bhaviripudi and Jing Kong

3 **Carbon Nanotube Field-Effect Transistors** ... 63
Ali Javey

4 **Measuring the AC Response of SWNT-FETs** ... 87
Islamshah Amlani

5 **Device Simulation of SWNT-FETs** ... 107
Jing Guo and Mark Lundstrom

6 **Carbon Nanotube Device Modeling and Circuit Simulation** ... 133
H.-S. Philip Wong, Albert Lin, Jie Deng, Arash Hazeghi, Tejas Krishnamohan and Gordon Wan

7 **Performance Modeling for Carbon Nanotube Interconnects** ... 163
Azad Naeemi and James D. Meindl

8 **Chemical Sensing with SWNT FETs** ... 191
Kyeong-Jae Lee and Jing Kong

9 **Single–Walled Carbon Nanotubes for High Performance Thin Film Electronics** ... 211
Qing Cao, Coskun Kocabas, Matthew A. Meitl, Seong Jun Kang, Jang Ung Park and John A. Rogers

10 Circuits, Applications and Outlook 247
Ali Keshavarzi and Arijit Raychowdhury

Index 263

Contributors

Islamshah Amlani Motorola, Corporations, Tempe, AZ, USA

Sreekar Bhaviripudi Department of Materials Science and Engineering, Massachusetts Institute of Technology, Cambridge, MA, USA

Qing Cao Department of Chemistry, University of Illinois at Urbana-Champaign, Urbana, IL, USA

Jie Deng Department of Electrical Engineering, Stanford University, Stanford, CA, USA

Jing Guo Department of Electrical and Computer Engineering, University of Florida, Gainesville, FL, USA

Arash Hazeghi Department of Electrical Engineering, Stanford University, Stanford, CA, USA

Mario Hofmann Department of Electrical Engineering and Computer Science, Massachusetts Institute of Technology, Cambridge, MA, USA

Ali Javey Department of Electrical Engineering and Computer Science, University of California, Berkeley, CA, USA

Seong Jun Kang Department of Materials Science and Engineering, University of Illinois at Urbana-Champaign, Urbana, IL, USA

Ali Keshavarzi Intel Corporation, Hillsboro, OR, USA

Coskun Kocabas Department of Physics, University of Illinois at Urbana-Champaign, Urbana, IL, USA

Jing Kong Department of Electrical Engineering and Computer Science, Massachusetts Institute of Technology, Cambridge, MA, USA

Tejas Krishnamohan Department of Electrical Engineering, Stanford University, Stanford, CA, USA

Kyeong-Jae Lee Department of Electrical Engineering and Computer Science, Massachusetts Institute of Technology, Cambridge, MA, USA

Albert Lin Department of Electrical Engineering, Stanford University, Stanford, CA, USA

Mark Lundstrom Department of Electrical and Computer Engineering, Purdue University, West Lafayette, IN, USA

James D. Meindl Microelectronics Research Center, Georgia Institute of Technology, Atlanta, GA, USA

Matthew A. Meitl Department of Materials Science and Engineering, University of Illinois at Urbana-Champaign, Urbana, IL, USA

Azad Naeemi Microelectronics Research Center, Georgia Institute of Technology, Atlanta, GA, USA

Jang Ung Park Department of Materials Science and Engineering, University of Illinois at Urbana-Champaign, Urbana, IL, USA

Ji-Yong Park Ajou University, School of Information Technology, Suwon, Korea

H.-S. Philip Wong Department of Electrical Engineering, Stanford University, Stanford, CA, USA

Arijit Raychowdhury Intel Corporation, Hillsboro, OR, USA

John A. Rogers Department of Computer Science, University of Illinois at Urbana-Champaign, Urbana, IL, USA

Gordon Wan Department of Electrical Engineering, Stanford University, Stanford, CA, USA

Chapter 1
Band Structure and Electron Transport Physics of One-Dimensional SWNTs

Ji-Yong Park

The electronic band structures of single-walled carbon nanotubes (SWNTs) along with their small size and low dimension are responsible for their unique electrical transport properties. In this chapter, we summarize the electronic band structures of one-dimensional (1D) SWNTs and the various electrical transport properties associated with them.

1.1 Introduction to the Band Structures of SWNTs

In this section, we examine how band structures of SWNTs can be obtained from that of graphene, a two-dimensional (2D) layer of graphite, along with their experimental verifications and manifestations.

1.1.1 Electronic Band Structure of Graphene

A SWNT can be considered as a graphene sheet, rolled up to form a hollow cylinder. As we will see, understanding the band structure of graphene is essential for understating those of SWNTs. The lattice structure of graphene in real space consists of hexagonal arrangement of carbon atoms as shown in Fig. 1.1(a). Isolated carbon atoms have four valence electrons in $2s$, $2p_x$, $2p_y$, and $2p_z$ atomic orbitals. As carbon atoms form graphene, three atomic orbitals, $2s$, $2p_x$, and $2p_y$, are hybridized into three sp^2 orbitals in the same plane while the $2p_z$ orbital remains perpendicular to other orbitals. The hybridized orbitals are responsible for σ bonds between the adjacent carbon atoms and the $2p_z$ orbital results in π bonds out of the plane of graphene. Generally, electrical transport properties are determined by the electrons (holes) near the Fermi level, since only these electrons (holes) have easy access to

J.-Y. Park (✉)
Department of Physics and Division of Energy Systems Research, Ajou University, Suwon, 443-749, Korea

A. Javey, J. Kong (eds.), *Carbon Nanotube Electronics*, Series on Integrated Circuits and Systems, DOI 10.1007/978-0-387-69285-2_1,

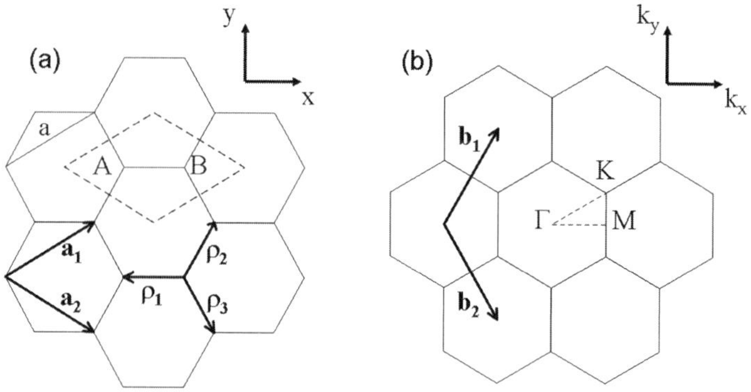

Fig. 1.1 (**a**) Real space representation of a graphene lattice. A unit cell is shown as a dashed rhombus with two carbon atoms (A and B). Unit vectors, $\mathbf{a}_1$ and $\mathbf{a}_2$, with length "a" are shown. Vectors $\boldsymbol{\rho}_j$ connect nearest neighbor carbon atoms. (**b**) Reciprocal space representation of a graphene lattice with two unit vectors $\mathbf{b}_1$ and $\mathbf{b}_2$. High-symmetry points (Γ, K, M) in a Brillouin zone are also shown

the unoccupied (occupied) states. Therefore, for the purpose of this chapter, we only consider the band structure near the Fermi level. In graphene, the π orbitals, which lie near the Fermi level, are responsible for the electrical transport properties by forming delocalized states.

The band structure of graphene derived from π orbitals can be calculated by the tight-binding approximations [1]. Graphene has a unit cell with two nonequivalent carbon atoms, A and B (all other atoms can be translated back into either of the two by a suitable combination of two unit vectors, $\mathbf{a}_1$ and $\mathbf{a}_2$), as shown in Fig. 1.1(a). The reciprocal lattice of graphene with unit vectors, $\mathbf{b}_1$ and $\mathbf{b}_2$, and high-symmetry points is shown in Fig. 1.1(b).

In order to find the band structure of the graphene π orbitals, we need to find the solutions of the Schrödinger equation

$$H\Psi = E\Psi, \tag{1.1}$$

where H is the Hamiltonian, Ψ is the total wave function, and E is the energy of electrons in the π orbitals of graphene. In a periodic system as in graphene, the total wave function can be constructed from a linear combination of Bloch functions u_i, which has a periodicity of the lattice. In the tight-binding approximation, u_i is represented by a linear combination of wave functions localized at each atom site, i.e., atomic wave functions. Since only the π orbitals that originate from the $2p_z$ orbital of each carbon atom are considered, Bloch function u_i for each atom can be constructed from $2p_z$ orbitals of atoms A and B as

$$u_{\mathrm{A(B)}} = \frac{1}{\sqrt{N}} \sum_{\mathrm{A(B)}} \mathrm{e}^{\mathrm{i}\mathbf{k}\cdot\mathbf{r}_{\mathrm{A(B)}}} X\left(\mathbf{r} - \mathbf{r}_{\mathrm{A(B)}}\right), \tag{1.2}$$

where $X(\mathbf{r})$ is the orbital $2p_z$ wave function for an isolated carbon atom. Then, Ψ in Eq. (1.1) can be written as follows:

$$\Psi = C_{\mathrm{A}} u_{\mathrm{A}} + C_{\mathrm{B}} u_{\mathrm{B}}. \tag{1.3}$$

By substituting Eq. (1.3) into (1.1), the Schrödinger equation can be solved in a matrix form as follows:

$$\begin{pmatrix} H_{\mathrm{AA}} & H_{\mathrm{AB}} \\ H_{\mathrm{BA}} & H_{\mathrm{BB}} \end{pmatrix} \begin{pmatrix} C_{\mathrm{A}} \\ C_{\mathrm{B}} \end{pmatrix} = E \begin{pmatrix} S_{\mathrm{AA}} & S_{\mathrm{AB}} \\ S_{\mathrm{BA}} & S_{\mathrm{BB}} \end{pmatrix} \begin{pmatrix} C_{\mathrm{A}} \\ C_{\mathrm{B}} \end{pmatrix}. \tag{1.4}$$

Here,

$$H_{ij} = \langle u_i \, |H| \, u_j \rangle, \;\; S_{ij} = \langle u_i \, | \, u_j \rangle. \tag{1.5}$$

For simplicity, the overlap between $2p_z$ wave functions of different atoms are neglected, i.e., $S_{\mathrm{AB}} = S_{\mathrm{BA}} = 0$. We can also see that $S_{\mathrm{AA}} = S_{\mathrm{BB}} = 1$ (normalized), then Eq. (1.4) is simplified to

$$\begin{pmatrix} H_{\mathrm{AA}} - E & H_{\mathrm{AB}} \\ H_{\mathrm{BA}} & H_{\mathrm{BB}} - E \end{pmatrix} \begin{pmatrix} C_{\mathrm{A}} \\ C_{\mathrm{B}} \end{pmatrix} = \begin{pmatrix} 0 \\ 0 \end{pmatrix}. \tag{1.6}$$

This matrix equation has a nontrivial solution only when

$$\begin{vmatrix} H_{\mathrm{AA}} - E & H_{\mathrm{AB}} \\ H_{\mathrm{BA}} & H_{\mathrm{BB}} - E \end{vmatrix} = 0. \tag{1.7}$$

Further, we can see that $H_{\mathrm{AA}} = H_{\mathrm{BB}}$ by symmetry of the graphene lattice (atoms A and B are not distinguishable) and $H_{\mathrm{AB}} = H_{\mathrm{BA}}{}^{*}$. Then, Eq. (1.7) leads to the solution

$$E = H_{\mathrm{AA}} \mp |H_{\mathrm{AB}}|. \tag{1.8}$$

H_{AA} ($=H_{\mathrm{BB}}$) can be calculated by inserting Eq. (1.2) into Eq. (1.5) as follows:

$$H_{\mathrm{AA}} = \frac{1}{N} \sum_{\mathrm{A}} \sum_{\mathrm{A}^*} \mathrm{e}^{\mathrm{i}\mathbf{k}\cdot(\mathbf{r}_{\mathrm{A}} - \mathbf{r}_{\mathrm{A}^*})} \int X^* (\mathbf{r} - \mathbf{r}_{\mathrm{A}}) \, H X (\mathbf{r} - \mathbf{r}_{\mathrm{A}^*}) \, \mathrm{d}\tau. \tag{1.9}$$

If we only consider the effects of the nearest neighbors, we need to evaluate Eq. (1.9) for each atom A (B) with three nearest neighbor B (A) atoms,

$$H_{\mathrm{AA}} = \int X^* (\mathbf{r} - \mathbf{r}_{\mathrm{A}}) \, H X (\mathbf{r} - \mathbf{r}_{\mathrm{A}}) \, \mathrm{d}\tau = E_0, \tag{1.10}$$

while

$$
\begin{aligned}
H_{\mathrm{AB}} =& \frac{1}{N}\sum_{\mathrm{A}}\sum_{\mathrm{B}} \mathrm{e}^{\mathrm{i}\mathbf{k}\cdot(\mathbf{r}_{\mathrm{A}}-\mathbf{r}_{\mathrm{B}})} \int X^{*}(\mathbf{r}-\mathbf{r}_{\mathrm{A}})\, H X(\mathbf{r}-\mathbf{r}_{\mathrm{B}})\,\mathrm{d}\tau \\
=& \frac{1}{N}\sum_{i} \mathrm{e}^{\mathrm{i}\mathbf{k}\cdot\rho_i} \int X^{*}(\mathbf{r})\, H X(\mathbf{r}-\rho_i)\,\mathrm{d}\tau,
\end{aligned}
\tag{1.11}
$$

where $\boldsymbol{\rho}_i$ is a vector connecting atom A to its three nearest neighbor B atoms (see Fig. 1.1(a)). By referring to the coordinate system of the graphene in Fig. 1.1(a),

$$
\begin{aligned}
H_{\mathrm{AB}} =& \left(\mathrm{e}^{\mathrm{i}\mathbf{k}\cdot\rho_1} + \mathrm{e}^{\mathrm{i}\mathbf{k}\cdot\rho_2} + \mathrm{e}^{\mathrm{i}\mathbf{k}\cdot\rho_3}\right) \int X^{*}(\mathbf{r})\, H X(\mathbf{r}-\rho_1)\,\mathrm{d}\tau \\
=& \gamma_0 \left(\mathrm{e}^{-\mathrm{i}k_x a/\sqrt{3}} + 2\mathrm{e}^{\mathrm{i}k_x a/2\sqrt{3}} \cos\left(\frac{k_y a}{2}\right)\right).
\end{aligned}
\tag{1.12}
$$

γ_0 is often called as the tight-binding integral or transfer integral which measures the strength of exchange interaction between nearest neighbor atoms. Then, from Eqs. (1.10) and (1.12), the energy dispersion in Eq. (1.8) can be calculated as follows:

$$
E = E_0 \mp \gamma_0 \left(1 + 4\cos\left(\frac{\sqrt{3}k_x a}{2}\right)\cos\left(\frac{k_y a}{2}\right) + 4\cos^2\left(\frac{k_y a}{2}\right)\right)^{1/2}. \tag{1.13}
$$

In Eq. (1.13), negative sign denotes valence bands of graphene formed by bonding π orbitals, while positive sign represents conduction bands formed by antibonding π^* orbitals. The dispersion relation in Eq. (1.13) is plotted in Fig. 1.2 along high-symmetry points in the reciprocal space with $E_0 = 0$. The surface and contour plots of the energy dispersion are also shown in Fig. 1.3(a) and (b), respectively. The main feature of the energy dispersion of graphene is the six K points at the corners of the Brillouin zone, where the conduction and valence bands meet so that the bandgap is zero only at these points. Also note that the two K points (K_1 and K_2) are nonequivalent due to symmetry (they originated from two nonequivalent atoms in the real space unit cell). The circular contour around each K point in Fig. 1.3(b) indicates the conic shape of dispersion near each K point.

The density of states (DOS) in graphene can be derived from the energy dispersion relation and it is found to be zero at the Fermi level [2]. Along with the zero bandgap, this is why graphene is a zero bandgap semiconductor. The slope of the conic shape dispersion near K points is proportional to the Fermi velocity of electrons in graphene, $v_{\mathrm{F}} = 8\times10^5$ m/s [3], as follows:

$$
\left.\frac{\mathrm{d}E}{\mathrm{d}k}\right|_{\mathrm{K}} = \frac{\sqrt{3}}{2} a\gamma_0 = \hbar v_{\mathrm{F}}. \tag{1.14}
$$

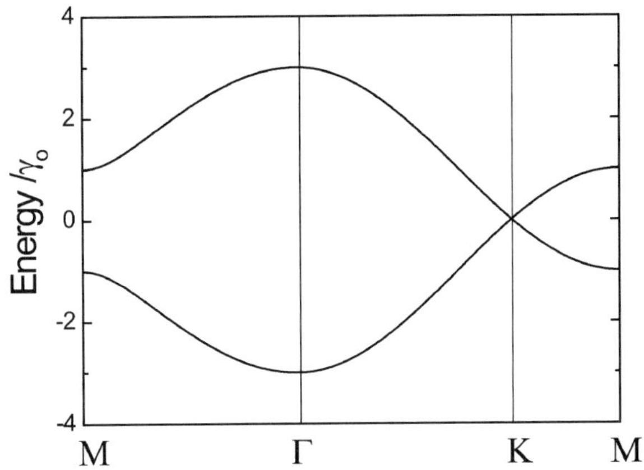

Fig. 1.2 Energy dispersion of graphene along high-symmetry points as indicated in Fig. 1.1(b)

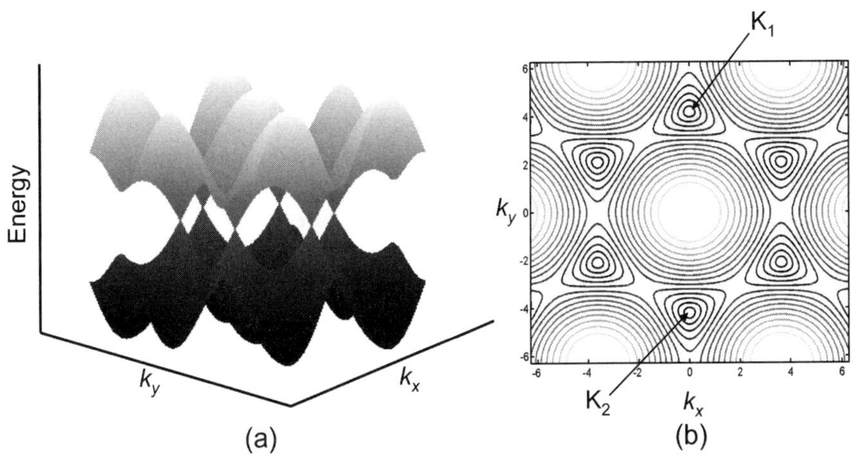

Fig. 1.3 (**a**) Surface plot and (**b**) contour plot of the energy dispersion in graphene as given by Eq. (1.13). Note that there are six K points where the band gap becomes zero. Of the six K points, only two are nonequivalent, denoted by K_1 and K_2

Before moving on to the SWNT case, it is appropriate to check the validities of some approximations and simplifications made for the tight-binding approximation presented here. In deriving the energy dispersion in Eq. (1.13), two main assumptions were made. First, the overlap integrals S_{AB} and S_{BA} between carbon atoms A and B are neglected in Eq. (1.4). Second, only the nearest neighbor interactions are considered in evaluating Eq. (1.11). In the literature [4], tight-binding calculations of graphene with more relaxed parameters were performed and compared to the ab initio calculations. As shown in Fig. 1.4, the energy dispersion obtained in Eq. (1.13) tends to deviate further from the first principle result far away from K points. It is found that the inclusion of a small overlap integral ($S_{AB} < 0.1$) and the third nearest neighbor interactions are needed to obtain a better fit [4]. However, since we are

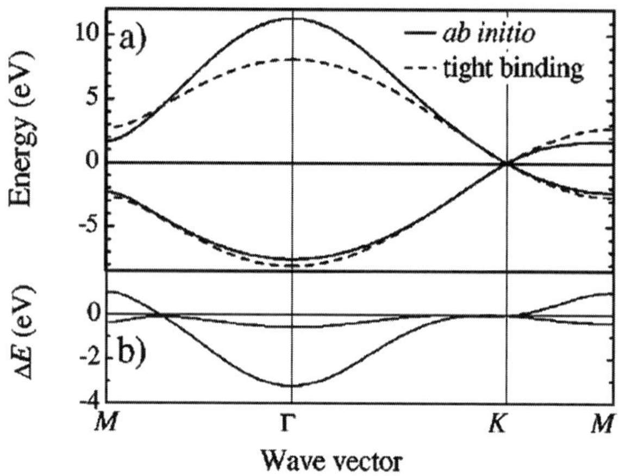

Fig. 1.4 (**a**) Energy dispersion of graphene calculated by first principle (*solid lines*) and nearest neighbor tight-binding method as in this chapter (*dashed lines*). (**b**) Difference between the energy dispersions calculated in (a). Reprinted figure with permission from [4]. Copyright 2002 by the American Physical Society

only concerned with the energy dispersion near the K points, the results obtained here with some simplifications should still be a good approximation for the purpose of understanding electron transport properties.

1.1.2 Band Structure of SWNT from Graphene

A SWNT can be uniquely defined by a chiral vector, $\mathbf{C} = n_1\mathbf{a}_1 + n_2\mathbf{a}_2$, where n_1 and n_2 are integers and $\mathbf{a}_1$ and $\mathbf{a}_2$ are the unit vectors of the graphene lattice as shown in Fig. 1.5. A SWNT is formed by rolling a graphene sheet in such a way that two carbon atoms pointed by **C** coincide. With wrapping indices, n_1 and n_2, SWNTs can be uniquely defined and described. SWNTs which are described by wrapping indices (n, n) and $(n, 0)$, are called armchair and zigzag SWNTs, respectively.

Since a SWNT is a rolled-up sheet of graphene, the band structure can be constructed from that of graphene by imposing an appropriate boundary condition around the SWNT. If we consider a SWNT as an infinitely long cylinder, there are two wave vectors associated with it. The wave vector $\mathbf{k}_{||}$, which is parallel to the SWNT axis, is continuous since the SWNT is assumed to be infinitely long, while the wave vector $\mathbf{k}_\perp$, which is along the circumference of a SWNT, should satisfy a periodic boundary condition (i.e., the wave function repeats itself as it rotates 2π around a SWNT),

$$\mathbf{k}_\perp \cdot \mathbf{C} = \pi d k_\perp = 2\pi m, \tag{1.15}$$

where d is the diameter of a SWNT and m is an integer. This boundary condition leads to quantized values of allowed $k_\perp$ for SWNTs. Then, the 1D band structure of SWNTs can be obtained from cross-sectional cutting of the energy dispersion of 2D graphene with these allowed $k_\perp$ states as shown in Fig. 1.6(a). This is called zone-folding scheme of obtaining the band structure of SWNTs. Each cross-sectional cutting gives rise to a 1D subband. Therefore, the 1D band structures of SWNTs are

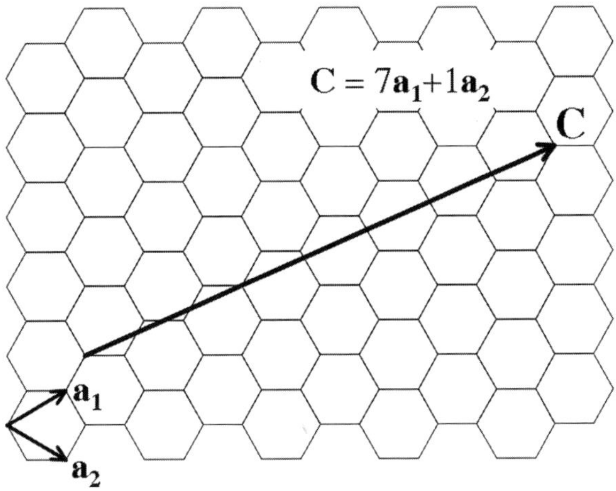

Fig. 1.5 Representation of a SWNT by a chiral vector, **C** $= n_1\mathbf{a}_1 + n_2\mathbf{a}_2$. In this figure, the wrapping index of the SWNT described by **C** is (7, 1). $\mathbf{a}_1$ and $\mathbf{a}_2$ are unit vectors of the underlying graphene lattice

determined by the spacing between allowed $k_\perp$ states and their angles with respect to the surface Brillouin zone of graphene, which is set by the diameter and chirality of SWNTs, i.e., wrapping indices. Especially, the band structure near the Fermi level, most relevant for transport properties, is given by allowed $k_\perp$ states that are closest to the K points. When the allowed $k_\perp$ states pass directly through the K points as in Fig. 1.6(b), the energy dispersion shows two linear bands crossing at the Fermi level without a bandgap. However, if the allowed $k_\perp$ states miss the K points as in Fig. 1.6(c), then there are two parabolic 1D bands with an energy bandgap. Therefore, we can expect two different kinds of SWNTs depending on the wrapping indices, metallic SWNTs without a bandgap as in Fig. 1.6(b) and semiconducting SWNTs with a bandgap as in Fig. 1.6(c). In this section, we will investigate the 1D subbands closest to the K points for zigzag SWNTs. Readers are referred to literatures for more general treatments of this subject [5–7].

Zigzag SWNTs, represented by wrapping index (n, 0), can be either metallic or semiconducting as will be shown below. Since the circumference is na ($\mathbf{C} = n\mathbf{a}_1$), the boundary condition in Eq. (1.15) becomes

$$k_x na = 2\pi m. \tag{1.16}$$

When n is a multiple of 3 ($n = 3q$, where q is an integer), there is an allowed k_x that coincides with a K point, which is at $(0, 4\pi/3a)$. By substitution,

$$k_x = \frac{2\pi m}{na} = \frac{3Km}{2n} = \frac{Km}{2q}. \tag{1.17}$$

Then, there is always an integer m $(= 2q)$ that makes k_x pass through K points so that these kinds of SWNTs (with $n = 3q$) are always metallic without a bandgap as shown in Fig. 1.6(b).

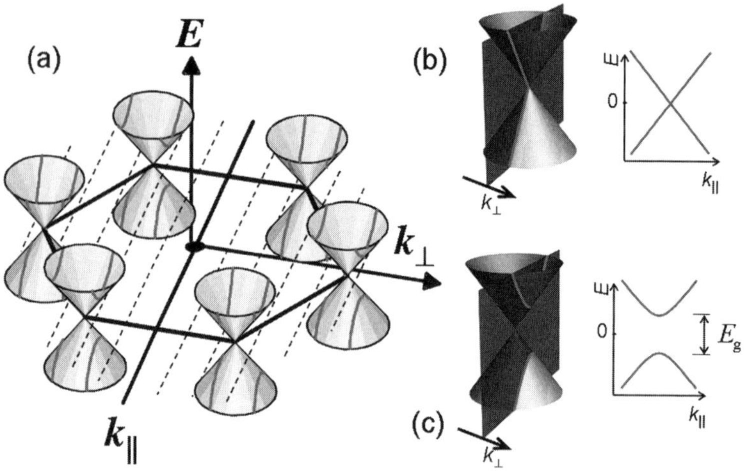

Fig. 1.6 (**a**) A first Brillouin zone of graphene with conic energy dispersions at six K points. The allowed $k_\perp$ states in a SWNT are presented by dashed lines. The band structure of a SWNT is obtained by cross-sections as indicated. Zoom-ups of the energy dispersion near one of the K points are schematically shown along with the cross-sections by allowed $k_\perp$ states and resulting 1D energy dispersions for (**b**) a metallic SWNT and (**c**) a semiconducting SWNT. Adapted from [8]

There are two cases when n is not a multiple of 3. If $n = 3q+1$, we can find k_x closest to K point at $m = 2q+1$ (see Fig. 1.6(c)).

$$k_x = \frac{2\pi m}{na} = \frac{3Km}{2n} = \frac{3K\,(2q-1)}{2\,(3q+1)} = K + \frac{K}{2}\frac{1}{3q+1}. \tag{1.18}$$

Similarly, for $n = 3q–1$, the allowed k_x closest to K is when $m = 2q–1$,

$$k_x = \frac{2\pi m}{na} = \frac{3Km}{2n} = \frac{3K\,(2q-1)}{2\,(3q-1)} = K - \frac{K}{2}\frac{1}{3q-1}. \tag{1.19}$$

In these two cases, allowed k_x misses K point by

$$\Delta k_x = \frac{K}{2}\frac{1}{3q \pm 1} = \frac{2}{3}\frac{\pi}{na} = \frac{2}{3}\frac{\pi}{\pi d} = \frac{2}{3d}. \tag{1.20}$$

Therefore, the smallest misalignment between an allowed k_x and a K point is inversely proportional to the diameter. Then, from the slope of a cone near K points (see Eq. (1.14)), the bandgap E_g is given by

$$E_g = 2 \times \left(\frac{\partial E}{\partial k}\right) \times \frac{2}{3d} = 2\hbar v_F \left(\frac{2}{3d}\right) \approx 0.7\,\text{eV}/d(\text{nm}). \tag{1.21}$$

Therefore, semiconducting SWNTs ($d = 0.8$–3 nm) have a bandgap in the order of 0.2–0.9 eV, inversely proportional to the diameter. Similar treatment for armchair SWNTs (n, n) leads to the conclusion that they are always metallic.

Extending the above results, SWNTs with wrapping indices (n_1, n_2) can be categorized into three different kinds based on p, which is the remainder when the difference between wrapping indices n_1 and n_2 is divided by 3 (i.e., $n_1 - n_2 = 3q + p$) [5, 6]:

(i) $p = 0$; metallic with linear subbands crossing at the K points.
(ii) $p = 1, 2$; semiconducting with a bandgap, $E_g \sim 0.7$ eV/d (nm).

In 1D metals, composed of chains of atoms or molecules, bond alternations (dimerizations) occur which open up a bandgap, a phenomenon known as Peierls distortion [9]. This is why 1D conducting polymers are not metallic at room temperature without doping. Although metallic SWNTs are 1D metals, they consist of rigid interwoven carbon bonds which cannot be easily distorted to open a bandgap. Therefore, Peierls distortion is not expected in SWNTs [5, 6, 10].

1.1.3 Deviation from Simple Zone-Folding Tight-Binding Picture

In the zone-folding derivation of the electronic band structure of SWNTs in Section 1.1.2, SWNTs were essentially treated as graphene sheets with a periodic boundary condition. In doing so, it is assumed that π orbitals are still orthogonal to σ orbitals in the graphene plane so that we can treat π orbitals independently as in the graphene case. This assumption, however, is not valid for SWNTs due to the curvature which mixes π and σ orbitals. This leads to hybridization between π and σ orbitals, and the degree of hybridization becomes larger as the diameter of a SWNT gets smaller. This π–σ hybridization effect has been considered and calculated in the literature [6, 11, 12]. The main result is that a small bandgap opens up in metallic SWNTs, except in the armchair nanotubes due to the symmetry. This is a secondary effect as the bandgap goes as $1/d^2$, ranging from a few meV to tens of meV for a diameter of ~3 nm or smaller as shown in Fig. 1.7.

1.1.4 Density of States in SWNTs

The DOS is an actual physical quantity measured in many measurements. The DOS, $n(E)$, with 1D subbands $\varepsilon(k)$, can be calculated as [13]

$$n(E) = \frac{\partial N(E)}{\partial E} = \frac{2}{l} \sum_i \int \mathrm{d}k \delta\left(k - k_i\right) \left|\frac{\partial \varepsilon}{\partial k}\right|^{-1}, \tag{1.22}$$

Fig. 1.7 The calculated curvature-induced bandgaps for metallic SWNTs. The bandgap is still zero for armchair SWNTs. Reprinted figure with permission from [11]. Copyright 1997 by the American Physical Society

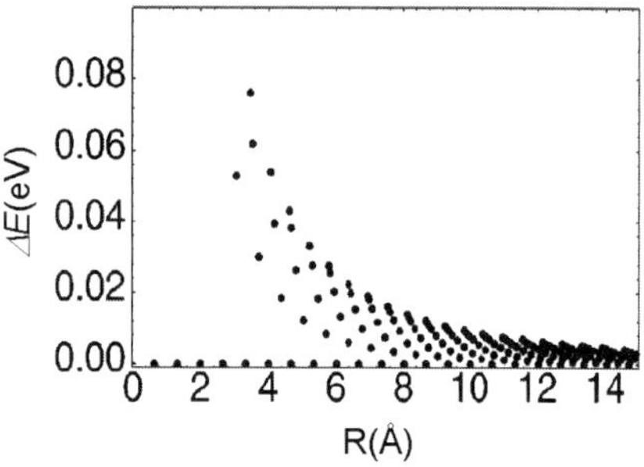

where k_i are the roots of the equation $E - \varepsilon(k_i) = 0$, l is the length of the 1D Brillouin zone, and $N(E)$ is the total number of electron states per unit cell below a given energy E. For parabolic 1D subbands such as those of SWNTs or free electrons, the resulting DOS in Eq. (1.22) is proportional to $1/E^{1/2}$ for each subband and diverges for each onset of a subband, giving rise to the so-called van Hove singularity in 1D systems. This happens in 1D, while DOS goes to zero as $E^{1/2}$ in 3D and over series of steps at each onset of a subband in 2D systems [14]. On the other hand, the metallic bands at K points can be approximated by linear dispersions as discussed in Section 1.1.2, so they lead to finite, constant DOS. Equation (1.22) can be solved for SWNTs as [13]

$$n(E) = \frac{2\sqrt{3}}{\pi^2}\frac{d}{\gamma_0 D}\sum_{m=-\infty}^{\infty} g\,(E, \varepsilon_m), \tag{1.23}$$

where

$$g\,(E, \varepsilon_m) = \begin{cases} |E|\,/\sqrt{E^2 - \varepsilon_m^2}, & |E| > |\varepsilon_m| \\ 0, & |E| < |\varepsilon_m| \end{cases}. \tag{1.24}$$

$g(E, \varepsilon_m)$ becomes divergent whenever $E = \varepsilon_m$, corresponding to van Hove singularity and $g(E, 0) = 1$. These lead to 1D density of states for SWNTs as shown in Fig. 1.8(a) and (b) for semiconducting and metallic SWNTs, respectively. Note that while DOS is zero at K points in graphene (zero bandgap), it is finite for metallic SWNTs due to the 1D characteristics of SWNTs. Zero bandgap and finite DOS at the Fermi level entitles metallic SWNTs as truly metallic, unlike graphene.

1.1.5 Experimental Verifications of the Band Structure of SWNTs

In this section, three experimental techniques verifying the electronic band structure of SWNTs will be discussed. They are (1) scanning tunneling microscopy/spectroscopy (STM/STS), (2) electrical transport measurements, and (3) optical measurements, such as photoluminescence (PL).

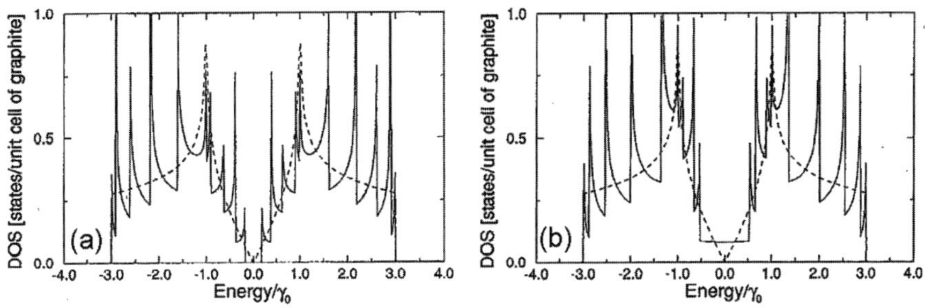

Fig. 1.8 The electronic density of states for a (**a**) (10, 0) semiconducting and (**b**) (9, 0) metallic SWNT. *Dotted lines* are for the density of states of graphene. Note that the density of states at the Fermi level is zero for graphene. Reprinted with permission from [2]. Copyright 1992, American Institute of Physics

The most direct verification of the electronic band structure of SWNTs was obtained by STM/STS. In STM, an atomically sharp and conducting tip is brought close to the sample surface, and atomic scale images can be obtained by monitoring the tip–sample distance while maintaining a constant tunneling current [15]. In STS, the tunneling current between the tip and the sample is measured as a function of voltage bias between them. The first derivative of the tunneling current with respect to the bias is proportional to the local DOS of the sample surface [16, 17]. In this way, STS can directly map the DOS of SWNTs, which was first demonstrated in 1998 [18, 19], as shown in Fig. 1.9. The STS spectra clearly show zero and finite DOS at the Fermi level for semiconducting and metallic SWNTs, respectively, which confirms the prediction in Section 1.1.4. Peaks associated with van Hove singularities at higher energies are also evident in the STS spectra as shown in Fig. 1.9. With STM, the atomic structure of SWNTs can be resolved so that the direct assignment of wrapping indices (n_1, n_2) is possible, which confirms the relation between wrapping indices and the band structure as described in Section 1.1.2.

Small bandgaps in metallic SWNTs due to the curvature-induced mixing of π–σ orbitals as discussed in Section 1.1.3 and even in armchair SWNT bundles (due to tube–tube interaction) were also confirmed by STS measurements [20] as shown in Fig. 1.10.

In order to perform electrical transport measurements of SWNTs, a three terminal device configuration is often used. Two electrodes directly contact an individual SWNT (source and drain electrodes), and there is often a third electrode, capacitively coupled to the nanotube through an insulating dielectric layer (gate electrode, as in transistors). Figure 1.11 shows a schematic diagram of a back-gated SWNT device. Typical transport measurements involve source/drain current (I_{DS}) measurements as a function of either source/drain (V_{DS}) or gate (V_{GS}) voltage. Unlike STS, this setup involves two junctions between the sample (SWNT) and the probes (source and drain), which makes it harder to associate the measurement results with simple physical quantities such as DOS.

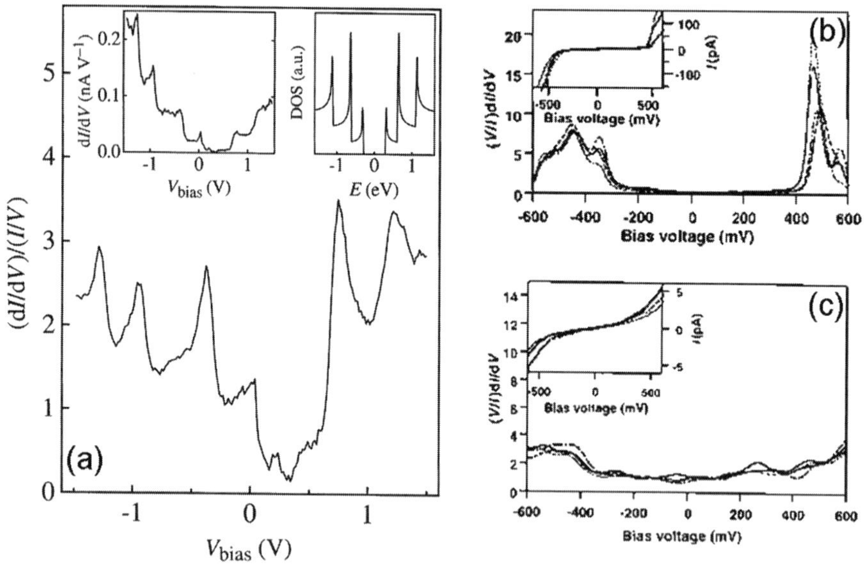

Fig. 1.9 (**a**)–(**c**) all are STS spectra. (**a**) and (**b**) are for semiconducting SWNTs. (**c**) is for a metallic SWNT. (dI/dV)/(I/V) is proportional to DOS in SWNTs. Note that DOS is zero for semiconducting SWNTs and finite for metallic SWNTs. Also, the van Hove singularities are evident in the STS spectra. Reprinted by permission from Macmillan Publishers Ltd: Nature [18, 19]. Copyright 1998

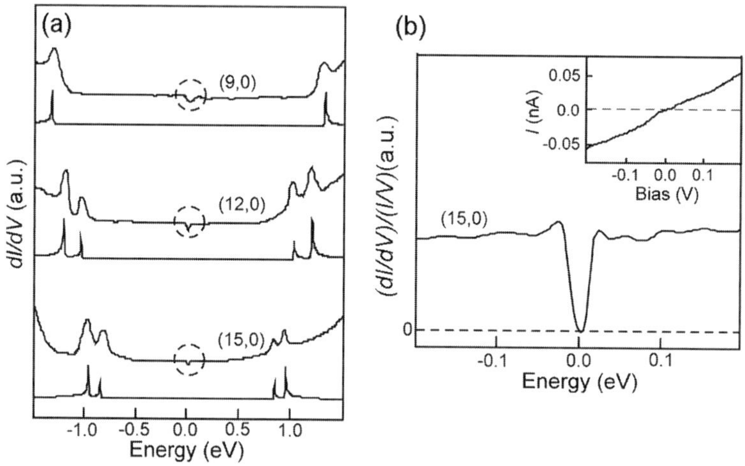

Fig. 1.10 (**a**) STS spectra of three different zigzag SWNTs. (**b**) Normalized conductance (proportional to DOS) for a (15, 0) zigzag SWNT. Note the small bandgap in this case. Reprinted with permission from [20]. Copyright 2001 AAAS

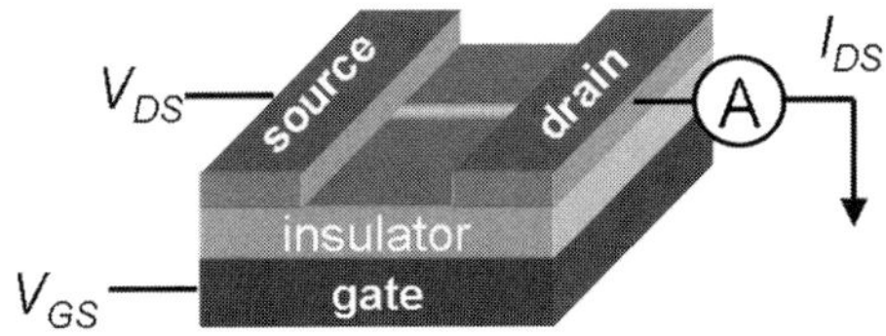

Fig. 1.11 A schematic of a typical setup for electrical transport measurements of SWNTs. Typically, I_{DS} as a function of V_{DS} or V_{GS} is measured for transport experiments

The first electrical transport measurements of individual SWNTs were reported in the late 1990s for both metallic and semiconducting SWNTs [21–24]. It was found that I_{DS}–V_{GS} characteristics of SWNTs depend strongly on whether they are semiconducting or metallic, as depicted in Fig. 1.12(a) and (b). The gate voltage (V_{GS}) shifts the Fermi level of the SWNTs while the quasi-Fermi levels of source and drain metal electrodes are fixed, since the DOS in a bulk metal electrode is much larger. For semiconducting SWNTs, this changes the relative positions of the SWNT bandgap to the Fermi levels of the source and drain, which modulates I_{DS} significantly as shown in Fig. 1.12(a). The current can be suppressed down to zero by V_{GS}. But for metallic SWNTs, there is no bandgap and the DOS is finite (constant) up to the next 1D subbands so that there is almost no change in I_{DS} as a function of V_{GS} as in Fig. 1.12(b). The large modulation of the I_{DS} by V_{GS} in semiconducting SWNTs suggests that they can work as field effect transistors (FETs). Since its first demonstrations [22, 24], SWNT-based FETs have extensively been studied for possible applications in electronic devices and are subjects of later chapters in this book. I_{DS}–V_{GS} characteristics as in Fig. 1.12(c) are also often observed and they have been attributed to the metallic SWNTs with small bandgaps (for instance, due to the curvature-induced σ–π mixing) [25]. Note that due to much smaller bandgaps, the current is modulated by gate voltage but it is not reduced to zero, unlike the semiconducting SWNTs. These three kinds of I_{DS}–V_{GS} characteristics in various SWNTs confirm the existence of different kinds of SWNTs as discussed in previous sections. In most cases, the electrical transport measurements are limited to the first 1D subbands near the Fermi level, since the typical gate voltage range is not

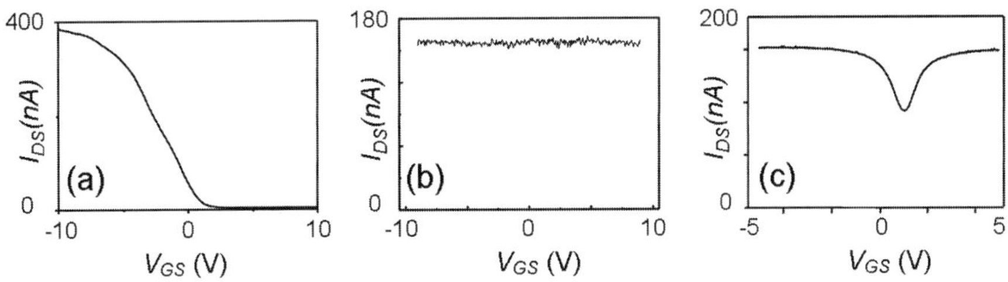

Fig. 1.12 Current (I_{DS}) as a function of the gate voltage (V_{GS}) for a (**a**) semiconducting, (**b**) metallic, and (**c**) semiconducting with a small bandgap SWNTs. V_{DS} = 10 mV for all data

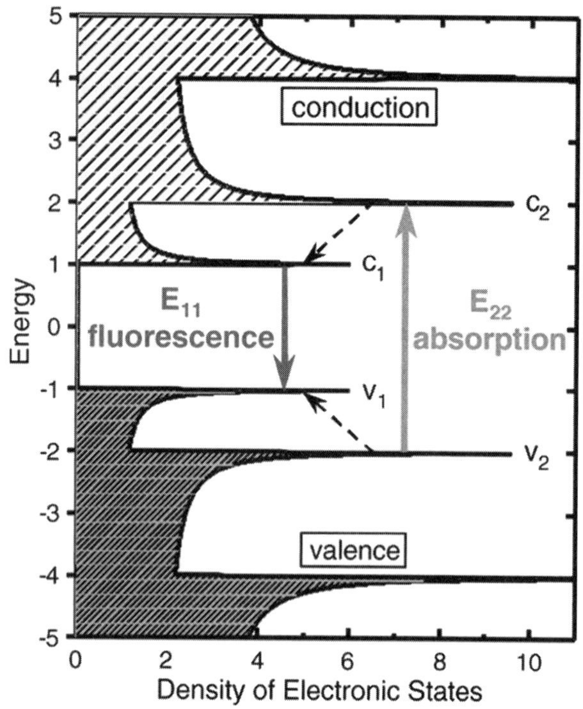

Fig. 1.13 A schematic DOS for a semiconducting SWNT. Optical excitations are marked by *solid arrows* and nonradiative relaxations are marked by *dashed arrows*. Reprinted with permission from [29]. Copyright 2002 AAAS

large enough to allow transport through higher 1D subbands.[1] Therefore, the existence of higher 1D subbands with singularities are not easily seen in the transport measurements as in STS measurements [26].

The optical characterizations by Raman spectroscopy [27], optical absorption [28], photoluminescence (PL) [29, 30], and Rayleigh scattering [31] have been very powerful tools in elucidating many interesting properties of SWNTs. In particular, the optical absorption and PL spectra are directly correlated to the electronic band structure of SWNTs (see Fig. 1.13). Due to the enhanced DOS at each onset of 1D subbands (van Hove singularities), strong optical absorption occurs when the energy of incident photons correspond to the energy differences of the subbands (see Fig. 1.13). This results in "bands" of strong absorption in UV–Vis–NIR absorption spectra [28]. With suspension of individual SWNTs (as oppose to bundles) in solution, better resolved absorption spectra along with stronger PL were obtained [29, 30]. The PL signal is due to the light emission resulting from recombination of electrons and holes at the band edge (see Fig. 1.13). Therefore, from PL spectra, the bandgap energies of semiconducting SWNTs can be obtained [30]. More detailed PL measurements demonstrated the dependence of the spectra on the diameter and

[1]Much higher barriers between metal electrodes and higher subbands in SWNTs also suppress transports through these 1D subbands in SWNTs.

chirality of SWNTs, enabling wrapping index assignments [29] while resolving 1D subbands.

Notably, excitons (electron–hole pairs) with large binding energies are formed by optical excitations in SWNTs, and they also participate in various optical processes so that considerations of exciton effects beyond the simple one electron band structures presented here are necessary to fully understand optical measurements [32–35].

1.2 Quantum Transport in SWNTs

As discussed in Section 1.1, SWNTs are either metallic or semiconducting with discrete 1D subbands. SWNTs exhibit rich quantum transport phenomena as a result of their electronic band structure as well as their small size and reduced dimension. In this section, quantum transport properties of SWNTs will be briefly reviewed and discussed.

1.2.1 Quantum Conductance in 1D Systems

Due to the 1D structure of SWNTs, only a small number of subbands participate in the electrical transport under a given bias voltage. The transport through a finite number of 1D subbands is well described by the Landauer formula [36–38] and will be discussed in this section. We will consider a 1D system with one parabolic subband and a ballistic transport. The conductance between two electron reservoirs through such a system is depicted in Fig. 1.14. Under thermal equilibrium (no bias applied to reservoirs), there will be equal number of electrons moving to the left and right; therefore, there is no net current (Fig. 1.14(a)). The reservoirs at both ends can thermalize the entering electrons to their own electrochemical potential (Fermi level). As a small bias $-V$ is applied to the right reservoir with respect to the left (Fig. 1.14(b)), the quasi-Fermi level of the right reservoir is moved up by eV, resulting in a net current flow.

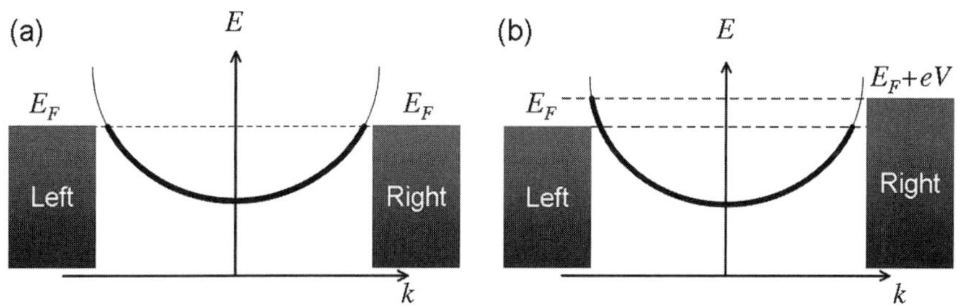

Fig. 1.14 Energy dispersion of a 1D subband with the Fermi level of the charge reservoirs (**a**) at equilibrium (no bias) and (**b**) when small bias is applied between the charge reservoirs

The net current is given by

$$I = \Delta nev = \frac{DeV}{2}ev = \frac{2}{hv}ve^2 V = \frac{2e^2}{h}V, \tag{1.25}$$

where Δn is the excess electron density, e is the electron charge, v is the charge velocity, D is the 1D DOS (of free electrons), and $D/2$ is used since only the left moving electrons contribute to the current. From Eq. (1.25), we can see that current depends only on the voltage and is independent of carrier velocity in 1D system (assuming ballistic transport). Then, the two-terminal conductance I/V and resistance V/I of an ideal 1D system with one subband and no scattering can be calculated from Eq. (1.25) as follows:

$$G_{\mathrm{Q}} = \frac{2e^2}{h}, \quad R_{\mathrm{Q}} = \frac{h}{2e^2} = 12.9\,\mathrm{k}\Omega. \tag{1.26}$$

We can see that a 1D channel (transport mode) with perfect transmission (ballistic conduction) has a finite conductance and resistance, which are called the conductance quantum G_{Q} and resistance quantum R_{Q}, respectively. The total current I_{T}, carried by multiple 1D channels, is the current per channel multiplied by the number of channels N as follows:

$$I_{\mathrm{T}} = N\frac{2e^2}{h}V. \tag{1.27}$$

By incorporating transmission probability $T_i(E_{\mathrm{F}})$ of electrons in each channel, the effect of carrier scattering can be incorporated into the formula as follows:

$$I_{\mathrm{T}} = \frac{2e^2}{h}V\sum_i T_i. \tag{1.28}$$

For ballistic transmission, $\Sigma T = N$, recovering Eq. (1.27).

This equation is often called the Landauer Formula. In metallic SWNTs, there are two 1D subbands (arising from 2-fold band degeneracy) at the Fermi level, participating in the electrical transport at low bias. If the transport is ballistic, the expected resistance for a SWNT is $R_{\mathrm{Q}}/2 \sim 6.5$ kΩ.

1.2.2 Quantum Transport in SWNTs

SWNT is a low dimensional system in which various quantum mechanical effects have been observed. In this section, transport properties of SWNTs as quantum systems having dimensions of 0 and 1 will be discussed.

As the scale of a material is reduced to 0D, discrete energy levels arise due to the quantum confinement effect. Such structures are known as artificial atoms or

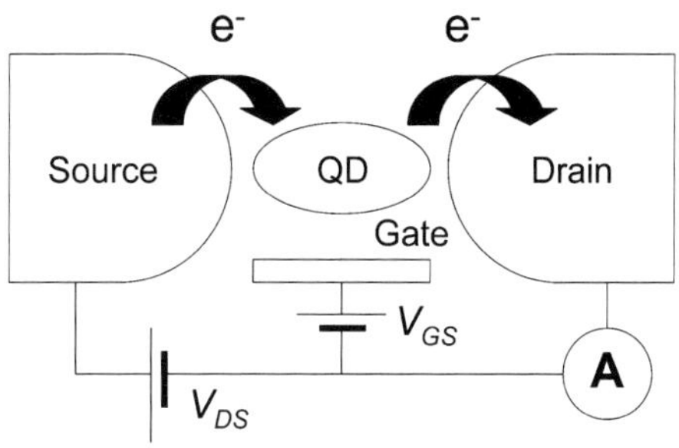

Fig. 1.15 A schematic of a QD coupled to electrodes, source, drain, and gate. Typical measurement setup is also shown

quantum dots (QDs). Many mesoscopic or nanoscale systems such as semiconductor nanocrystals, metal nanoparticles, lithographically defined small islands on semiconductor heterostructures, and SWNTs with tunnel barriers are examples of QDs [39, 40]. Before discussing transport properties of SWNT QDs, the general properties of QDs will briefly be introduced [41].

When a QD is weakly coupled to the source/drain electrodes (Fig. 1.15), it acts as an island for electrons to hop in and out.

QDs have two different energy scales. The first is the discrete electron energy level spacing due to the confinement of electrons inside a QD. The second is the electrostatic charging energy required when adding or removing electrons from the system. Changing the number of electrons in the QD by one requires an electrostatic energy $U_c = e^2/C$, where C is the total capacitance of the system. Therefore, in order for an electron to tunnel from one electrode to the QD, it must pay an energy cost of the charging energy plus energy level spacing as

$$E_{\mathrm{add}} = U_c + \Delta E, \tag{1.29}$$

where $\Delta E = E_{n+1} - E_n$ is the electron energy level spacing between single electron levels. The relative electrochemical potentials of the electrodes and the QD determine the electron transport properties. Transport characteristics related to single electrons are observable when the energy scale, U_c or ΔE, is larger than the thermal energy, $k_B T$. Otherwise, the discrete levels are smeared out due to the thermal fluctuations (note that the width of Fermi–Dirac distribution of electrons at the Fermi level is proportional to $k_B T$). Also, the coupling between the QD and the electrodes should be weak for tunneling to occur. This condition is typically given by the resistance of the coupling, R_t, as $R_t >> h/e^2$ [41].

When $U_c >> k_B T$ but $\Delta E << k_B T$, only the charging energy becomes important and the QD can be thought to have discrete charge states separated by E_{add} ($=U_c$) as shown in Fig. 1.16(a). With a small bias V_{DS} applied between the source and the drain, the transport through the QD is suppressed when there is no available charge states in between the Fermi levels of the electrodes (Fig. 1.16(a)). This is called the Coulomb blockade. Since the electrochemical potential of the QD can be shifted up and down by the gate voltage V_{GS}, transport will occur when the electrochemical

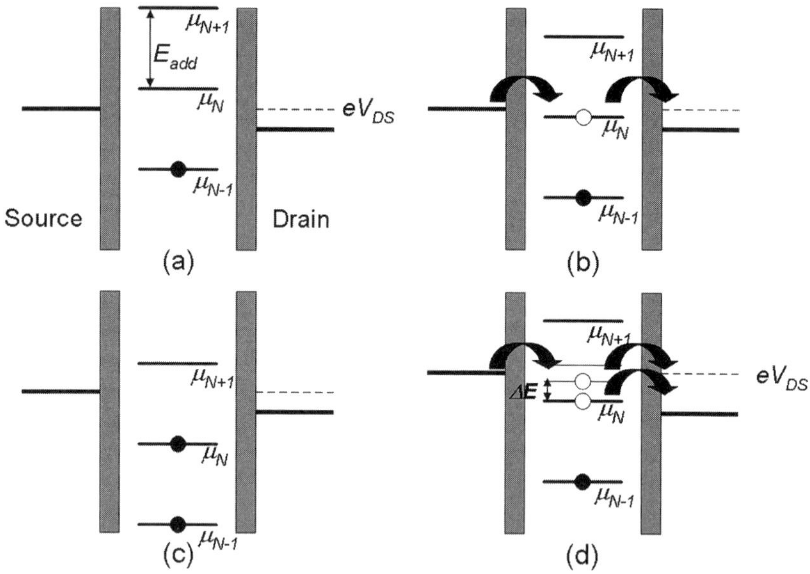

Fig. 1.16 A schematic energy level diagrams for a QD coupled to source and drain electrodes. Each charge state in a QD is denoted as μ_N separated by E_{add}. The charge states are out of alignment with the Fermi levels of electrodes in (**a**) and (**c**), while in alignment in (**b**), which results in single electron flows. (**d**) When the single electron energy levels are discrete, more single electron energy levels are available for electron transport as the bias is increased for a given charge state

potential of the N charge state falls between Fermi levels of the source and drain electrodes as in Fig. 1.16(b). Then, an electron can hop from the source electrode to the QD and from the QD to the drain electrode, resulting in a net flow of current. This process repeats as the electrochemical potential of the QD is shifted up again as in Fig. 1.16(c).

Transport occurs whenever new charge state falls in between the Fermi levels of the electrodes, giving rise to periodic peaks in the current as a function of V_{GS}, as shown in Fig. 1.17(a). This is called the Coulomb oscillations and is a manifestation of single electron tunneling phenomenon in QDs. The charge state can also be accessed by changing V_{DS} while fixing the gate voltage. The overall effect of V_{DS} and V_{GS} on the transport through a QD can be captured by plotting differential conductance as functions of both gate and bias voltage in a so-called Coulomb diamond plot as shown in Fig. 1.17(b). Inside each diamond-shaped region, the number of electrons is fixed due to the Coulomb blockade. The Coulomb oscillation plot, such as in Fig. 1.17(a), can be obtained by taking a cross-section along $V_{\mathrm{DS}} \sim 0$ V in the Coulomb diamond plot as in Fig. 1.17(b).

When $U_{\mathrm{c}} >> k_{\mathrm{B}}T$ and $\Delta E >> k_{\mathrm{B}}T$, beside the charging energy, single electron energy levels are also important in determining the transport properties (typically $U_{\mathrm{c}} >> \Delta E$). This effect is usually observed by sweeping V_{DS} as a function of V_{GS} since, for a given charge state, increasing the bias voltage leads to additional single

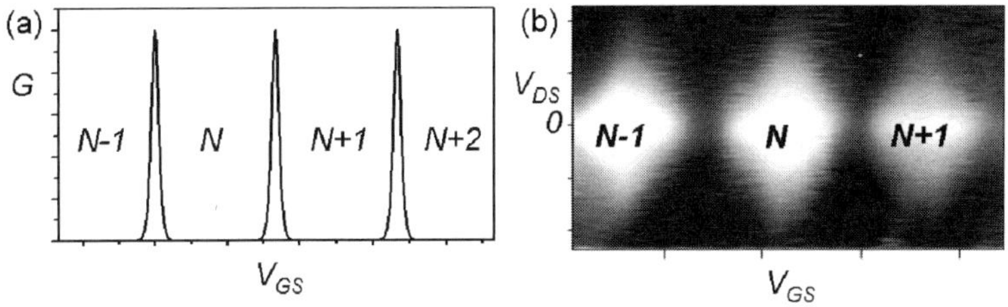

Fig. 1.17 (**a**) Coulomb oscillations as a function of V_{GS}. Each peak in the conductance is due to the single electron transport as schematically shown in Fig. 1.16(b). (**b**) An example of Coulomb diamond plot which includes the effect of V_{DS} in single electron transport through a QD. Differential conductance is zero inside each diamond shaped region

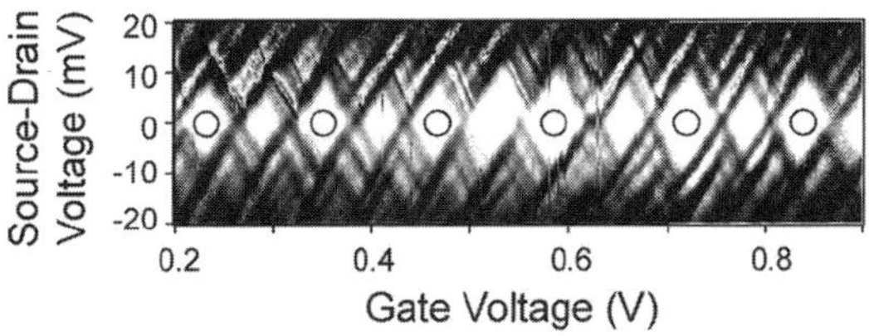

Fig. 1.18 A differential conductance map showing excited levels as extra lines parallel to diamonds. Reprinted with permission from [42]. Copyright 2006, American Institute of Physics

electron levels (excited states) for transport as shown in Fig. 1.16(d). The transport through the excited levels are shown as extra lines parallel to the edge of the Coulomb diamonds in differential conductance maps as shown in Fig. 1.18. This is called the excitation level spectroscopy.

When a short SWNT (typically $L < 1$ μm) is placed between source/drain electrodes with tunneling interfaces and capacitively coupled to a gate electrode, it can act as a QD. In Section 1.1.2, the $k_{||}$ along the axis of SWNTs is assumed to be continuous due to the infinite length of a SWNT. However, when the length of a SWNT becomes finite, $k_{||}$ is also quantized, leading to discrete energy levels as in a QD. By assuming electrons confined within a metallic SWNT of length L by infinite barriers (such as two electrodes with tunneling barriers at both ends), the energy level spacing ΔE can be estimated as

$$\Delta E = \frac{\partial E}{\partial k_{||}} \frac{\Delta k_{||}}{2} = \hbar v_{\mathrm{F}} \frac{\pi}{2L} = \frac{h v_{\mathrm{F}}}{4L} \approx \frac{0.8\,\mathrm{meV}}{L(\mu\mathrm{m})}, \tag{1.30}$$

where the factor 2 comes from the existence of two degenerate 1D subbands at the Fermi level of metallic SWNTs. The second energy scale is the charging energy U_{c}, which can be roughly estimated with a gate capacitance of a cylindrical object with radius r and length L, on a Si/SiO_2 (thickness h and $\varepsilon \sim 3.9$) substrate (assuming a back-gate geometry as shown in Fig. 1.11 for 500 nm-thick SiO_2 and a SWNT with $r = 1$ nm) by [43]

$$U_c = \frac{e^2}{C} \approx \frac{e^2}{2\pi\varepsilon\varepsilon_0 L/\ln(2h/r)} \approx \frac{5\,\mathrm{meV}}{L(\mu\mathrm{m})}, \tag{1.31}$$

assuming the capacitance between the SWNT and the source/drain electrodes is small compared to the gate capacitance. From Eqs. (1.30) and (1.31), we can see that the charging energy is about six times larger than the energy level spacing in SWNT QDs, independent of the length of the SWNT. Compared to QDs defined on 2D electron gases of semiconductor heterostructures, the energy level spacing is large and independent of the number of electrons for metallic SWNTs [41]. Due to the small size of a SWNT, the spin-orbit coupling in a SWNT QD is small for a magnetic field perpendicular to the SWNT axis, which makes the spin-related effect more pronounced. These factors along with the simple band structure at the Fermi level make metallic SWNTs a model system to study QD phenomena.

Indeed, SWNT QDs have been extensively studied. Early studies [21, 23] showed the charging and excited energy levels in SWNT QDs similar to Fig. 1.18. Since then, various interesting quantum effects in SWNT QDs, such as the Zeeman splitting [44, 45], Kondo effect [46], shell filling [47–49], double QDs in a SWNT [50], electron–hole symmetry in a semiconducting SWNT QD [51], and the coupling of suspended SWNT QDs with phonon modes [52], have been observed and studied. This is not an exhaustive list, and still many interesting and new QD physics of SWNTs are being investigated.

In our discussions so far, the SWNT is assumed to be weakly coupled to the contacts and act as an isolated island (QD) for electrons. When the contacts become nearly transparent (i.e., nearly ohmic but still with small scattering) and the single electron charging becomes negligible, the SWNT recover its identity as a 1D quantum wire of electron waveguide. In this case, the Coulomb interactions become negligible, and the quantum interference of the two propagating wave modes caused by electron scattering at the SWNT-metal contacts become dominant. This interference of quantum mechanical electron waves shows up in differential conductance maps as crisscross patterns as in Fig. 1.19, which demonstrates the quantum mechanical wave nature of electrons and ballistic transport in SWNTs [53, 54].

Electron transport in 1D systems, such as SWNTs, are expected to be affected by strong electron–electron (e–e) interactions. However, the Fermi liquid theory which is suitable for e–e interactions in higher dimensions is no longer applicable. The 1D transport is typically described and studied within the framework of Tomonaga–Luttinger liquid theory [55, 56]. SWNTs are expected to be a good model system for studying Luttinger liquid phenomenon and have been a subject of many theoretical studies [57, 58]. Theoretical predictions of power-law dependence of conductance on bias or temperature and zero bias anomaly have been reported for SWNTs [59–61]. In order to verify the Tomonaga–Luttinger liquid theory by transport measurements, the SWNTs need to be connected by metal contact electrodes with tunneling barriers as in the SWNT QDs [59–61]. However, depending on the temperature and length, the transport can occur in the Tomonaga–Luttinger liquid regime, the Coulomb-blockade regime, or the crossover between the two regimes, which makes the interpretation of experimental results difficult and gives

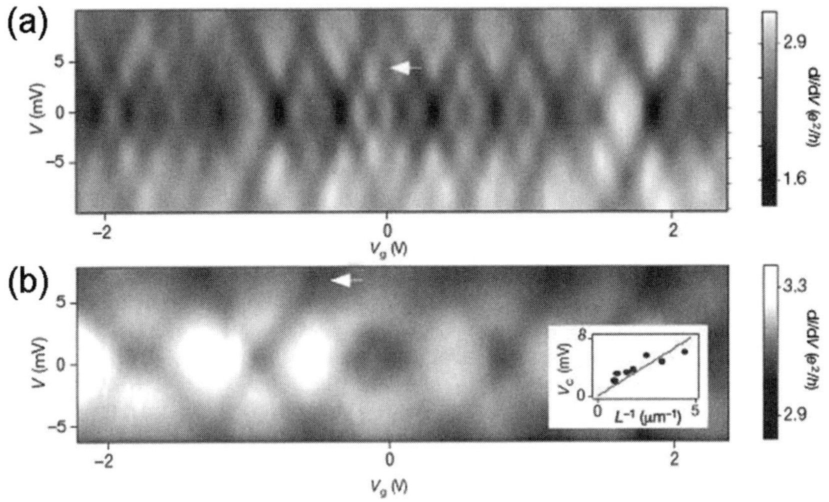

Fig. 1.19 Differential conductance maps in the quantum interference regime for a (**a**) 530 nm- and (**b**) 220 nm-long SWNT. The differential conductance never reduces to zero unlike in Fig. 1.18. Reprinted by permission from Macmillan Publishers Ltd: Nature [54]. Copyright 2001

uncertainty to experimentally measured Luttinger parameters [62]. More detailed experimental and theoretical works are needed to shed light on the precise nature of e–e interactions in SWNTs.

1.3 Modifications to the Band Structure

In Section 1.1.2, the 1D electronic band structure of SWNTs was obtained from "cutting" the graphene band structure with discrete $k_\perp$ states imposed by the periodic boundary condition. External perturbations, such as magnetic field, electric field, or mechanical deformation which affect the periodic boundary condition, will induce relative displacements between the allowed $k_\perp$ states and K points, leading to the modification of the band structure of SWNTs. In this section, we will examine how the electronic band structure and the transport properties of a SWNT are modified by these external perturbations.

1.3.1 External Fields

Under the influence of an applied magnetic field, electrons acquire additional phase factors (i.e., Aharonov–Bohm effect) proportional to the magnetic flux enclosed by the electron paths [63, 64]. Then, the boundary condition (Eq. (1.15)) along the circumference of a SWNT can be generalized to include this phase shift due to a magnetic field as

$$\mathbf{k}_\perp \cdot \mathbf{C} + 2\pi \frac{\phi}{\phi_0} = 2\pi m, \tag{1.32}$$

where ϕ is the magnetic flux enclosed by a SWNT and $\phi_o = h/e$ is the magnetic flux quantum. By comparing this generalized boundary condition with the one without magnetic field in Eq. (1.15), we can see that the allowed $k_\perp$ states will now shift by $(2/d)(\phi/\phi_0)$. Since ϕ is maximum when the magnetic field is parallel to the SWNT, the shift is largest when the field is parallel to the SWNT axis. For metallic SWNTs, this shift results in the deviation of the allowed $k_\perp$ states from the K points by $(2/d)(\phi/\phi_0)$ (they were originally passing through the K points), inducing metal–semiconductor transition with a bandgap change ΔE_g, which can be estimated from Eq. (1.21) as

$$\Delta E_g \sim 2 \times \left(\frac{\partial E}{\partial k}\right) \times \Delta k = 2\hbar v_F \left(\frac{2}{d}\right)\left(\frac{\phi}{\phi_o}\right) = 2\hbar v_F \left(\frac{2}{d}\right)\left(\frac{e\pi d^2 B}{4h}\right) = \frac{e v_F d B}{2}, \tag{1.33}$$

where B is the applied magnetic field. The bandgap change expected from this equation is about 1 meV per 1 Tesla for a $d = 2.5$ nm SWNT. In semiconducting SWNTs, two 1D subbands are degenerate (with the same bandgap) near the two nonequivalent K points under no magnetic field as schematically shown in Fig. 1.20(a). As an external magnetic field shifts the allowed $k_\perp$ states by $(2/d)(\phi/\phi_O)$ in the same direction as shown in Fig. 1.20(b), the bandgap near K_1 decreases while the bandgap increases near K_2. The overall result is the lifting of the band degeneracy for semiconducting SWNTs as shown in Fig. 1.20(b). The effect of an external magnetic field on the band structure of SWNTs were theoretically examined [65, 66], and the expected bandgap modulation of different SWNT chiralities were calculated as a function of external magnetic field for different angles as shown in Fig. 1.21.

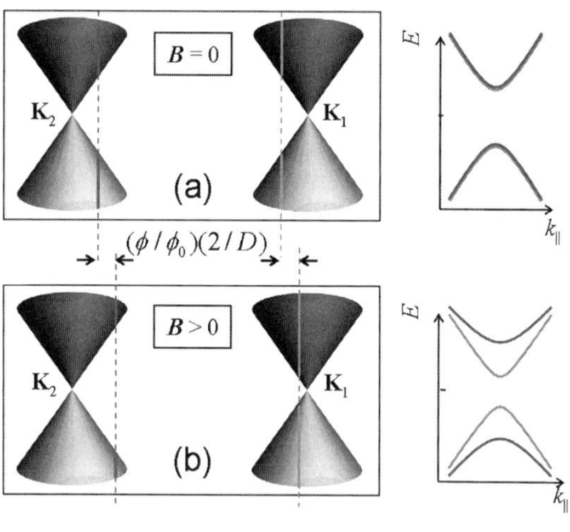

Fig. 1.20 Relative positions of the allowed $k_\perp$ states and K points and the resulting 1D subbands of a semiconducting SWNT (**a**) without and (**b**) with an external magnetic field. Note that the 1D subbands are not degenerate anymore under a B-field. Adapted from [8]

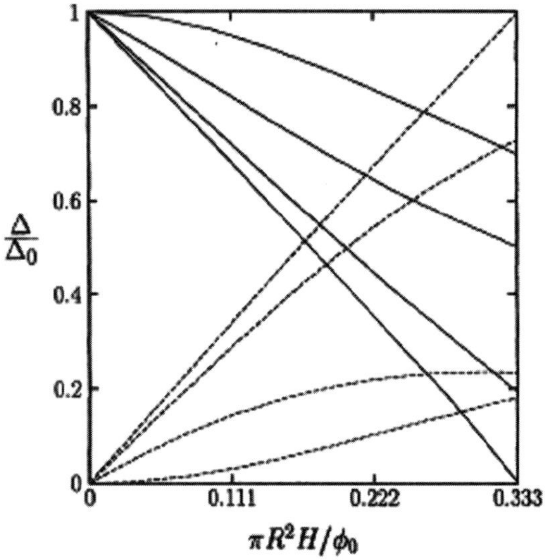

Fig. 1.21 The energy gap changes of semiconducting (*solid curve*) and metallic (*dashed curve*) SWNTs as a function of a magnetic field for different angles θ, between the SWNT and the external magnetic field. $\theta = \pi/2, \pi/3, \pi/6, 0$ (top-down for *solid* and bottom-up for *dashed curves*). The change is largest when the magnetic field is parallel to the SWNT. Reprinted figure with permission from [66]. Copyright 1995 by the American Physical Society

Experimental works have confirmed the theoretical predictions of the magnetic field effects on transport properties. For instance, the Aharonov–Bohm effect in SWNTs was experimentally observed [63, 64]. Also, the bandgap modulation due to a magnetic field was observed [67, 68]. In this experiment, the conductance modulation due to the applied magnetic field and the temperature was fitted by thermally activated transport of carriers, consistent with the expected theory (Fig. 1.22). Notably, suspended SWNTs provide a defect-free environment for carrier transport as the lack of substrate interactions preserve the pristine and intrinsic nature of 1D SWNTs. Therefore, for many of these transport physics experiments, the use of suspended nanotubes is necessary.

In the case of an applied electric field, a perturbation in the form of electrostatic potential to the Hamiltonian of electrons in a SWNT is expected. For example, homogeneous electric field E, perpendicular to SWNTs (defined as y axis) modifies the Hamiltonian as

$$H = H_0 - eEy. \tag{1.34}$$

where H_0 is the Hamiltonian under equilibrium conditions. The presence of perturbing potential can lead to the mixing of states due to symmetry breaking. This results in the bandgap modulation of semiconducting and metallic SWNTs except for armchair tubes, similar to the curvature-induced state mixing as considered in Section 1.1.3 [69–72]. The calculation results for a (15, 0) small bandgap and a (17, 0) semiconducting zigzag SWNT under a homogeneous transverse electric field are shown in Fig. 1.23(a) and (b), respectively, which show a bandgap opening and closing depending on the strength of the transverse electric field. The

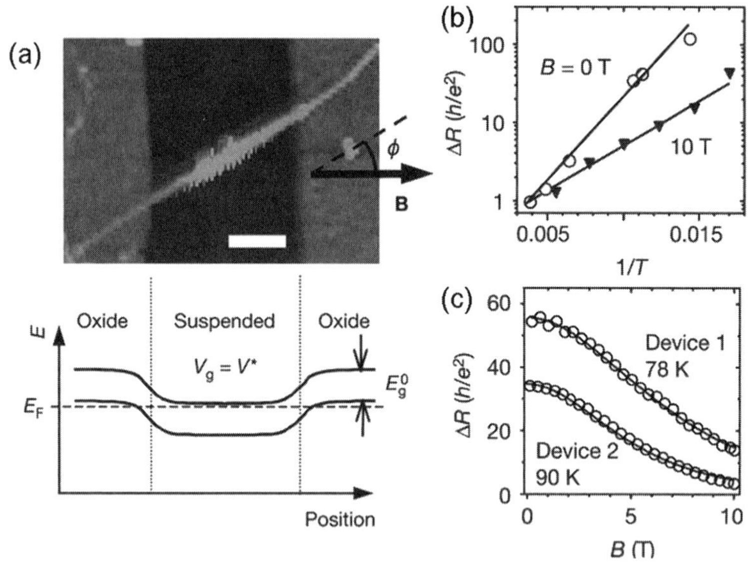

Fig. 1.22 Conductance through a partially suspended, small band gap SWNT is measured as a function of temperature and external magnetic field. (**a**) An atomic force microscope image of a SWNT and corresponding energy band diagram. (**b**) Resistance change as a function of inverse temperature. (**c**) Resistance change as a function of applied magnetic field. Reprinted with permission from Macmillan Publishers Ltd: Nature [68]. Copyright 2004

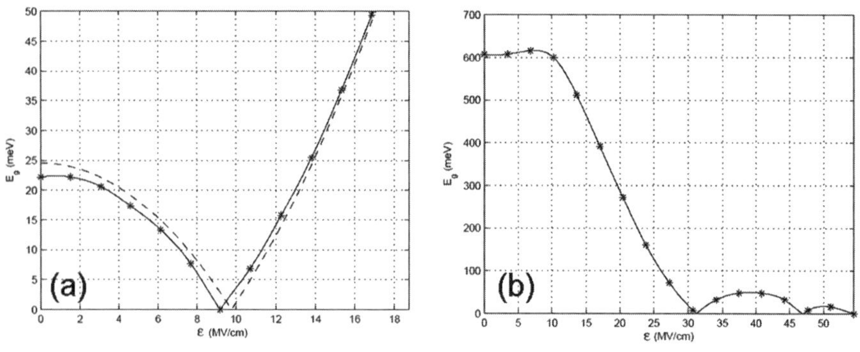

Fig. 1.23 The band gap modulation of SWNTs as a function of transverse electric field strength for (**a**) a (15, 0) small band gap SWNT and (**b**) a (17, 0) semiconducting SWNT. Adapted from [72]

magnitude of the electric field that is expected to give rise to a significant bandgap change is in the order of 0.1–1 V/Å, which is attainable for SWNT devices with ultrathin (1–10 nm) gate dielectrics [73]. The armchair SWNTs remain metallic with no bandgap opening from the curvature or transverse electric field, since the mirror symmetry is still preserved under these perturbations. But the symmetry can still be broken if inhomogeneous transverse electric field is applied [73] or if the armchair SWNT contains defects [74, 75], in which case a bandgap can be developed or electrical switching becomes possible.

1.3.2 Mechanical Deformation

Various mechanical deformations also result in changes in the band structure of SWNTs. Deformations, such as tensile or compressive strains, flattening, and twist, basically lead to deformation of the unit cell of SWNTs. Then, depending on the symmetry of the original SWNT, the relative position of the K points and the allowed $k_\perp$ states changes, giving rise to the modification of the electronic band structure of SWNTs [11, 76–83].

In this section, the effects of axial tensile strain on zigzag SWNTs will be briefly discussed [83]. The tensile strain can be modeled as a uniform stretching of the unit cell in SWNTs as shown in Fig. 1.24(a) and (b). The bond lengths between the nearest neighbor carbon atoms are now different, which invalidates the assumption of the same transfer integrals for the three nearest carbon atom pairs that was used to obtain Eq. (1.12). Then, the energy dispersion near the K points can be calculated based on the stretched-bond lengths between the nearest neighbor carbon atoms in a similar way as in Section 1.1.1. By writing σ as the strain and ν as the Poisson ratio (ratio of the transverse contracting strain to the elongation strain), the vectors connecting the three nearest neighbor carbon atoms are now changed to (Fig. 1.24(b))

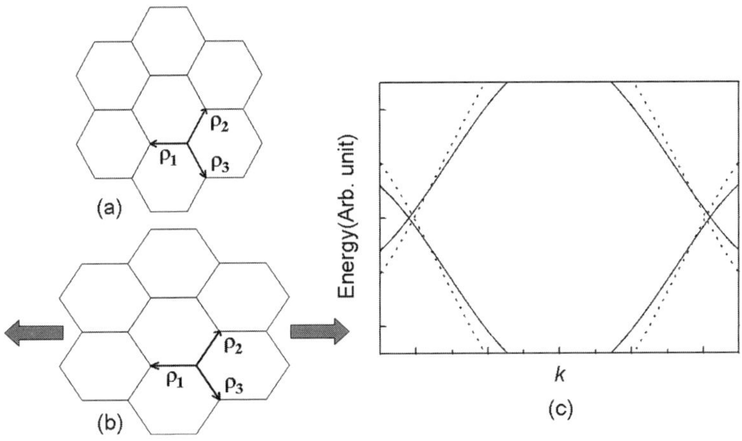

Fig. 1.24 The lattice structure of graphene (**a**) without and (**b**) with tensile strain. (**c**) The resulting band structure along the Γ–K direction for unstrained (*solid curve*) and strained (*dotted curve*) graphene

$$\begin{aligned}\boldsymbol{\rho}_1 &= -\frac{1}{\sqrt{3}}a\,(1+\sigma)\,\mathbf{x},\\ \boldsymbol{\rho}_{2,3} &= \frac{1}{2\sqrt{3}}a\,(1+\sigma)\,\mathbf{x} \pm \frac{1}{2}a(1-\sigma\nu)\mathbf{y}.\end{aligned} \tag{1.35}$$

The non-diagonal matrix element in Eq. (1.11), by allowing different transfer integrals between the nearest neighbor atoms, now becomes

$$\begin{aligned}H_{\mathrm{AB}} &= \frac{1}{N}\sum_i \mathrm{e}^{\mathrm{i}\mathbf{k}\cdot\rho_i}\int X^*(\mathbf{r})\,H\,X(\mathbf{r}-\rho_i)\,\mathrm{d}\tau\\ &= \gamma_1\mathrm{e}^{\mathrm{i}\mathbf{k}\cdot\rho_1} + \gamma_2\mathrm{e}^{\mathrm{i}\mathbf{k}\cdot\rho_2} + \gamma_3\mathrm{e}^{\mathrm{i}\mathbf{k}\cdot\rho_3}\\ &= \gamma_1\mathrm{e}^{-\mathrm{i}k_x a(1+\sigma)/\sqrt{3}} + 2\gamma_2\mathrm{e}^{\mathrm{i}k_x a(1+\sigma)/2\sqrt{3}}\cos\left(\frac{k_y a\,(1-\sigma\nu)}{2}\right),\end{aligned} \tag{1.36}$$

where γ_i refers to a transfer integral for each pair of atoms and $\gamma_2 = \gamma_3$. The transfer integrals γ_i of strained bonds are related to the transfer integral γ_0 of unstrained bonds as $\gamma_i = \gamma_0(a_0/a_i)^2$, where a_0 and a_i are the bond lengths of unstrained and strained bonds, respectively [83]. Then, the energy dispersion in Eq. (1.8) becomes

$$E = E_0 \mp \gamma_1\left(\begin{array}{l}1 + 4\dfrac{\gamma_2}{\gamma_1}\cos\left(\dfrac{\sqrt{3}k_x a\,(1+\sigma)}{2}\right)\cos\left(\dfrac{k_y a\,(1-\sigma\nu)}{2}\right)\\ +4\left(\dfrac{\gamma_2}{\gamma_1}\right)^2\cos^2\left(\dfrac{k_y a\,(1-\sigma\nu)}{2}\right)\end{array}\right)^{1/2} \tag{1.37}$$

The energy dispersion is plotted along the Γ–K direction without strain ($\sigma = 0$, $\gamma_1 = \gamma_2$) and with strain ($\sigma = 0.2$, ν=0.2) in Fig. 1.24(c). We can see that the bands still cross at the Fermi level with no bandgap, but the crossing points (K points) shift.

The band structure of SWNTs under strain can be obtained by performing zone folding on the band structure of the stretched graphene. Zigzag SWNTs $(n, 0)$ with $n = 3q$ and $n = 3q \pm 1$ will be considered here. The periodic boundary condition in Eq. (1.15) still applies, but with a reduced diameter of $(1\text{–}\sigma\nu)d$. The K points shift according to Eq. (1.37), while the allowed $k_\perp$ states also shift due to the reduced diameters as applied to the boundary conditions in Eq. (1.15). These relative shifts result in changes in the band structure of SWNTs. For metallic zigzag SWNTs with $n = 3q$, similar treatments as the one shown in Section 1.1.1 lead to a relative displacement of the $k_\perp$ state nearest to K points by an amount of

$$\Delta k = \frac{1}{a}\sqrt{3}\,(1+\nu)\,\sigma. \tag{1.38}$$

For $n = 3q \pm 1$, similarly

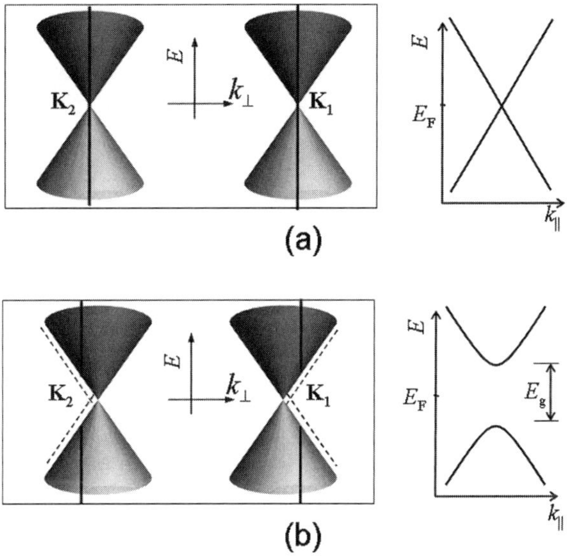

Fig. 1.25 The band structure of a metallic SWNT (**a**) without and (**b**) with a tensile strain with the resulting 1D subbands shown on the right side. Bandgap opens in metallic SWNTs due to the deformation. Adapted from [8]

$$\Delta k = \frac{1}{a}\sqrt{3}\,(1+\nu)\,\sigma \pm \frac{2}{3d}. \tag{1.39}$$

The effects of the displacement are schematically shown in Fig. 1.25 for a metallic SWNT. We can see that the strain induces a bandgap for metallic zigzag SWNTs, and the size of the bandgap is determined by Δk. For semiconducting SWNTs, the bandgap can either increase ($n = 3q+1$) or decrease ($n = 3\,q-1$) due to the relative displacement of the $k_\perp$ states with respect to the K points. In contrast to the magnetic field effects as in Fig. 1.20, the 1D subbands are still degenerate after deformation in this case.

By considering both uniaxial and torsional strains, Yang and Han [83] derived a general formula of bandgap variation (ΔE_{gap}) for all kinds of SWNTs due to different kinds of strains:

$$\Delta E_{\text{gap}} = \text{sgn}(2q+1)3t_0[(1+\nu)\,\sigma\,\cos 3\theta + \gamma\,\sin 3\theta], \tag{1.40}$$

where q is from $n_1 - n_2 = 3p+q$, ν is the Poisson's ratio, θ is the SWNT chiral angle, σ is the uniaxial strain, and γ is the torsional strain. We can see that the bandgap modulation due to the uniaxial strain is largest for zigzag SWNTs and zero for armchair SWNTs, whereas the modulation due to the torsional strain (twist) is largest for armchair SWNTs and zero for zigzag SWNTs. According to Eq. (1.40), a uniaxial strain as small as 1% can induce a bandgap modulation as large as ~100 meV.

Other mechanical deformations like radial deformation and bending were also shown to induce similar changes in the band structures, such as bandgap

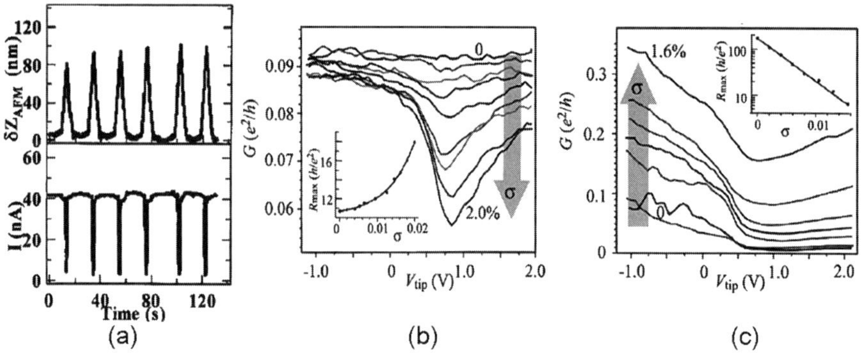

Fig. 1.26 Strong influence of mechanical strain on the transport properties of SWNTs. Suspended SWNTs are pushed down by an AFM tip. (**a**) Current decrease associated with an AFM tip pushing on a suspended CNT, (**b**) bandgap enhancement with strain, and (**c**) bandgap decrease with strain. Reprinted figures with permission from [85, 86]. Copyright 2003 by the American Physical Society

opening/closing and shifts of the K points, demonstrating the sensitivity of SWNTs to various mechanical deformations [76, 78, 81].

There are experimental works relating the mechanical deformations to changes in the transport properties of SWNTs [84–86]. In these experiments, an atomic force microscope (AFM) tip was used to exert tensile strain on suspended SWNTs by directly [84, 86] or indirectly [85] pushing on the SWNTs while the SWNT conductance was simultaneously monitored. The conductance of a SWNT was found to change (see Fig. 1.26) with the strain, and it was attributed to the strong local bonding deformations (from sp^2 to sp^3) due to an AFM tip [84] or bandgap modulations due to stretching [85, 86] as discussed in this section, possibly with extra contribution from other electromechanical deformations [85]. In obtaining Fig. 1.26(b) and (c), an AFM tip is used to induce tensile strains on suspended SWNTs while electrically gating them simultaneously. The measurements demonstrated the bandgap opening in metallic SWNTs (Fig. 1.26(b)) and bandgap narrowing in certain semiconducting SWNTs with increasing strain (Fig. 1.26(c)), manifested by changes in the conductance as a function of the gate voltage, confirming the theoretical predictions in Eq. (1.40). In these experiments, local deformations under the AFM tip or near the contact electrodes can also contribute to the changes in the transport properties, which may explain some discrepancies found in the experiments. Other experiments, such as STM [87] and PL measurements [88], also showed evidence of bandgap modification due to strain.

These theoretical and experimental results indicate that the transport properties of SWNTs are in general very sensitive to various mechanical deformations. Due to this strong sensitivity to deformations, SWNTs may find applications as mechanical transducers. Also, given the strong van der Waals interaction between SWNTs and the substrate surfaces where they lie on, local deformations due to the interaction with the substrate may account for some of the transport properties reported in the literatures. There are reports showing clean or ideal transport properties when SWNTs are suspended, not in contact with the substrates [68, 89].

1.4 Electron Transport Properties of SWNTs

In previous sections, we explored the band structure of nanotubes and its modification due to external perturbations. Once the band structure was determined, the transport through each 1D subband was assumed to be ballistic, without any scattering. But in the real world, there are various sources of carrier scattering in SWNTs that affect the transport characteristics.

Resistance in solid systems is due to the scattering of charge carriers as they travel through the system. In typical solid state systems, main origins of scatterings (thereby resistance) are static defects (imperfections, impurities, and so on) and phonons (lattice vibration). For example, in typical semiconductors, scattering with phonons is dominant at high temperatures while impurity scattering prevails at low temperatures [90, 91]. In this section, we will consider these two scattering mechanisms in SWNTs. Related to the scattering, charge carrier mobility is an important performance parameter for electronic devices, and it will also be discussed in this section.

1.4.1 Scatterings in SWNTs

Charge carrier scattering in 1D systems is quite different from the one in higher dimensions in a sense that carriers can only forward or backward scatter while charge carriers can be scattered into many different directions in higher dimensions. Additionally, metallic SWNTs have a very limited momentum space (as represented by a small number of 1D subbands) available for the backscattering process while satisfying both momentum and energy conservations. This contributes to the suppression of backscattering and results in a very long mean free path in metallic SWNTs, even at room temperature [92].

In general, there are two types of scattering processes in solid systems. One is the scattering by static potentials such as impurities which does not change the energy of the particles being scattered. The other is the scattering by time-varying potentials such as phonons which results in the inelastic scattering with energy as well as momentum changes. In quantum mechanics, scattering involves transition of a particle from one state to the other in the presence of these scattering potentials. In order to assess the effects of the scattering on the transport properties, we need to find the transition rate in the presence of a scattering potential, which is a measure of the time between subsequent scatterings. Fermi's golden rule is used for this purpose. In Fermi's golden rule, the transition (scattering) rate W can be calculated by

$$W_{\mathrm{fi}} = \frac{2\pi}{\hbar} \left| V_{\mathrm{fi}} \right|^2 \delta \left(E_{\mathrm{f}} - E_{\mathrm{i}} \right), \tag{1.41}$$

where V_{fi} is the matrix element connecting the final state to the initial state through a scattering potential. Then, the problem becomes finding the scattering matrix for each specific scatterer.

Scattering by defects or impurities is due to scattering potentials which are generally static in time. The scattering potential can be a screened or unscreened Coulomb potential due to charged impurities or a more complex short-range potential due to neutral defects. Although SWNTs retain high crystallinity, they inevitably have different kinds of structural defects such as vacancies or pentagon–heptagon pairs and impurities such as foreign atoms or chemicals deposited or formed on the surface. Despite these imperfections, metallic SWNTs are expected to have a long mean free path, even at room temperature, and this is attributed to the suppressed back scatterings due to the high symmetry of SWNTs (especially armchair SWNTs) [93, 94]. Metallic SWNTs have two degenerate 1D subbands crossing at the Fermi level. Each subband corresponds to a different moving direction for electrons (left-going vs. right-going) and backscattering requires electrons move from one subband to the other. Due to symmetry, these two crossing 1D subbands are orthogonal to each other and do not mix. As a result, backscattering in metallic SWNTs is significantly suppressed [93]. As long as the defects do not significantly perturb the band symmetry of SWNTs, this argument holds and as a result, metallic SWNTs have a large mean free path. It was also pointed out that the wave functions of SWNTs near the Fermi level are delocalized and extend around their circumferences. These delocalized wave functions will experience an averaged defect potential over the circumference, which reduces scattering [94]. The effects of various defects on transport properties of SWNTs have been a subject of many theoretical studies [95–98]. Defects such as boron and nitrogen impurities and pentagon–heptagon pairs are expected to affect the electrical transport properties of metallic SWNTs, but only far away from the Fermi level [95] as shown in Fig. 1.27(a). On the other hand, vacancies which have a short range potential seem to affect the transport near the Fermi level [95, 96] as shown in Fig. 1.27(b). Experimentally, defects are quite well characterized by STM measurements where atomic scale features can be resolved, and STS measurements can be acquired exactly at the defect sites [99–103]. However,

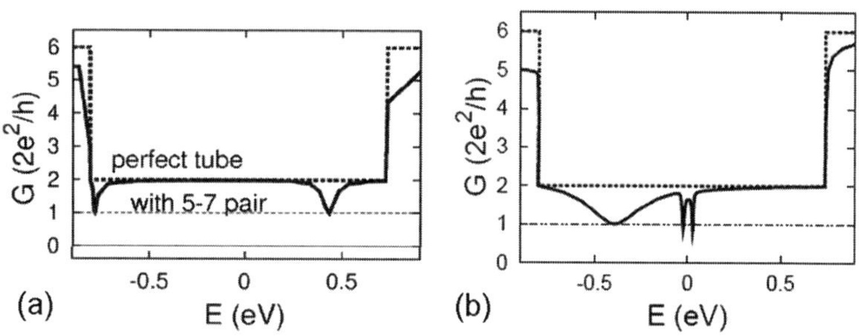

Fig. 1.27 Conductance of a (10, 10) armchair SWNT (**a**) with a pentagon–heptagon pair defect and (**b**) a point vacancy, based on an ab initio pseudopotential method. Reprinted figures with permission from [95]. Copyright 2000 by the American Physical Society

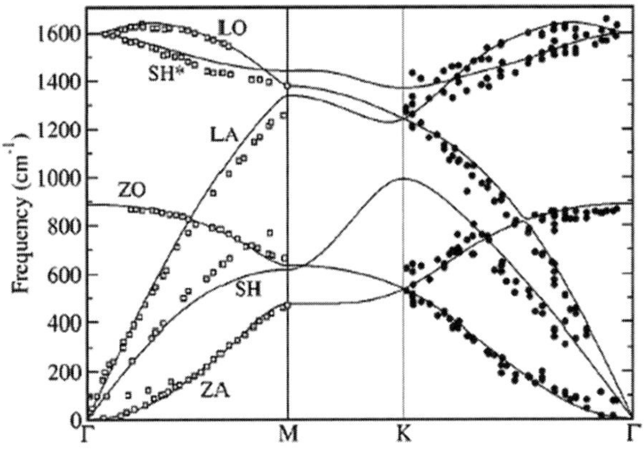

Fig. 1.28 Phonon dispersion relation of graphene. A reprinted figure with permission from [105]. Copyright 2003 by the American Physical Society

correlations between the specific types of defects and the actual transport properties are difficult to study experimentally since the atomic structure of defects cannot be resolved for the nanotube devices (SWNTs are on insulating substrates, which are not accessible with an STM). There is a report on the creation of vacancy defects with ion irradiation and its influence on the nanotube resistance; however, more detailed experimental study of such defects on transport properties is still needed [104].

As scattering due to defects are suppressed in SWNTs, the main source of scattering is believed to be phonons, especially at high temperatures. Various phonon modes exist in SWNTs (Fig. 1.28) and they have been extensively studied both theoretically and experimentally, especially with Raman spectroscopy [27, 105].

Phonon scattering is an inelastic scattering process which requires the momentum and energy conservations by electrons and phonons combined. Limited momentum space of SWNTs along with the symmetry requirements leave only three possible electron–phonon backscattering processes satisfying the momentum and energy conservations, which are shown in Fig. 1.29(a)–(c). The first one is scattering by low energy acoustic phonons which involves a small momentum and energy changes (Fig. 1.29(a)). The other two scattering processes are due to the high energy optical and zone boundary phonons which require large energy changes

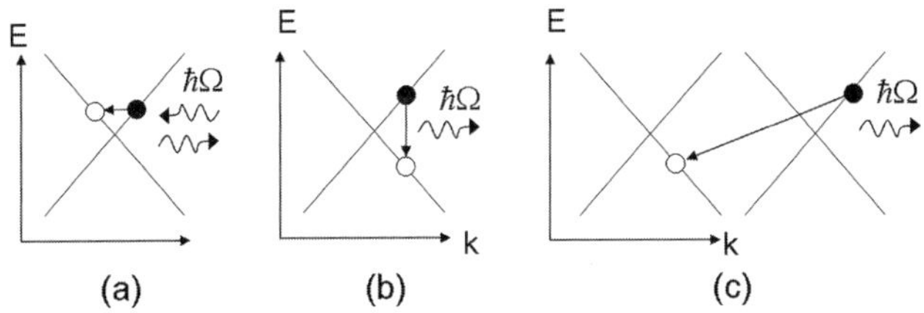

Fig. 1.29 Allowed electron backscattering processes in a metallic SWNT by (**a**) acoustic phonons, (**b**) optical phonons, and (**c**) zone boundary phonons with energy $\hbar\Omega$

(150–180 meV) with small and large momentum changes, respectively (Fig. 1.29(b) and (c)). Phonon scattering involves either emission or absorption of the phonons by the traveling charge carriers. At low V_{DS}, acoustic phonon scattering (Fig. 1.29(a)) is the only available scattering process since electrons do not have enough energy to interact with the high-energy phonons. In this regime, resistance is inversely proportional to the temperature since the scattering rate is determined by the number of phonons available and the occupation of acoustic phonons is proportional to the temperature [3]. Due to their high energies (compared to the room temperature, $k_B T{\sim}25$ meV), the optical and zone boundary phonons are not expected to be present at room temperature so that the scattering in this case corresponds to only the emission of phonons by electrons (Fig. 1.29(b) and (c)).

At high electric fields (high V_{DS}), electrons can acquire enough energy to emit optical and zone boundary phonons, which results in the backscattering of electrons. As long as the mean free path for high energy phonon scattering is much smaller than the SWNT length, we can assume that electrons backscatter instantly when they acquire enough energy to emit high-energy phonons. Then the steady state population of electrons moving in one direction have an energy difference of $\Delta E = \hbar\Omega$ (corresponding to the phonon energy) in respect to the electrons moving in the opposite direction. As a result, the net current carried by the electrons is given by

$$I_0 = \left(4e/h\right)\hbar\Omega. \tag{1.42}$$

Since the energy of high energy phonon is $\hbar\Omega{\sim}0.16$ eV, $I_0{\sim}25$ μA, which can explain universal current saturation at 20–25 μA for long (>~100 nm) metallic SWNTs. This current saturation was first reported by Yao et al. as shown in Fig. 1.30(a) [106]. By fitting their result to the numerical calculations based on the Boltzmann transport equation, they found the mean free path for optical phonon scattering, $l_{pb} = 10$ nm. Later, two independent experiments [107, 108] systematically characterized the $I_{DS}-V_{DS}$ of metallic SWNTs of different lengths for low- and high-bias regimes. The length-dependence was acquired by measuring various SWNTs with different lengths and Pd ohmic metal contacts (Fig. 1.30(b), [107]) or by employing a gold-coated AFM tip contacting the SWNT at different points along the length (Fig. 1.30(c) [108]). As shown in Fig. 1.30(b) and (c), similar $I_{DS}-V_{DS}$ characteristics were observed from both experiments with the currents at high bias not saturating for short nanotubes (<~100 nm), but rather increasing almost linearly but at different slopes from the low-bias regime. The non-saturating, high-bias behavior can be understood by the fact that the lengths of the short nanotubes approach the mean free path for high energy phonon scattering, and therefore not all electrons lose their energy to optical and zone boundary phonons. The resistivity ρ of a 1D, metallic SWNT can be written as a function of scattering mean free path ℓ by extending the Landauer formula, Eq. (1.25) as [14]

$$\rho = \left(h/4e^2\right)\left(1/\ell\right). \tag{1.43}$$

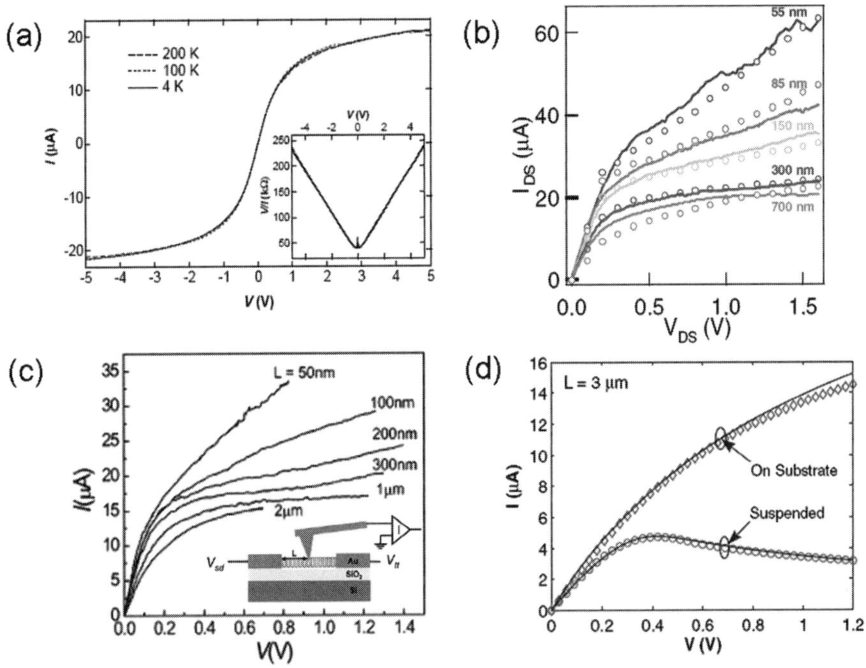

Fig. 1.30 Electron transport through a metallic SWNT from low- to high-bias regime. (**a**) Current saturation at high bias for a 1 μm long CNT. The inset is a plot of $R = V/I$ vs. bias. (**b**), (**c**) Current characteristics of metallic SWNTs with various lengths. The inset is a schematic of the measurements. (**d**) Current characteristics for metallic SWNTs on the substrate and suspended. Reprinted figures with permission from [106–109]. Copyright 2000, 2004, 2005 by the American Physical Society and Copyright 2004 American Chemical Society

Then, the mean free paths for various phonon scattering processes (such as low vs. high-energy phonons) can be deduced by measuring the differential resistance at different bias regimes from the data shown in Fig. 1.30(b) and (c) [107, 108]. Since we can attribute the low bias transport mainly to acoustic phonon scattering and high-bias transport to high-energy optical and zone boundary phonons, the measured mean free path from each bias regime can be regarded as the mean free path for each kind of phonon scattering. From these measurements, the acoustic phonon scattering mean free path is found to be ~300 nm [107] and 1.6 μm [108], respectively. The high energy phonon scattering mean free paths are found in the range between 10 and 15 nm [106–108]. The mean free path for high energy phonon scattering is found to be much shorter than the one for acoustic phonons, indicating much higher electron–phonon coupling strength. The large discrepancies for mean free paths at low bias can be due to different kinds of SWNTs measured in different experiments or due to the presence of additional scattering centers like defects which also contribute to the scattering process. The current carried by a short metallic SWNT can exceed 60 μA as seen in Fig. 1.30(b). On the other hand, much smaller maximum current was found in suspended SWNTs, as shown in Fig. 1.30(d). This is due to the self-heating effects induced by insufficient heat dissipation through the substrate, showing the importance of heat sinking for higher current carrying applications [109].

Theoretically, mean free paths for electron–phonon scattering in SWNTs can be calculated from Fermi's golden rule as in Eq. (1.41) with the electron–phonon coupling potentials calculated using tight-binding method or density functional theory. The resulting mean free path values for scattering with high-energy optical and zone boundary phonons vary depending on the calculated electron–phonon coupling strength, ranging from $\sim$30 to 150 nm [107, 108, 110–112], much longer than experimentally determined values of 10–15 nm [106–108]. The difference was later attributed to non-equilibrium high-energy phonon populations and local heating effects due to the fact that hot phonon generation time is much faster than the thermalization time for high energy-phonons, which gives rise to both phonon emission and absorption with electron back scatterings [109, 110, 113, 114]. The effect is more pronounced for suspended nanotubes, but it still affects the nanotubes on substrates due to the poor heat dissipation capability of the commonly used SiO_2 substrates. In semiconducting SWNTs, the main scattering mechanisms are also expected to be low- and high-energy phonons and have been theoretically studied [112, 115–119]. Since induced charge density varies with the gate voltage in semiconducting SWNTs, the scattering rate also depends on the gate voltage and the onset of additional subbands [112, 115, 118, 119]. Also, in CNTFETs, another source of scattering could be the metal contacts, depending heavily on the Schottky nature of the metal interfaces as further discussed in Chapter 3 [116, 117].

1.4.2 Carrier Mobility in SWNTs

One of the important parameters characterizing the transport properties of a semiconducting material is its carrier mobility. Mobility measures how fast the charge carriers respond to an external electric field. It is defined as

$$\mu = \frac{v}{E} = \frac{e\tau}{m^*} = \frac{\sigma}{ne}, \tag{1.44}$$

where v is the drift velocity of charge carriers, E is the applied electric field, σ is the conductivity, and ne is the charge density. Carrier mobility in a FET channel is an important parameter that determines the performance of the device, especially frequency response and ON current delivering capability. Mobility of a semiconducting material is typically measured by two methods. One method utilizes the Hall Effect (thus it is called the Hall mobility), and the other is based on the analysis of the transfer characteristics of the FETs (called effective and field-effect mobility). Since the Hall mobility cannot be measured for 1D systems, the effective or field-effect mobility is typically measured and reported for semiconducting SWNTs [120]. The effective mobility μ_{eff} and the field-effect mobility μ_{FE}, in the linear region (low-bias regime) are defined as

$$\mu_{\mathrm{eff}} = \frac{L}{C_{\mathrm{G}}} \frac{G}{|V_{\mathrm{G}} - V_{\mathrm{T}}|}, \quad \mu_{\mathrm{FE}} = \frac{L}{C_{\mathrm{G}}} \left| \frac{\partial G}{\partial V_{\mathrm{G}}} \right|, \tag{1.45}$$

where C_{G} is the gate capacitance per unit length, V_{G} is the gate voltage, V_{T} is the threshold voltage, and L is the length of a device. Note that the use of Eq. (1.45) for evaluation of mobility applies only to channel-controlled MOSFET-like CNTFETs, and not to the Schottky-barrier controlled FETs. The experimentally reported mobility values for CNTFETs at room temperature have been $>\sim 4000\ \mathrm{cm^2/Vs}$, with the highest reported value of $\sim 100{,}000\ \mathrm{cm^2/Vs}$ [121]. In comparison, the mobility of a typical Si MOSFET is $\sim 1000\ \mathrm{cm^2/Vs}$. The field effect mobility of a semiconducting SWNT FET can be estimated based on the band structure of semiconducting SWNTs [122]. The energy dispersion of the lowest 1D subbands of a semiconducting SWNT can be written as

$$E = \pm\sqrt{(m^* v_{\mathrm{F}}^2)^2 + (\hbar k v_{\mathrm{F}})^2}, \tag{1.46}$$

where $v_{\mathrm{F}} = 8 \times 10^5$ m/s is the Fermi velocity of electrons in SWNTs. We can see that the energy dispersion represented by Eq. (1.46) approaches that of a metallic SWNT ($E = \pm\ v_{\mathrm{F}} k$) when k becomes large; therefore, a semiconducting SWNT will behave similarly to a metallic one high gate voltages are applied (the ON state). By noting that the induced charge density in a semiconducting SWNT is given by $C_{\mathrm{G}}|V_{\mathrm{G}} - V_{\mathrm{T}}|$, the conductance and field-effect mobility as a function of the gate voltage can be derived as [122]

$$G(V_{\mathrm{G}}) = \frac{4e^2}{h} \frac{\ell_0}{L} \frac{(\Delta V_{\mathrm{G}}/a)^2}{1 + (\Delta V_{\mathrm{G}}/a)^2}, \quad \mu_{\mathrm{FE}} = \frac{e\tau_0}{m^*} \frac{(\Delta V_{\mathrm{G}}/a)}{\left(1 + (\Delta V_{\mathrm{G}}/a)^2\right)^2}, \tag{1.47}$$

where $\Delta V_{\mathrm{G}} = |V_{\mathrm{G}} - V_{\mathrm{T}}|$, $a = 8e/3\pi d C_{\mathrm{G}}$, l_0 and τ_0 are the mean free path and scattering time at high energies, and V_{T} is the FET threshold voltage. Equation (1.47) shows that conductance G saturates at high-gate voltage and μ_{FE} peaks at a gate voltage near V_{T} as shown in Fig. 1.31. At low V_{DS}, the dominant scattering is due to the acoustic phonons, similar to the metallic SWNTs. From the fact that scattering time is proportional to the diameter of a SWNT and is inversely proportional to the temperature, the peak mobility, μ_{peak}, and maximum conductance, G_{max}, of a semiconducting SWNT as can be shown to be

$$G_{\mathrm{max}} = \frac{4e^2 v_0}{h\alpha L} \frac{d}{T}, \quad \mu_{\mathrm{peak}} = 0.48 \frac{e v_0}{\hbar\alpha} \frac{d^2}{T}, \tag{1.48}$$

where α is a proportionality factor. The experimental results have also confirmed the diameter and temperature dependence predicted by Eq. (1.48) [122].

A more detailed theoretical calculation based on the multiband Boltzmann equation obtained an almost similar functional form as Eq. (1.48) [112]. The experimentally measured effective mobility values scatter a lot as mentioned before and they

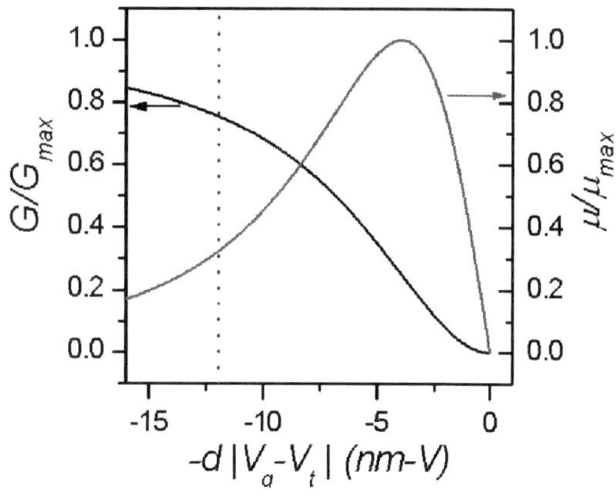

Fig. 1.31 Plot of conductance and mobility of a SWNT FET from Eq. (1.48). A reprinted figure with permission from [122]. Copyright 2005 by the American Physical Society

are also different from theoretically estimated values [112]. The effects of contact characteristics (ohmic, Schottky barrier, or tunneling) and inhomogeneous response to the gate voltage [122] may account for the variations and discrepancies seen in the different measurements.

1.5 Summary

In this chapter, we reviewed the electronic band structure of SWNTs and the associated electrical transport characteristics. Essential features of the nanotube band structure can be obtained by zone folding of graphene. SWNTs can be either metallic with linear dispersions and finite DOS near the Fermi level or semiconducting with a direct bandgap and zero DOS near the Fermi level, depending on their chirality and diameter. Due to the 1D nature of SWNTs, singularities known as van Hove singularities are formed in the DOS. As a low-dimensional system, SWNTs exhibit rich quantum transport characteristics, such as single electron effect, electron interference, and Luttinger liquid characteristics. Modifications to the electronic band structure of SWNTs due to various perturbations such as external fields or mechanical deformations can dramatically affect the transport properties. Charge carrier scattering processes responsible for the resistance in SWNTs are also discussed, emphasizing the role of phonon and defect scattering on the transport characteristics.

References

1. P. R. Wallace, "The band theory of graphite," *Physical Review*, vol. 71, pp. 622–634, 1947.
2. R. Saito, M. Fujita, G. Dresselhaus, and M. S. Dresselhaus, "Electronic structure of chiral graphene tubules," *Applied Physics Letters*, vol. 60, pp. 2204–2206, 1992.
3. C. L. Kane, E. J. Mele, R. S. Lee, J. E. Fischer, P. Petit, H. Dai, A. Thess, R. E. Smalley, A. R. M. Verschueren, S. J. Tans, and C. Dekker, "Temperature-dependent resistivity of single-wall carbon nanotubes," *Europhysics Letters*, vol. 41, pp. 683–688, 1998.

4. S. Reich, J. Maultzsch, C. Thomsen, and P. Ordejón, "Tight-binding description of graphene," *Physical Review B*, vol. 66, p. 035412, 2002.
5. R. Saito, G. Dresselhaus, and M. S. Dresselhaus, *Physical Properties of Carbon Nanotubes*. Imperial College Press, 1998.
6. R. Saito, M. Fujita, G. Dresselhaus, and M. S. Dresselhaus, "Electronic structure of graphene tubules based on C_{60}," *Physical Review B*, vol. 46, pp. 1804–1811, 1992.
7. J.-C. Charlier, X. Blase, and S. Roche, "Electronic and transport properties of nanotubes," *Reviews of Modern Physics*, vol. 79, pp. 677–656, 2007.
8. E. Minot, "Tuning the Band Structure of Carbon Nanotubes." Ph.D. thesis, Cornell University, 2004.
9. R. E. Peierls, *Quantum Theory of Solids*. London: Oxford University Press, 1955.
10. J. W. Mintmire, B. I. Dunlap, and C. T. White, "Are fullerene tubules metallic?," *Physical Review Letters*, vol. 68, pp. 631–634, 1992.
11. C. L. Kane and E. J. Mele, "Size, shape, and low energy electronic structure of carbon nanotubes," *Physical Review Letters*, vol. 78, pp. 1932–1935, 1997.
12. A. Kleiner and S. Eggert, "Curvature, hybridization, and STM images of carbon nanotubes," *Physical Review B*, vol. 64, p. 113402, 2001.
13. J. W. Mintmire and C. T. White, "Universal density of states for carbon nanotubes," *Physical Review Letters*, vol. 81, pp. 2506–2509, 1998.
14. P. L. McEuen, "Nanostructures," in *Introduction to Solid State Physics*, 8th ed, C. Kittel, Ed. John Wiley and Sons, 2005.
15. G. Binning, H. Rohrer, C. Gerber, and E. Weibel, "Surface studies by scanning tunneling microscopy," *Physical Review Letters*, vol. 49, pp. 57–60, 1982.
16. R. M. Feenstra and P. Mårtensson, "Fermi-level pinning at the Sb/GaAs(110) surface studied by scanning tunneling spectroscopy," *Physical Review Letters*, vol. 61, pp. 447–450, 1988.
17. N. D. Lang, "Spectroscopy of single atoms in the scanning tunneling microscope," *Physical Review B*, vol. 34, pp. 5947–5950, 1986.
18. T. W. Odom, J.-L. Huang, P. Kim, and C. M. Lieber, "Atomic structure and electronic properties of single-walled carbon nanotubes," *Nature*, vol. 391, pp. 62–64, 1998.
19. J. W. G. Wildöer, L. C. Venema, A. G. Rinzler, R. E. Smalley, and C. Dekker, "Electronic structure of atomically resolved carbon nanotubes," *Nature*, vol. 391, pp. 59–62, 1998.
20. M. Ouyang, J.-L. Huang, C. L. Cheung, and C. M. Lieber, "Energy gaps in 'Metallic' single-walled carbon nanotubes," *Science*, vol. 292, pp. 702–705, 2001.
21. M. Bockrath, D. H. Cobden, P. L. McEuen, N. G. Chopra, A. Zettl, A. Thess, and R. E. Smalley, "Single-electron transport in ropes of carbon nanotubes," *Science*, vol. 275, pp. 1922–1925, 1997.
22. R. Martel, T. Schmidt, H. R. Shea, T. Hertel, and P. Avouris, "Single- and multi-wall carbon nanotube field-effect transistors," *Applied Physics Letters*, vol. 73, pp. 2447–2449, 1998.
23. S. J. Tans, M. H. Devoret, H. Dai, A. Thess, R. E. Smalley, L. J. Geerligs, and C. Dekker, "Individual single-wall carbon nanotubes as quantum wires," *Nature*, vol. 386, pp. 474–477, 1997.
24. S. J. Tans, A. R. M. Verschueren, and C. Dekker, "Room-temperature transistor based on a single carbon nanotube," *Nature*, vol. 393, pp. 49–52, 1998.
25. C. Zhou, J. Kong, and H. Dai, "Intrinsic electrical properties of individual single-walled carbon nanotubes with small band gaps," *Physical Review Letters*, vol. 84, pp. 5604–5607, 2000.
26. J. Appenzeller, J. Knoch, M. Radosavljevic, and P. Avouris, "Multimode transport in Schottky-barrier carbon-nanotube field-effect transistors," *Physical Review Letters*, vol. 92, p. 226802, 2004.
27. M. S. Dresselhaus and P. C. Eklund, "Phonons in carbon nanotubes," *Advances in Physics*, vol. 49, pp. 705–814, 2000.
28. H. Kataura, Y. Kumazawa, Y. Maniwa, I. Umezu, S. Suzuki, Y. Ohtsuka, and Y. Achiba, "Optical properties of single-qall carbon nanotubes," *Synthetic Metals*, vol. 103, pp. 2555–2558, 1999.

29. S. M. Bachilo, M. S. Strano, C. Kittrell, R. H. Hauge, R. E. Smalley, and R. B. Weisman, "Structure-assigned optical spectra of single-walled carbon nanotubes," *Science*, vol. 298, pp. 2361–2366, 2002.
30. M. J. O'Connell, S. M. Bachilo, C. B. Huffman, V. C. Moore, M. S. Strano, E. H. Haroz, K. L. Rialon, P. J. Boul, W. H. Noon, C. Kittrell, J. Ma, R. H. Hauge, R. B. Weisman, and R. E. Smalley, "Band gap fluorescence from individual single-walled carbon nanotubes," *Science*, vol. 297, pp. 593–596, 2002.
31. M. Y. Sfeir, F. Wang, L. Huang, C.-C. Chuang, J. Hone, S. P. O'Brien, T. F. Heinz, and L. E. Brus, "Probing electronic transitions in individual carbon nanotubes by Rayleigh scattering," *Science*, vol. 306, pp. 1540–1543, 2004.
32. T. Ando, "Excitons in carbon nanotubes," *Journal of the Physical Society of Japan*, vol. 66, pp. 1066–1073, 1997.
33. S. Reich, C. Thomsen, and J. Robertson, "Exciton resonances quench the photoluminescence of zigzag carbon nanotubes," *Physical Review Letters*, vol. 95, p. 077402, 2005.
34. P. T. Araujo, S. K. Doorn, S. Kilina, S. Tretiak, E. Einarsson, S. Maruyama, H. Chacham, M. A. Pimenta, and A. Jorio, "Third and fourth optical transitions in semiconducting carbon nanotubes," *Physical Review Letters*, vol. 98, pp. 067401–067404, 2007.
35. F. Wang, G. Dukovic, L. E. Brus, and T. F. Heinz, "The optical resonances in carbon nanotubes arise from excitons," *Science*, vol. 308, pp. 838–841, 2005.
36. S. Datta, *Electronic Transport in Mesoscopic Systems*. Cambridge University Press, 1995.
37. Y. Imry, *Introduction to Mesoscopic Physics*. New York: Oxford University Press, 1997.
38. R. Landauer, "Conductance from transmission: common sense points," *Physica Scripta*, vol. T42, p. 110, 1992.
39. L. Kouwenhoven and C. Marcus, "Quantum dots," *Physics World*, pp. 35–39, 1998.
40. M. A. Kastner, "Artificial Atoms," *Physics Today*, vol. 46, pp. 24–31, 1993.
41. L. P. Kouwenhoven, C. M. Marcus, P. L. McEuen, S. Tarucha, R. M. Westervelt, and N. S. Wingreen, "Electron transport in quantum dots," in *Mesoscopic Electron Transport*, L. P. Kouwenhoven, G. Schon, and L. L. Sohn, Eds., Kluwer, 1997.
42. K. Ishibashi, S. Moriyama, D. Tsuya, T. Fuse, and M. Suzuki, "Quantum-dot nanodevices with carbon nanotubes," *Journal of Vacuum Science & Technology A*, vol. 24, pp. 1349–1355, 2006.
43. J. Nygård, D. H. Cobden, M. Bockrath, P. L. McEuen, and P. E. Lindelof, "Electrical transport measurements on single-walled carbon nanotubes," *Applied Physics A: Materials Science & Processing*, vol. 69, pp. 297–304, 1999.
44. D. H. Cobden, M. Bockrath, P. L. McEuen, A. G. Rinzler, and R. E. Smalley, "Spin splitting and even–odd effects in carbon nanotubes," *Physical Review Letters*, vol. 81, pp. 681–684, 1998.
45. S. J. Tans, M. H. Devoret, R. J. A. Groeneveld, and C. Dekker, "Electron–electron correlations in carbon nanotubes," *Nature*, vol. 394, pp. 761–764, 1998.
46. J. Nygård, D. H. Cobden, and P. E. Lindelof, "Kondo physics in carbon nanotubes," *Nature*, vol. 408, pp. 342–346, 2000.
47. D. H. Cobden and J. Nygård, "Shell filling in closed single-wall carbon nanotube quantum dots," *Physical Review Letters*, vol. 89, p. 046803, 2002.
48. W. Liang, M. Bockrath, and H. Park, "Shell filling and exchange coupling in metallic single-walled carbon nanotubes," *Physical Review Letters*, vol. 88, p. 126801, 2002.
49. S. Moriyama, T. Fuse, M. Suzuki, Y. Aoyagi, and K. Ishibashi, "Four-electron shell structures and an interacting two-electron system in carbon-nanotube quantum dots," *Physical Review Letters*, vol. 94, p. 186806, 2005.
50. N. Mason, M. J. Biercuk, and C. M. Marcus, "Local gate control of a carbon nanotube double quantum dot," *Science*, vol. 303, pp. 655–658, 2004.
51. P. Jarillo-Herrero, S. Sapmaz, C. Dekker, L. P. Kouwenhoven, and H. S. J. van der Zant, "Electron–hole symmetry in a semiconducting carbon nanotube quantum dot," *Nature*, vol. 429, pp. 389–392, 2004.

52. S. Sapmaz, P. Jarillo-Herrero, Y. M. Blanter, C. Dekker, and H. S. J. van der Zant, "Tunneling in suspended carbon nanotubes assisted by longitudinal phonons," *Physical Review Letters*, vol. 96, p. 026801, 2006.
53. J. Kong, E. Yenilmez, T. W. Tombler, W. Kim, H. Dai, R. B. Laughlin, L. Liu, C. S. Jayanthi, and S. Y. Wu, "Quantum interference and ballistic transmission in nanotube electron waveguides," *Physical Review Letters*, vol. 87, p. 106801, 2001.
54. W. Liang, M. Bockrath, D. Bozovic, J. H. Hafner, M. Tinkham, and H. Park, "Fabry-Perot interference in a nanotube electron waveguide," *Nature*, vol. 411, pp. 665–669, 2001.
55. J. M. Luttinger, "An exactly soluble model of a many-fermion system," *Journal of Mathematical Physics*, vol. 4, pp. 1154–1162, 1963.
56. S.-I. Tomonaga, "Remarks on Bloch's method of sound waves applied to many-fermion problems," *Progress of Theoretical Physics*, vol. 5, pp. 544–569, 1950.
57. R. Egger and A. O. Gogolin, "Effective low-energy theory for correlated carbon nanotubes," *Physical Review Letters*, vol. 79, pp. 5082–5085, 1997.
58. C. Kane, L. Balents, and M. P. A. Fisher, "Coulomb interactions and mesoscopic effects in carbon nanotubes," *Physical Review Letters*, vol. 79, pp. 5086–5089, 1997.
59. M. Bockrath, D. H. Cobden, J. Lu, A. G. Rinzler, R. E. Smalley, L. Balents, and P. L. McEuen, "Luttinger-liquid behaviour in carbon nanotubes," *Nature*, vol. 397, pp. 598–601, 1999.
60. B. Gao, A. Komnik, R. Egger, D. C. Glattli, and A. Bachtold, "Evidence for Luttinger-liquid behavior in crossed metallic single-wall nanotubes," *Physical Review Letters*, vol. 92, p. 216804, 2004.
61. Z. Yao, H. W. C. Postma, L. Balents, and C. Dekker, "Carbon nanotube intramolecular junctions," *Nature*, vol. 402, pp. 273–276, 1999.
62. S. Bellucci, J. González, and P. Onorato, "Crossover from the Luttinger-liquid to coulomb-blockade regime in carbon nanotubes," *Physical Review Letters*, vol. 95, p. 186403, 2005.
63. A. Bachtold, C. Strunk, J.-P. Salvetat, J.-M. Bonard, L. Forró, T. Nussbaumer, and C. Schönenberger, "Aharonov–Bohm oscillations in carbon nanotubes," *Nature*, vol. 397, pp. 673–675, 1999.
64. S. Zaric, G. N. Ostojic, J. Kono, J. Shaver, V. C. Moore, M. S. Strano, R. H. Hauge, R. E. Smalley, and X. Wei, "Optical signatures of the Aharonov–Bohm phase in single-walled carbon nanotubes," *Science*, vol. 304, pp. 1129–1131, 2004.
65. H. Ajiki and T. Ando, "Magnetic properties of carbon nanotubes," *Journal of the Physical Society of Japan*, vol. 62, pp. 2470–2480, 1993.
66. J. P. Lu, "Novel magnetic properties of carbon nanotubes," *Physical Review Letters*, vol. 74, pp. 1123–1126, 1995.
67. U. C. Coskun, T.-C. Wei, S. Vishveshwara, P. M. Goldbart, and A. Bezryadin, "h/e magnetic flux modulation of the energy gap in nanotube quantum dots," *Science*, vol. 304, pp. 1132–1134, 2004.
68. E. D. Minot, Y. Yaish, V. Sazonova, and P. L. McEuen, "Determination of electron orbital magnetic moments in carbon nanotubes," *Nature*, vol. 428, pp. 536–539, 2004.
69. Y.-H. Kim and K. J. Chang, "Subband mixing rules in circumferentially perturbed carbon nanotubes: Effects of transverse electric fields," *Physical Review B*, vol. 64, p. 153404, 2001.
70. Y. Li, S. V. Rotkin, and U. Ravaioli, "Electronic response and bandstructure modulation of carbon nanotubes in a transverse electrical field," *Nano Letters*, vol. 3, pp. 183–187, 2003.
71. J. O'Keeffe, C. Wei, and K. Cho, "Bandstructure modulation for carbon nanotubes in a uniform electric field," *Applied Physics Letters*, vol. 80, pp. 676–678, 2002.
72. D. Gunlycke, C. J. Lambert, S. W. D. Bailey, D. G. Pettifor, G. A. D. Briggs, and J. H. Jefferson, "Bandgap modulation of narrow-gap carbon nanotubes in a transverse electric field," *Europhysics Letters*, vol. 73, pp. 759–764, 2006.
73. S. V. Rotkin and K. Hess, "Possibility of a metallic field-effect transistor," *Applied Physics Letters*, vol. 84, pp. 3139–3141, 2004.
74. Y.-W. Son, J. Ihm, M. L. Cohen, S. G. Louie, and H. J. Choi, "Electrical Switching in Metallic Carbon Nanotubes," *Physical Review Letters*, vol. 95, p. 216602, 2005.

75. J.-Y. Park, "Electrically tunable defects in metallic single-walled carbon nanotubes," *Applied Physics Letters*, vol. 90, p. 023112, 2007.
76. L. F. Chibotaru, S. A. Bovin, and A. Ceulemans, "Bend-induced insulating gap in carbon nanotubes," *Physical Review B*, vol. 66, p. 161401, 2002.
77. Y. N. Gartstein, A. A. Zakhidov, and R. H. Baughman, "Mechanical and electromechanical coupling in carbon nanotube distortions," *Physical Review B*, vol. 68, p. 115415, 2003.
78. O. Gülseren, T. Yildirim, S. Ciraci, and Ç. Kılıç, "Reversible band-gap engineering in carbon nanotubes by radial deformation," *Physical Review B*, vol. 65, p. 155410, 2002.
79. R. Heyd, A. Charlier, and E. McRae, "Uniaxial-stress effects on the electronic properties of carbon nanotubes," *Physical Review B*, vol. 55, pp. 6820–6824, 1997.
80. S. Ogata and Y. Shibutani, "Ideal tensile strength and band gap of single-walled carbon nanotubes," *Physical Review B*, vol. 68, p. 165409, 2003.
81. A. Rochefort, D. R. Salahub, and P. Avouris, "The effect of structural distortions on the electronic structure of carbon nanotubes," *Chemical Physics Letters*, vol. 297, pp. 45–50, 1998.
82. L. Yang, M. P. Anantram, J. Han, and J. P. Lu, "Band-gap change of carbon nanotubes: Effect of small uniaxial and torsional strain," *Physical Review B*, vol. 60, pp. 13874–13878, 1999.
83. L. Yang and J. Han, "Electronic structure of deformed carbon nanotubes," *Physical Review Letters*, vol. 85, pp. 154–157, 2000.
84. T. W. Tombler, C. Zhou, L. Alexseyev, J. Kong, H. Dai, L. Liu, C. S. Jayanthi, M. Tang, and S.-Y. Wu, "Reversible electromechanical characteristics of carbon nanotubes under local-probe manipulation," *Nature*, vol. 405, pp. 769–772, 2000.
85. J. Cao, Q. Wang, and H. Dai, "Electromechanical properties of metallic, quasimetallic, and semiconducting carbon nanotubes under stretching," *Physical Review Letters*, vol. 90, p. 157601, 2003.
86. E. D. Minot, Y. Yaish, V. Sazonova, J.-Y. Park, M. Brink, and P. L. McEuen, "Tuning carbon nanotube band gaps with strain," *Physical Review Letters*, vol. 90, p. 156401, 2003.
87. J. Lee, H. Kim, S. J. Kahng, G. Kim, Y. W. Son, J. Ihm, H. Kato, Z. W. Wang, T. Okazaki, H. Shinohara, and Y. Kuk, "Bandgap modulation of carbon nanotubes by encapsulated metallofullerenes," *Nature*, vol. 415, pp. 1005–1008, 2002.
88. D. Karaiskaj, C. Engtrakul, T. McDonald, M. J. Heben, and A. Mascarenhas, "Intrinsic and extrinsic effects in the temperature-dependent photoluminescence of semiconducting carbon nanotubes," *Physical Review Letters*, vol. 96, pp. 106805–4, 2006.
89. J. Cao, Q. Wang, and H. Dai, "Electron transport in very clean, as-grown suspended carbon nanotubes," *Nature Materials*, vol. 4, pp. 745–749, 2005.
90. J. H. Davies, *The Physics of Low-Dimensional Semiconductors: An Introduction*. Cambridge: Cambridge University Press, 1998.
91. S. M. Sze, *Physics of Semiconductor Devices*. Wiley-Interscience, 1981.
92. P. L. McEuen, M. S. Fuhrer, and P. Hongkun, "Single-walled carbon nanotube electronics," *IEEE Transactions on Nanotechnology*, vol. 1, pp. 78–85, 2002.
93. T. Ando and T. Nakanishi, "Impurity scattering in carbon nanotubes: absence of back scattering," *Journal of the Physical Society of Japan*, vol. 67, pp. 1704–1713, 1998.
94. C. T. White and T. N. Todorov, "Carbon nanotubes as long ballistic conductors," *Nature*, vol. 393, pp. 240–242, 1998.
95. H. J. Choi, J. Ihm, S. G. Louie, and M. L. Cohen, "Defects, quasibound states, and quantum conductance in metallic carbon nanotubes," *Physical Review Letters*, vol. 84, pp. 2917–2920, 2000.
96. M. Igami, T. Nakanishi, and T. Ando, "Conductance of carbon nanotubes with a vacancy," *Journal of the Physical Society of Japan*, vol. 68, pp. 716–719, 1999.
97. T. Kostyrko, M. Bartkowiak, and G. D. Mahan, "Reflection by defects in a tight-binding model of nanotubes," *Physical Review B*, vol. 59, pp. 3241–3249, 1999.
98. P. L. McEuen, M. Bockrath, D. H. Cobden, Y.-G. Yoon, and S. G. Louie, "Disorder, pseudospins, and backscattering in carbon nanotubes," *Physical Review Letters*, vol. 83, pp. 5098–5101, 1999.

99. M. Ishigami, H. J. Choi, S. Aloni, S. G. Louie, M. L. Cohen, and A. Zettl, "Identifying defects in nanoscale materials," *Physical Review Letters*, vol. 93, p. 196803, 2004.
100. H. Kim, J. Lee, S. J. Kahng, Y. W. Son, S. B. Lee, C. K. Lee, J. Ihm, and Y. Kuk, "Direct observation of localized defect states in semiconductor nanotube junctions," *Physical Review Letters*, vol. 90, p. 216107, 2003.
101. S. Lee, G. Kim, H. Kim, B.-Y. Choi, J. Lee, B. W. Jeong, J. Ihm, Y. Kuk, and S.-J. Kahng, "Paired gap states in a semiconducting carbon nanotube: Deep and shallow levels," *Physical Review Letters*, vol. 95, p. 166402, 2005.
102. M. Ouyang, J.-L. Huang, C. L. Cheung, and C. M. Lieber, "Atomically resolved single-walled carbon nanotube intramolecular junctions," *Science*, vol. 291, pp. 97–100, 2001.
103. L. C. Venema, J. W. Janssen, M. R. Buitelaar, J. W. G. Wildöer, S. G. Lemay, L. P. Kouwenhoven, and C. Dekker, "Spatially resolved scanning tunneling spectroscopy on single-walled carbon nanotubes," *Physical Review B*, vol. 62, pp. 5238–5244, 2000.
104. C. Gómez-Navarro, P. J. De Pablo, J. Gómez-Herrero, B. Biel, F. J. Garcia-Vidal, A. Rubio, and F. Flores, "Tuning the conductance of single-walled carbon nanotubes by ion irradiation in the Anderson localization regime," *Nature Materials*, vol. 4, pp. 534–539, 2005.
105. O. Dubay and G. Kresse, "Accurate density functional calculations for the phonon dispersion relations of graphite layer and carbon nanotubes," *Physical Review B*, vol. 67, p. 035401, 2003.
106. Z. Yao, C. L. Kane, and C. Dekker, "High-field electrical transport in single-wall carbon nanotubes," *Physical Review Letters*, vol. 84, pp. 2941–2944, 2000.
107. A. Javey, J. Guo, M. Paulsson, Q. Wang, D. Mann, M. Lundstrom, and H. Dai, "High-field quasiballistic transport in short carbon nanotubes," *Physical Review Letters*, vol. 92, p. 106804, 2004.
108. J.-Y. Park, S. Rosenblatt, Y. Yaish, V. Sazonova, H. Üstünel, S. Braig, T. A. Arias, P. W. Brouwer, and P. L. McEuen, "Electron–phonon scattering in metallic single-walled carbon nanotubes," *Nano Letters*, vol. 4, pp. 517–520, 2004.
109. E. Pop, D. Mann, J. Cao, Q. Wang, K. Goodson, and H. Dai, "Negative differential conductance and hot phonons in suspended nanotube molecular wires," *Physical Review Letters*, vol. 95, p. 155505, 2005.
110. M. Lazzeri, S. Piscanec, F. Mauri, A. C. Ferrari, and J. Robertson, "Electron transport and hot phonons in carbon nanotubes," *Physical Review Letters*, vol. 95, p. 236802, 2005.
111. G. D. Mahan, "Electron-optical phonon interaction in carbon nanotubes," *Physical Review B*, vol. 68, p. 125409, 2003.
112. V. Perebeinos, J. Tersoff, and P. Avouris, "Electron–phonon interaction and transport in semiconducting carbon nanotubes," *Physical Review Letters*, vol. 94, p. 086802, 2005.
113. M. A. Kuroda, A. Cangellaris, and J.-P. Leburton, "Nonlinear transport and heat dissipation in metallic carbon nanotubes," *Physical Review Letters*, vol. 95, p. 266803, 2005.
114. M. Lazzeri and F. Mauri, "Coupled dynamics of electrons and phonons in metallic nanotubes: Current saturation from hot-phonon generation," *Physical Review B*, vol. 73, p. 165419, 2006.
115. H. C. d'Honincthun, S. Galdin-Retailleau, J. Sée, and P. Dollfus, "Electron–phonon scattering and ballistic behavior in semiconducting carbon nanotubes," *Applied Physics Letters*, vol. 87, p. 172112, 2005.
116. J. Guo, "A quantum-mechanical treatment of phonon scattering in carbon nanotube transistors," *Journal of Applied Physics*, vol. 98, p. 063519, 2005.
117. J. Guo and M. Lundstrom, "Role of phonon scattering in carbon nanotube field-effect transistors," *Applied Physics Letters*, vol. 86, p. 193103, 2005.
118. G. Pennington and N. Goldsman, "Semiclassical transport and phonon scattering of electrons in semiconducting carbon nanotubes," *Physical Review B*, vol. 68, p. 045426, 2003.
119. A. Verma, M. Z. Kauser, and P. P. Ruden, "Ensemble Monte Carlo transport simulations for semiconducting carbon nanotubes," *Journal of Applied Physics*, vol. 97, p. 114319, 2005.

120. T. Dürkop, B. M. Kim, and M. S. Fuhrer, "Properties and applications of high-mobility semiconducting nanotubes," *Journal of Physics: Condensed Matter*, vol. 16, pp. R553–R580, 2004.
121. T. Dürkop, S. A. Getty, E. Cobas, and M. S. Fuhrer, "Extraordinary mobility in semiconducting carbon nanotubes," *Nano Letters*, vol. 4, pp. 35–39, 2004.
122. X. Zhou, J.-Y. Park, S. Huang, J. Liu, and P. L. McEuen, "Band structure, phonon scattering, and the performance limit of single-walled carbon nanotube transistors," *Physical Review Letters*, vol. 95, p. 146805, 2005.

Chapter 2
Direct Synthesis and Integration of SWNT Devices

Mario Hofmann, Sreekar Bhaviripudi and Jing Kong

2.1 Introduction

The unique properties of carbon nanotubes (as mentioned in Chapter 1) have attracted enormous attention during the past two decades. Significant progress and understanding have been made in this field, owing to the successful development of effective strategies for the fabrication of CNT-based electronic devices. Nevertheless, many challenges still need to be overcome, such as simple and reliable control of single-walled nanotube (SWNT) diameter, chirality, length, and orientation for the large-scale integration of nanotube devices and circuits.

In this chapter, we will describe the past progress and on-going efforts on the various aspects of SWNT synthesis and integration for electronic devices. There have been two parallel approaches for making SWNTs electronic devices: One is to produce bulk amount of SWNT materials first, and then followed by purification of the material and dispersion into solutions. After that the SWNTs are deposited on the substrate for device fabrications [1]. Significant progress has been made to address several issues involved in this process, such as separating the semiconducting and metallic nanotubes in solutions using various techniques [2–5] and controllable deposition using techniques such as dielectrophoresis [6, 7] or molecular recognition [8, 9]. With this method, since the substrate does not need to experience the high temperature of the SWNT synthesis, nanotube devices can be made on any substrate, such as flexible plastics, for a wide range of applications. However, as the post-growth manipulations (i.e. purification and solution–dispersions, etc.) tend to create damages in SWNT lattices, this method has been most successful with thin film SWNT devices and will be discussed in depth in Chapter 9. High performance SWNT FET devices have been mainly fabricated by the second route, which is to directly synthesis and integrate SWNTs. The second route will be the focus of this chapter.

M.Hofmann (✉)
Department of Electrical Engineering and Computer Science,
Massachusetts Institute of Technology, Cambridge,
MA, 02139, USA

A. Javey, J. Kong (eds.), *Carbon Nanotube Electronics*, Series on Integrated Circuits and Systems, DOI 10.1007/978-0-387-69285-2_2,

The direct synthesis of SWNTs from controlled locations on the surface of the substrate is achieved by depositing catalyst materials at lithographically defined locations on the substrate and performing chemical vapor deposition (CVD) synthesis with these catalyst-containing substrates. To fabricate SWNT FETs, metal electrodes can be put down either before or after the CVD growth of SWNTs. The additional challenge of this approach, as compared with the initially described one, is that the substrate will need to experience the high temperature associated with the CVD process, and thus will give rise to limitations on the applications. As a result, large effort has been made to reduce the synthesis temperatures. This will be discussed in Section 2.4.

The chapter is organized as follows. Section 2.2 gives a brief overview of the CVD method with discussions of several detailed aspects. Section 2.3 explains the various efforts in controlling the synthesis of SWNTs for device integration, such as controlling the diameter, position, and orientation of the nanotubes. Section 2.4 briefly introduces the on-going effort of integrating SWNTs with Si MOS circuits to obtain hybrid devices. Lastly Section 2.5 provides a summary and discussion about the existing challenges.

2.2 CVD Synthesis

2.2.1 The Method

Chemical vapor deposition has been a classical method to produce carbon materials, such as diamond films, carbon fibers [10, 11], filaments [12–14], and nanotubes [15]. For SWNT synthesis, it has been found that in most methods, metal catalyst nanoparticles are essential. Figure 2.1depicted a schematic setup for CVD synthesis with a typical tube furnace.

The process involves heating a catalyst material to high temperatures (~1000°C) and flowing a hydrocarbon gas (or other carbon containing gases, such as CO) through the furnace. The catalyst material contains nanoparticles of metal, and it is understood that the catalyst assists the hydrocarbon gas to decompose and carbon

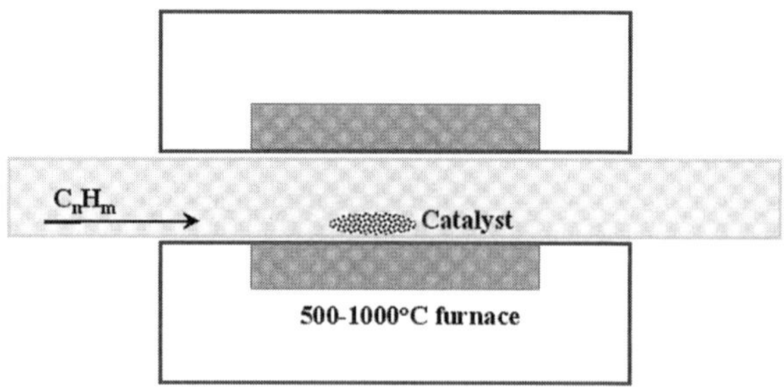

Fig. 2.1 Schematic setup of chemical vapor deposition synthesis

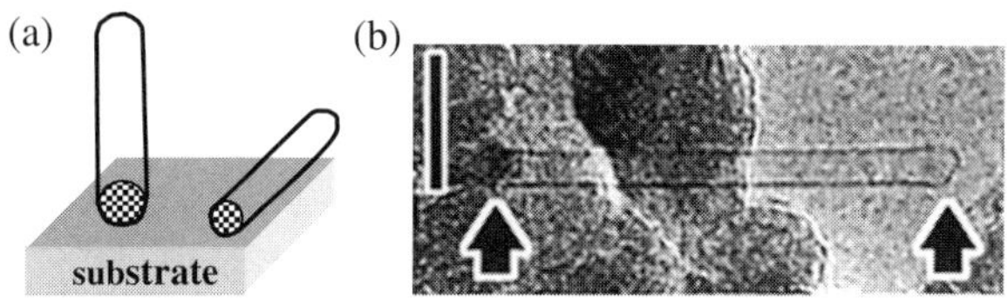

Fig. 2.2 (**a**) A schematic illustration of SWNTs grown from catalyst nanoparticles. (**b**) TEM image of a SWNT grown from a Fe nanoparticle. (Reprinted with permission from [16]. Copyright 2001 American Chemical Society.) Scale bar is 10 nm

will dissolve in the nanoparticles and precipitate out once the carbon-metal solution become supersaturated [12, 15]. The precipitation of carbon from the metal nanoparticles leads to the formation of tubular carbon with sp^2 structure on the sidewall and a hemispherical end cap. This kind of structure is energetically favored since the graphite basal plane is a low-energy form and the end cap also avoids dangling bond of open edges. Figure 2.2(a) is a schematic diagram showing SWNTs protruding from isolated metal nanoparticles on a flat substrate and Fig. 2.2(b) shows a transmission electron microscope (TEM) image of a nanotube growing from an Fe nanoparticle [16].

2.2.2 Direct Incorporation with the Device Fabrication Process

The understanding of the above described nanotube growth mechanism leads to the important conclusion that if the SWNT growth initiates from the catalyst particle, by controlling its position on the substrate, the location of the nanotube can be controlled. Therefore, an effective strategy can be developed for the fabrication of nanotube devices, in which all the steps are compatible with the semiconductor manufacturing processes.

Figure 2.3 is a schematic diagram of an examplary fabrication process flow. The process involves three steps of either electron beam lithography (EBL) or photolithography. Si wafers with thermally grown SiO_2 are typically used as substrates. In the first step, alignment markers are patterned by lithography and lifted-off (Fig. 2.3(a)). In the second lithography step, locations for the catalyst are defined on the substrates by opening up windows in the resist (Fig. 2.3(b)). Then catalyst can be deposited either by drop casting a liquid solution [16, 17] or by depositing multilayer metal catalyst thin films by evaporation or sputtering [18]. After lifting-off the resist, catalyst is deposited only at specific locations on the substrates (Fig.2.3(c)). Nanotubes can then be synthesized by CVD (Fig. 2.3(d)) [19–22]. The last lithography step, following CVD synthesis, is to place electrical contact pads over the nanotubes (Fig. 2.3(e)).

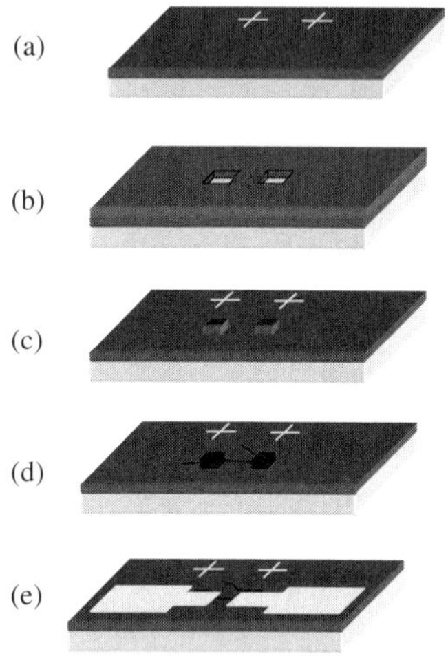

Fig. 2.3 Schematic of an example fabrication process. (**a**) Patterning of alignment markers on the substrate; (**b**) patterning of catalyst windows in resist; (**c**) catalyst deposition followed by resist lift-off; (**d**) CVD synthesis; (**e**) final device with metallic electrode

2.2.3 SWNT Synthesis on Metal Electrodes

By controlling the catalyst location via lithographical patterning, SWNTs can be directly grown on various substrates. A particular example is substrates that contain trenches, so that suspended nanotubes devices can be fabricated. There has been a lot of interest in studying these suspended nanotubes, since their intrinsic properties will not be perturbed by substrate interactions. The growth of SWNTs across trenches is straightforward; however, if the trench is wider than $\sim$1 μm, SWNTs can hardly survive the fabrication steps afterwards, such as resist spin-coating or solvent-drying at the end of the lift-off [23, 24]. The suspended nanotubes are either dragged to the bottom of the trench by the capillary force upon solvent drying or are broken during the spin-coating step. One alternative solution to overcome this problem is to reverse the procedure in Fig. 2.3(d) and (e), i.e., to directly grow SWNTs on the metal electrodes.

Synthesizing SWNTs on metal electrodes is more challenging than growth on insulating substrates due to the following reasons: (1) Most of the metal (particularly transition metal) surfaces play an active role in absorbing hydrocarbon gas molecules and are assisting the dissociation of these molecules. Therefore, the optimal CVD condition for SWNT synthesis is quite different from that without the presence of these metal electrodes. (2) Certain commonly used metals, such as gold, silver and copper, have a relatively low melting point ($\sim$1000°C); at the high temperatures during CVD synthesis these metal films break up and form small grains and lose their conductivity. (3) Since the CVD environment usually has an abundance of H_2, some metals form volatile metal hydrides at high temperatures, and become partially etched and highly resistive after synthesis [24]. Therefore, only a limited number of metal materials are compatible with the nanotube CVD process. In addition, a large variety of metals form oxides on their surface in air and thus

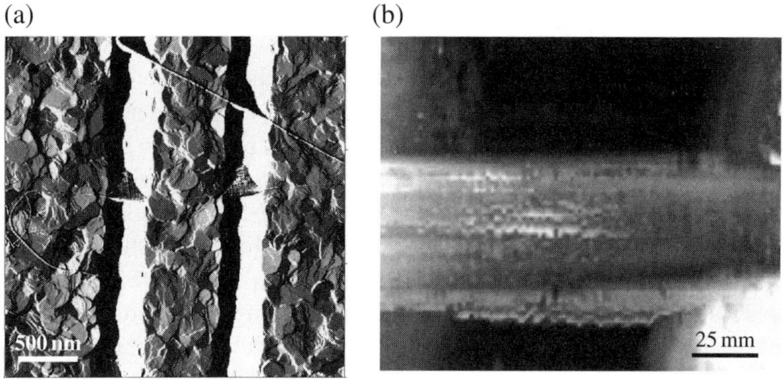

Fig. 2.4 (**a**) Atomic force microscope (AFM) image of SWNTs directly grown on Pt substrate with trenches. (**b**) STM image of a SWNT across the trench. Figure reprinted from [27]

can not be used. Nevertheless, through numerous research efforts these difficulties have been overcome. It was found that Molybdenum (Mo) is a high melting point metal (2617°C) which is also compatible with the CVD synthesis conditions. A reducing environment is required (this can be ensured by flowing H_2 during the heating-up and cooling-down stages) to prevent the oxidation of Mo. After the CVD, care needs to be taken to avoid device degradation in air. SWNT devices with resistances in the range of tens to hundreds of kΩ can be routinely fabricated. Apart from Mo, Platinum (Pt) has also been identified as a good electrode material suitable for nanotube growth [25, 26]. It has a melting point of 1772°C, high enough to withstand the temperatures of CVD. The reduction environment of the process gives rise to a surface with atomically flat grains as shown in Fig. 2.4(a). This enables the scanning tunneling microscope (STM) imaging of suspended SWNTs directly grown on Pt substrates (Fig. 2.4(b)). Phonon-assisted tunneling was observed from the scanning tunneling spectroscopy (STS) analysis of these samples [27]. More recently Au has been successfully used as an electrode material as well [28]. In order to prevent the Au electrodes from forming disconnected grains, thick films were used (~200 nm) with a CVD temperature of ~800°C.

2.2.4 Lowering the Synthesis Temperature

The high temperature required by CVD limits the applications, especially the choice of the substrate materials. Efforts have been made to reduce the temperature in order to make the nanotube fabrication compatible with more substrates and applications. This has been carried out by two different approaches. Since, in CVD, the high temperature is mostly used for decomposing the carbon precursors, one approach uses additional means to assist the hydrocarbon gas decomposition; thus the substrates can be held at a relatively lower temperature. One example is hot filament

CVD (HFCVD) where a hot metal wire (usually tungsten, $>1000^\circ$C) is put above the substrate or at an upstream position of the gas flow so that the hydrocarbon gas molecules can be dissociated first by the hot filament before reaching the substrate. SWNTs have been synthesized with HFCVD using C_2H_2 source at a substrate temperature of 590°C [29]. Plasma-enhanced CVD (PECVD) has also been used for this purpose, where the hydrocarbon molecules are decomposed by a plasma source first, so that the substrate temperature can be lowered. SWNTs have been synthesized with PECVD at temperatures as low as 450°C [30]. More interestingly, it has been found that SWNTs synthesized by PECVD with CH_4 are ~90% semiconducting [31]. This is in sharp contrast to the 67% semiconducting population as observed by normal CVD growth. The latter is based on a 1:2 metal:semiconducting ratio when assuming that a uniform chirality distribution occurs in the growth of SWNTs.

The second approach resorts to nanofabrication techniques to build local resistive heaters on the substrate. Both SWNTs and MWNTs have been made this way [32–34]. Localized resistive heating at the catalyst sites allows controlled growth of nanotubes on metal electrodes. It was reported in [34] that the surrounding silicon chip reaches an average temperature of 60°C. Apart from the highly local control of the CNT growth, one additional advantage of this approach is that for normal CVD synthesis where both the whole substrate and the growth chamber heat up to 900°C, the processing time is usually on the order of one hour per synthesis. Most of the time is attributed to the heating and cooling of the synthesis chamber. In contrast, the nanotube growth using micro-fabricated heaters occurs in a room-temperature chamber and is significantly faster due to the much reduced thermal mass. This method has great potential for future integration of SWNT devices with CMOS circuits where the thermal budget is ~550°C. However, at present the quality of SWNTs grown by these approaches [34] still needs to be improved significantly.

2.3 Controlling the SWNT Growth

In order to achieve deterministic integration of SWNT devices, the ultimate goal in controlling nanotube synthesis includes defining the locations as well as orientations of SWNTs with nanometer accuracy, and their atomic structures such as diameter and chirality. The following subsections discuss the research efforts addressing these issues.

2.3.1 Location

For the ultimate large-scale integration in semiconductor industry, the locations of the SWNTs need to be accurately defined. Thus, it is highly desirable to have one SWNT from each catalytic site. This is challenging with current lithographical technology, since the finest resolution that can be achieved is ~10–20 nm which will still give rise to multiple nanoparticles (1–2 nm in size) at each site.

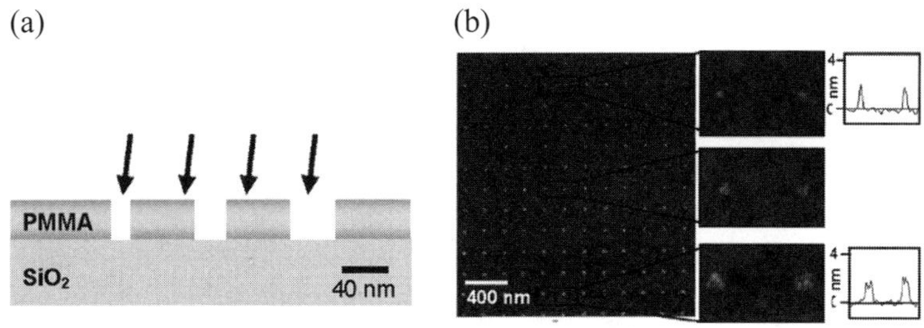

Fig. 2.5 (**a**) Schematic illustration of metal deposition into 20–50 nm holes in PMMA through angle evaporation. (**b**) AFM image of rows of single (*top right*) and multiple (*bottom right*) ~2.2 nm Co particles and topographic line scans. Reprinted with permission from [39]. Copyright 2005 American Chemical Society

The following sections show several example schemes to address this issue. One method [35] utilizes angle evaporation in combination with the high-resolution EBL patterning. As shown in Fig. 2.5(a), wells with a diameter d ~20 nm were first patterned by EBL on a polymethylmethacrylate (PMMA, 100 nm thick)-coated SiO_2 substrate. Thin films (nominally 2–20 Å) of metal (Co, Fe, Pt, etc.) were then evaporated at an angle (5–10°) with respect to the substrate normal. After PMMA lift-off and thermal annealing (700–900°C), discrete clusters were formed in an arrayed fashion (Fig. 2.5(b)). The number of particles per site depended on the size of the patterned-PMMA wells and the angle of evaporation (Fig. 2.5), and the diameter of the particles was controlled by the thickness of metal deposited. This approach affords arrays of various metal clusters with tunable diameters from tens of nanometers down to ~1–2 nm. CVD synthesis conditions were then tuned to have the majority of the nanoparticles each produce a SWNT. One more advantage of this method is that the catalyst particles are similar in size, which gives rise to a narrower SWNT diameter distribution.

Another method [36] mixed organo-metallic compounds into negative resist which was then patterned and annealed to form nanoparticles with diameters significantly smaller than the lithography limit. Figure 2.6 illustrates the process. First a high-resolution negative EBL resist consisting of only C, O, and H (p-methyl methylacetoxy calix[6]-arene, MC6) is mixed with Fe(III) acetylacetonate to obtain Fe-doped EBL resist. Exposure conditions were tuned to obtain resist dots with a typical diameter of 20 nm and a height of 20 nm. The substrate was then loaded in an evacuated chamber and annealed at 650–800°C for 5–30 min. The doped resist was transformed into amorphous carbon with metal compound particles inside. Oxygen plasma or heating in oxygen will remove the amorphous carbon and leave only Fe oxide particles in the originally defined locations. Particle size of 1.7 ± 0.6 nm was demonstrated, and it is obvious that by varying the doping concentration and dopant species, different metal particles of tunable sizes can be achieved; thus the diameter of SWNTs can be tuned in this way.

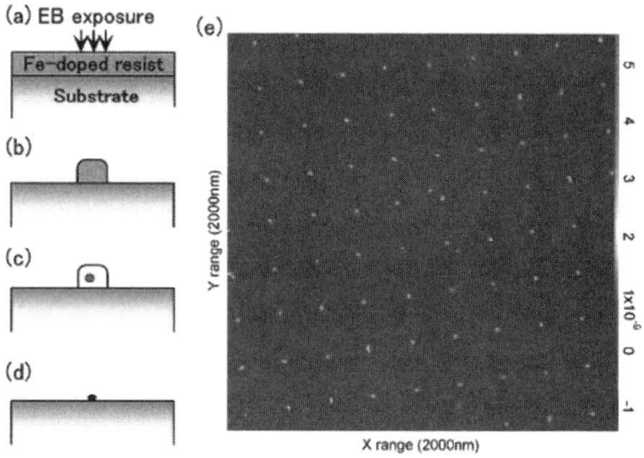

Fig. 2.6 (**a**)–(**d**) Schematic illustration of nanoparticle fabrication process. (**e**) AFM image of Fe nanoparticle array patterned at 200 nm pitch. Reprinted with permission from [40]

2.3.2 Orientation

During the past few years, significant progress has been made in controlling the orientation of SWNTs during the CVD synthesis. Three different approaches have been demonstrated, each working using different principles. Correspondingly, they have different requirements for the synthesis setup, or substrate structure.

The first one is the SWNT alignment through an in-situ electric field [16]. Since the polarizability along the tube axis $\alpha_{//}$ is much higher than that perpendicular to the tube axis $\alpha_{\perp}$ [16], in an electric field the torque acting on the nanotube dipole aligns it to the direction of the field. It was found that a field of 0.5–2 V/μm was large enough to overcome the thermal agitation at the CVD growth temperature and gas flow effect. Figure 2.7(a) shows arrays of SWNTs aligned with the field direction, in contrast to the randomly oriented nanotubes without an applied field (Fig. 2.7(b)).

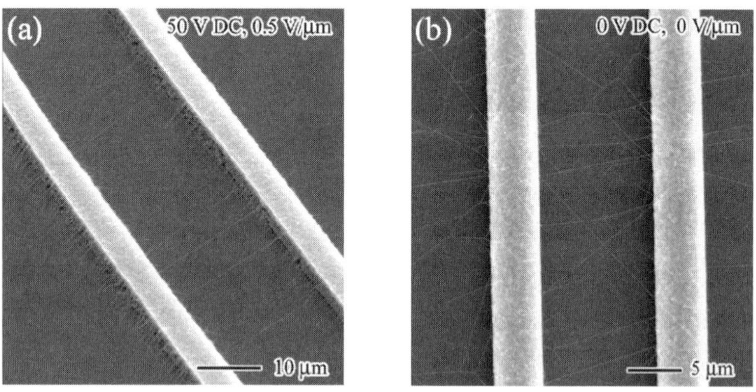

Fig. 2.7 (**a**) SEM image of suspended SWNTs aligned with the electric field direction in between two poly-Si electrodes. (**b**) Without the electric field, SWNTs grown in random orientations. Reused with permission from [41]. Copyright 2001, American Institute of Physics

SWNTs that are well aligned along the electric-field direction are clearly observed. This method requires fabrication of closely spaced (~100 μm) and large enough electrodes in order to generate the required strength of electric field. In addition, synthesis parameters need to be adjusted to avoid arcing between the electrodes at high temperature in the CVD gas environment.

The second approach aligns the nanotubes with the gas flow during the synthesis. This works particularly well with the ultralong SWNTs (millimeter up to centimeter) (Fig. 2.8(a)). It was found that there is a critical length that flow alignment can take effect [37]. This approach relies on the fact that nanotubes need to float above the substrate during the growth. It was found that the nanotube is lifted along their whole length by a thermal buoyancy induced gas motion ("Kite mechanism") [38, 39]. However with this method, at present the density of nanotubes are relatively low (~20 μm spacing) and the straightness of the nanotubes are not well under control. There can be many curly features along the nanotube locally (Fig. 2.8(c)), which are not fully understood at this stage. The ultimate reliable and controlled alignment of SWNT during growth requires the unraveling of the mechanism in this process. Nevertheless, this method is the easiest to implement and is suitable for any substrate. SWNT architectures with cross-bar geometry can also be achieved by growing two consecutive steps, which have many potential applications.

More recently, substrate-directed growth has been reported by several groups [40–42]. The alignment was observed for SWNTs grown on single crystalline quartz and sapphire substrates with certain particular crystal orientation. It is proposed that the step edges, resulting from the miscut of the substrates, are responsible for the alignment of SWNTs during growth. This result has led to the fabrication of well-aligned, high-coverage arrays of SWNTs in well-defined geometries for thin-film electronic devices. This will be discussed at length in Chapter 9.

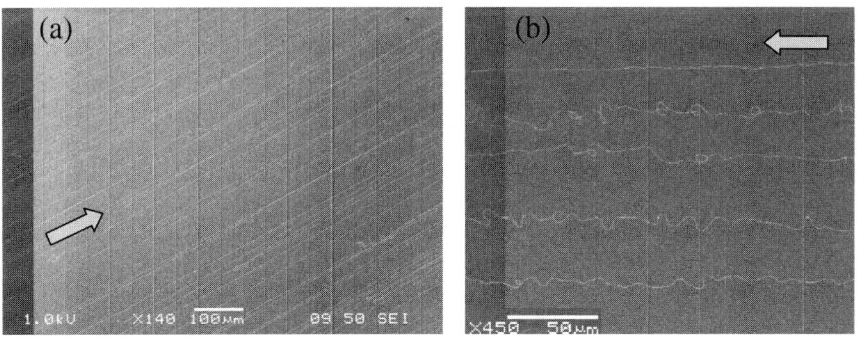

Fig. 2.8 (**a**) SEM image of ultralong SWNTs synthesized and aligned with the gas flow direction. (**b**) A higher magnification SEM image of the ultralong nanotubes indicating the wiggles along the nanotube. The green arrows in the figures indicate the gas flow direction

2.3.3 *Chirality*

As mentioned in Chapter 1, the electronic properties of SWNTs depend very sensitively on its structure (n, m chirality). Therefore, in order to incorporate SWNTs into large-scale integrated circuits, it is critical to obtain SWNTs of identical structures (or at least, very similar electrical properties). However, in most CVD synthesis, a large variety of nanotube structures are produced. By far, the most challenging topic in carbon nanotube research is to obtain SWNTs with a uniform chiral structure.

2.3.3.1 Narrowing the Diameter Distributions

Since the diameter of a nanotube is determined by its (n,m) chirality, and for a semiconducting nanotube, its bandgap is inversely proportional to its diameter [43], as a first step, a large amount of efforts have been devoted to the synthesis of nanotubes with narrower diameter distribution.

The strategy for diameter-controlled synthesis came from the understanding of the nanotube growth mechanism. It is shown [16, 44] that SWNTs originate from metal nanoparticles and the diameters of the nanoparticles dictate the diameters of the nanotube (Fig. 2.2(b)). Therefore, the goal of controlling the diameter of the SWNTs is translated into narrowing down the diameter distribution of the nanoparticles, which has been a heavily investigated topic for the nanoparticle/nanocrystal research field. Classical solution-based methods utilize specific protecting ligands and precise control of the reaction temperature during the particle precipitation to obtain a narrow distribution of nanoparticles [45]. Another route involves the use of a protein called ferritin, which has a spherical shell and can store up to 4500 iron atoms in the form of hydrous ferric oxide. The core of ferritin can be emptied to afford apoferritin, and subsequently a controllable number of metal ions, such as Fe or Co, can be placed into the core to synthesize nanoparticles with narrow size distributions. Nanoparticles with diameters in the range of 1–2 nm or 3–5 nm were obtained using ferritin [16]. More recently, block copolymer micelle method has been used to derive nanoparticles with narrower size distributions for SWNT synthesis [46, 47]. This method offers a very facile route for making nanoparticles of different kind, such as Fe, Co, Ni, and Au. In addition, since the block copolymer micelles self-assemble into regular arrays as they are deposited on a substrate (Fig. 2.9), well-organized patterns of nanoparticles can be obtained [48]. Thus the diameters of the SWNTs can be tuned by the sizes of these nanoparticles [47].

For the catalyst nanoparticles made by the aforementioned methods, even though they have a narrower distribution, after the CVD synthesis, the results are still a mixture of many different SWNTs. In fact, even from identical nanoparticles there are still several challenging issues that need to be tackled in order to realize the synthesis of SWNTs with the same structures. One is the aggregation of the nanoparticles on substrate. A previous investigation utilized identical molecular nanoclusters based on $[\mathrm{H}x\mathrm{PMo}_{12}\mathrm{O}_{40}\subset\mathrm{H4Mo^{VI}}_{72}\mathrm{Fe^{III}}_{30}(\mathrm{CH_3COO})_{15}\mathrm{O}_{254}(\mathrm{H_2O})_{98}]\cdot 60\mathrm{H_2O}$ as starting catalyst cores for the SWNT growth [49]. It was found that these

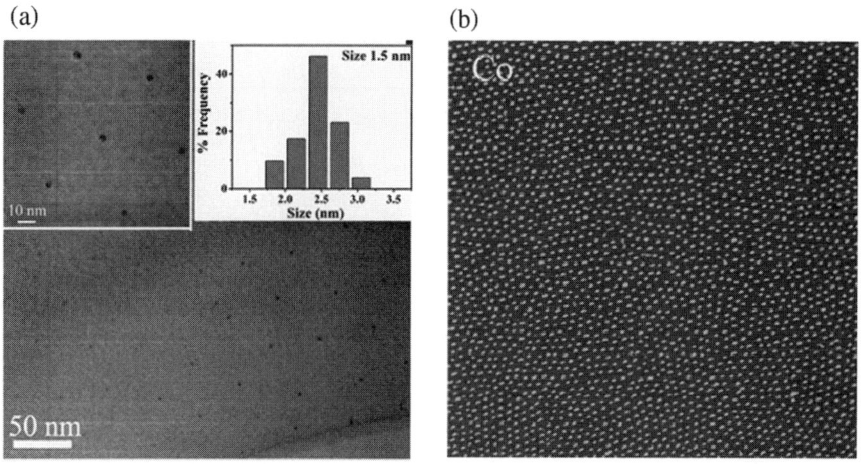

Fig. 2.9 (**a**) TEM images of Fe nanoparticles made by block copolymer micellar method. The upper right inset shows the diameter distribution. Reused with permission from [48]. Copyright IOP 2007. (**b**) AFM images of Co nanoparticles made by this approach. The area is 2.5 μm × 2.5 μm

nanoclusters tend to aggregate when they are deposited from solution onto the substrates; as a result, the nanotubes grown from these molecular nanocluster catalysts are not identical but still have a range of distribution in diameter. Another challenging issue is that during the pre-growth step, the nanoparticles could change their size due to various reasons, such as evaporation at high temperature, which again result in a spread of the diameter distribution. Finally, assuming the diameters of the catalyst nanoparticles can be well controlled, another question needs to be addressed as well, which is: do the synthesis conditions have any preference in determining the chirality of the nanotubes? The investigations into this issue have led to a series of chirality distribution analyses which will be discussed in the next section.

2.3.3.2 Chirality Distribution Analysis for Different CVD Processes

Before the chirality distribution analysis was possible for bulk amount SWNT materials, studies regarding the chirality preference of a specific synthesis process were derived based on the percentage of metallic and semiconducting nanotubes in the synthesized SWNTs. If there is a uniform (n,m) chirality distribution, from the "2n+m mod 3" rule as mentioned in Chapter 1, two-thirds (i.e., 67%) of the SWNTs should be semiconducting and one-third (i.e., 33%) of the SWNTs should be metallic (including "quasi"-metallic ones with a bandgap on the order of ~10 meV). In Ref. [31], electrical measurements were carried out on a few hundred SWNT devices made by different synthesis conditions for such a statistical analysis. The findings are summarized in Table 2.1, in which the percentage of the semiconducting and metallic SWNTs for different synthesis conditions are listed [31, 50]. It can

Table 2.1 Percentage of semiconducting and metallic SWNTs

Synthesis process	Catalyst	Semiconducting (%)
Normal thermal CVD with CH_4	Ferritin	62.5
Plasma-enhanced CVD with CH_4	Ferritin	89.3 ±2.3
HiPco with CO gas	$Fe(CO)_5$ or Ferrocene	61.0 ±7.6
Laser ablation	Ni/Co	30.0 ±6.0

be seen that the percentage of semiconducting nanotubes by certain synthesis conditions deviate far from 67%. This indicates that those synthesis conditions must have preferences on the chirality distributions. It should be noted that the plasma-enhanced CVD (PECVD) and the normal CVD in Table 2.1 use the same type of Ferritin nanoparticle catalyst [31, 50], but due to the differences in synthesis conditions, different chirality distributions (thus different metallicity types) have been obtained.

The efficient analysis of the chirality distribution for bulk amount of SWNTs from a particular synthesis process was enabled by the successful development of two types of optical spectroscopy techniques: resonant Raman spectroscopy (RRS) [51, 52] and Photoluminescence (PL) Excitation (PLE) technique [53, 54]. For RRS characterizations, intensity maps of the radial breathing mode (RBM) in SWNT Raman spectra are plotted as a function of the laser excitations (Fig. 2.10(a)). Since the different RBM peaks in the Raman spectra correspond to different (n,m) chirality nanotubes, from these intensity maps the chirality of the SWNTs within a bulk material can be identified. For quantitative analysis, theoretical calculations of the RRS cross section for each (n,m) species are needed, so that the intensities of the Raman peaks can be converted to the percentage information of that particular species [55]. In the PLE characterization experiments, SWNTs need to be isolated and dispersed in solutions, so that the quenching of the semiconducting SWNT PL signals by the metallic SWNTs can be prevented. Similar to the RRS characterization, in the PLE experiments intensity maps of PL peaks at each excitation wavelengths are obtained. These characteristic PLE peaks can be used to recognize the chirality of the SWNTs (Fig. 2.10(b)). Theoretical calculations are also required in order to use the intensity of the PLE peaks to derive the percentage of a particular (n,m) species. Due to the instrumental limitations, RRS and PLE can only be used to characterize SWNTs with diameters <2 nm, and for PLE characterization, only semiconducting SWNTs can be analyzed. Nevertheless, these two techniques have provided vast amount of information about the SWNT species and their synthesis process.

Studies using the PLE mapping have reported that different CVD processes give rise to different chirality distributions (the SWNTs were synthesized by the HiPco process in Fig. 2.10(b) and by alcohol CVD in Fig. 2.10(c)). This is consistent with the previous electrical characterization study [31]; furthermore, the specific species are being identified. A specific CVD synthesis process, called the CoMoCAT

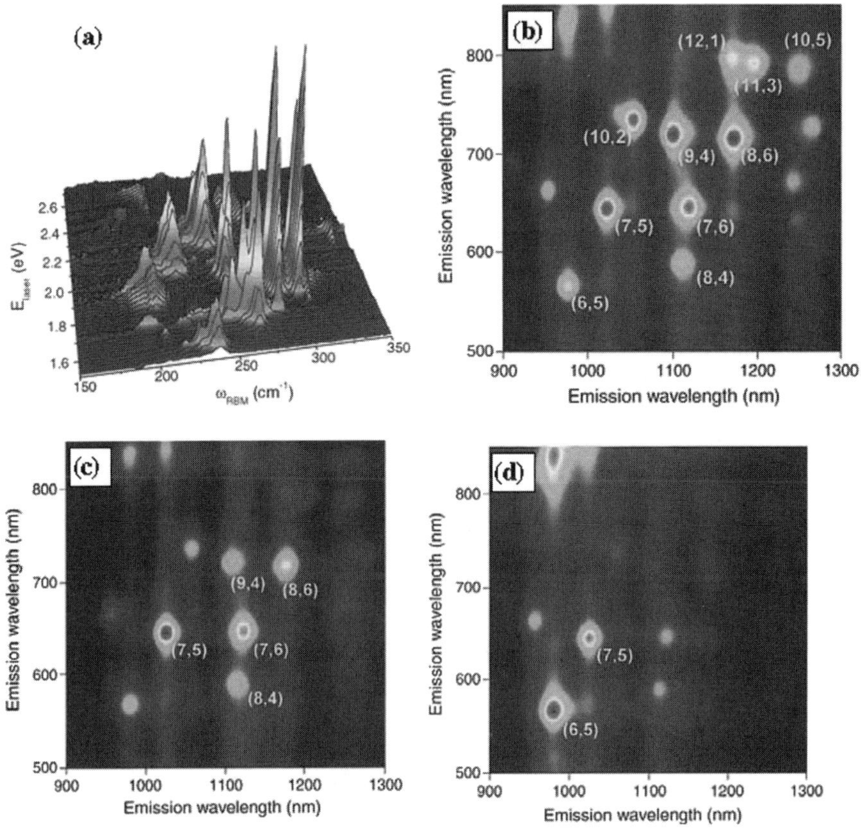

Fig. 2.10 (**a**) RRS intensity of SWNTs dispersed in solution as a function of resonant laser excitation (E_{laser}) and RBM frequency (ω_{RBM}). Reused with permission from [52]. Copyright 2004, American Institute of Physics. (**b**) PLE characterization of HiPco SWNTs. *Blue* corresponds to low intensity and *red* to high intensity. The identified (n,m) chiralities are labeled next to the PLE peaks. (**c**) PLE characterization of SWNTs grown by alcohol CVD at 850°C. (**d**) PLE characterization of SWNTs grown by the same alcohol CVD as in (c) but at 650°C. (b)–(d) are reused from [56]. Copyright 2004. With permission from Elsevier

process, was found to yield dominantly (6,5) and (7,5) species [53]. It was also found that by changing the synthesis conditions (such as temperature, feeding gas compositions, or type of catalyst support), the chirality distribution profile can be tailored [56, 57] (Fig. 2.10(c) and (d)). This indicates the possibility of chirality-selective synthesis by fine-tuning the CVD conditions in the future.

2.3.3.3 Selective Removal of the Metallic Nanotubes in FET Devices

In order to construct SWNT FETs with only semiconducting nanotubes, alternative efforts have also been made, for the past decade, to eliminate the metallic nanotubes while keeping the semiconducting ones intact a SWNT FET. The most straightforward method has been the selective electrical breakdown [31, 58] to remove the metallic SWNTs that are in parallel with the semiconducting ones. This is based on the fact that the semiconducting SWNTs can be turned off with a gate voltage, whereas at the same time the metallic nanotubes will still conduct. Therefore, if a large bias voltage is applied, the current will mainly go through the metallic nanotubes and the large current will heat up the SWNTs due to optical phonon scattering [59]. As a result, the metallic nanotubes will be burnt in air and the semiconducting nanotubes will be retained. However, in a SWNT network where the metallic and semiconducting SWNTs are physically contacted with each other, very often the semiconducting SWNTs are destroyed as well. In addition, this method is time consuming and is not suitable for large-scale device fabrication. A batch process is more preferred to remove only metallic SWNTs. It has been found that diazonium functionalization [4, 60] is preferential towards metallic SWNTs due to the difference in the availability of electrons near the Fermi level for metallic and semiconducting SWNTs. Metallic nanotubes can be effectively removed in SWNT network devices. This method requires very careful control on the concentration of the diazonium salt in the solution, because when the diazonium salt concentration is high, semiconducting SWNTs tend to be damaged as well. In addition, since the availability of electrons near the Fermi level depends on the doping level of the semiconducting SWNTs, which is in turn influenced by the previous treatment in the fabrication processes and the bandgap (thus the diameter) of each individual nanotube, a well-controlled selective removal using this method is challenging.

More recently, a methane plasma etching followed by an annealing process has been developed to selectively eliminate metallic SWNTs and retain semiconducting SWNTs that are grown on substrates. Etching of SWNTs in the methane plasma is attributed to hydrocarbonation, with SWNTs irreversibly etched [61] into hydrocarbon gas species as a result of reactions with neutral and positive ions of H and CH_3 species in a methane plasma [62]. It was found that the diameter is an important factor for the chemical reactivity of a SWNT toward hydrocarbonation. Smaller-diameter SWNTs are preferentially etched over larger ones because of the higher radius of curvature and higher strain in the C–C bonding configuration, as in other chemical reactions [63]. When the diameter of the SWNTs are in the range of 1.4–2 nm, a preferential etching of metallic over semiconducting ones were observed, which is consistent with first-principles calculations [64, 65] that the formation energies of same-diameter seminconducting SWNTs are lower than the metallic SWNTs because of the electronic energy gain resulting from the band gap opening [64] and the higher chemical reactivity of metallic SWNTs due to more abundant delocalized electronic states [65]. In this diameter range, although covalent functionalization to the semiconducting SWNTs also occurs under the plasma condition, an annealing step afterwards has been shown to be very effective in resuming the original SWNT structures. The retained SWNTs exhibit electrical properties similar to pristine materials. The dual effects of selective metal removal and diameter distribution narrowing combined with compatibility with microfabrication technology make the method promising for large-scale SWNT electronics.

2.4 Integration

Albeit the enormous effort of nanotube research up to date, the development of carbon nanotube electronics is still in its infancy compared with the Si technology. Nevertheless, as nanotube synthesis and fabrication procedures are compatible with semiconductor process technology, a hybrid technology which combines the two can be anticipated, where the advantages of both materials could be utilized. One example envisioned is a SWNT-based memory with a read-out circuitry based on silicon MOSFETs [66].

As an initial attempt toward this direction, a random access SWNT test chip was designed and fabricated which integrates the CVD grown SWNTs with an n-channel metal oxide semiconductor (NMOS) circuitry fabricated in a modified silicon integrated circuit process [67]. The chip is a simple switching network consisting of NMOS transistors, through which approximately 2003-terminal SWNT devices, obtained via growth from equally many catalyst sites, can be electrically accessed on an individual basis by using only 22 binary inputs. In this way, 2000 SWNT devices can be characterized directly, yielding a large sample size that enables statistical analysis of the device performance and/or the synthesis process. This work paves the way for the future integration of CNT devices within CMOS circuit.

2.5 Summary

In this chapter, various aspects of SWNT synthesis and device fabrication are discussed. Controlling nanotube synthesis with various strategies during CVD has enabled the integration of nanotube devices for fundamental characterization and potential applications. However, considerable challenges remain for the ultimate control of SWNTs with specific type/chirality being assembled at desired sites and orientations. In addition to it, there is a lack of detailed understanding of growth mechanism of carbon nanotubes [68]. Nevertheless, it can be expected that continuing progress in effective characterization techniques [54, 69, 70] will assist future understandings of the growth mechanism, and as a result new synthesis strategies may be developed upon this progress. If these efforts turn out to be successful in leading to the ultimate control in the nanotube synthesis and integration, revolutionary opportunities can be anticipated for the future of nanotube electronics.

Acknowledgments Most part of the work presented in this chapter was based on Dr. Jing Kong's thesis with Professor Hongjie Dai at Stanford University. We deeply appreciate the valuable advices and contribution from Prof. Dai. The authors would also like to thank Prof. Angela Belcher at MIT for her support of this work. This work was funded in part by the MARCO IFC Focus Research Center Program.

References

1. S. J. Tans, M. H. Devoret, H. Dai, A. Thess, R. E. Smalley, L. J. Geerligs, and C. Dekker, "Individual single-wall carbon nanotubes as quantum wires," *Nature*, vol. 386, pp. 474–477, 1997.
2. M. S. Arnold, A. A. Green, J. F. Hulvat, S. I. Stupp, and M. C. Hersam, "Sorting carbon nanotubes by electronic structure using density differentiation," *Nature Nanotechnology*, vol. 1, pp. 60–65, 2006.
3. M. Zheng, A. Jagota, M. S. Strano, A. P. Santos, P. Barone, S. G. Chou, B. A. Diner, M. S. Dresselhaus, R. S. McLean, G. B. Onoa, G. G. Samsonidze, E. D. Semke, M. Usrey, and D. J. Walls, "Structure-based carbon nanotube sorting by sequence-dependent DNA assembly," *Science*, vol. 302, pp. 1545–1548, 2003.
4. M. S. Strano, C. A. Dyke, M. L. Usrey, P. W. Barone, M. J. Allen, H. W. Shan,C.Kittrell, R. H. Hauge, J. M. Tour, and R. E. Smalley, "Electronic structure control of single-walled carbon nanotube functionalization," *Science*, vol. 301, pp. 1519–1522, 2003.
5. D. Chattopadhyay, L. Galeska, and F. Papadimitrakopoulos, "A route for bulk separation of semiconducting from metallic single-wall carbon nanotubes," *Journal of the American Chemical Society*, vol. 125, pp. 3370–3375,2003.
6. R. Krupke, F. Hennrich, H. von Lohneysen, and M. M. Kappes, "Separation of metallic from semiconducting single-walled carbon nanotubes," *Science*, vol. 301, pp. 344–347, 2003.
7. A. Vijayaraghavan, S. Blatt, D. Weissenberger, M. Oron-Carl, F. Hennrich, D. Gerthsen, H. Hahn, and R. Krupke, "Ultra-large-scale directed assembly of single-walled carbon nanotube devices," *Nano Letters*, vol. 7, pp. 1556–1560, 2007.
8. K. Keren, R. S. Berman, E. Buchstab, U. Sivan, and E. Braun, "DNA-templated carbon nanotube field-effect transistor," *Science*, vol. 302, pp. 1380–1382, 2003.
9. Y. H. Wang, D. Maspoch, S. L. Zou, G. C. Schatz, R. E. Smalley, and C. A. Mirkin, "Controlling the shape, orientation, and linkage of carbon nanotube features with nano affinity templates," *Proceedings of the National Academy of Sciences of the United States of America*, vol. 103, pp. 2026–2031, 2006.
10. G. G. Tibbetts, "Vapor-grown carbon fibers: Status and prospects," *Carbon*, vol. 27, pp. 745–747, 1989.
11. G. G. Tibbetts, "Lengths of carbon fibers grown from iron catalyst particles in natural gas," *Journal of Crystal Growth*, vol. 73, pp. 431–438, 1985.
12. G. G. Tibbetts, "Why are carbon filaments tubular?" *Journal of Crystal Growth*, vol. 66, pp. 632–638, 1984.
13. G. G. Tibbetts, M. G. Devour, and E. J. Rodda, "An adsorption–diffusion isotherm and its application to the growth of carbon filaments on iron catalyst particles," *Carbon*, vol. 25, pp. 367–375, 1987.
14. R. T. K. Baker, "Catalytic growth of carbon filaments," *Carbon*, vol. 27, pp. 315–323, 1989.
15. R. T. K. Baker and N. M. Rodriguez, "Catalytic Growth of Carbon Nanofibers and Nanotubes," presented at Symposium of the Materials Research Society, 1994.
16. Y. M. Li, W. Kim, Y. G. Zhang, M. Rolandi, D. W. Wang, and H. J. Dai, "Growth of single-walled carbon nanotubes from discrete catalytic nanoparticles of various sizes," *Journal of physical chemistry B*, vol. 105, pp. 11424–11431, 2001.
17. J. Kong, A. M. Cassell, and H. Dai, "Chemical vapor deposition of methane for single-walled carbon nanotubes," *Chemical Physics Letters*, vol. 292, pp. 567–574, 1998.
18. L. Delzeit, B. Chen, A. Cassell, R. Stevens, C. Nguyen, and M. Meyyappan, "Multilayered metal catalysts for controlling the density of single-walled carbon nanotube growth," *Chemical Physics Letters*, vol. 348, pp. 368–374, 2001.
19. J. Kong, H. Soh, A. Cassell, C. F. Quate, and H. Dai, "Synthesis of individual single-walled carbon nanotubes on patterned silicon wafers," *Nature*, vol. 395, p. 878, 1998.

20. J. Y. Park, S. Rosenblatt, Y. Yaish, V. Sazonova, H. Ustunel, S. Braig, T. A. Arias, P. W. Brouwer, and P. L. McEuen, "Electron–phonon scattering in metallic single-walled carbon nanotubes," *Nano Letters*, vol. 4, pp. 517–520, 2004.
21. B. Zheng, Y. Li, and J. Liu, "CVD synthesis and purification of single-walled carbon nanotubes on aerogel-supported catalyst," *Applied Physics a-Materials Science & Processing*, vol. 74, pp. 345–348, 2002.
22. Y. Murakami, Y. Miyauchi, S. Chiashi, and S. Maruyama, "Characterization of single-walled carbon nanotubes catalytically synthesized from alcohol," *Chemical Physics Letters*, vol. 374, pp. 53–58, 2003.
23. T. Tombler, C. Zhou, L. Alexeyev, J. Kong, H. Dai, L. Liu, C. Jayanthi, M. Tang, and S. Y. Wu, "Reversible nanotube electro-mechanical characteristics under local probe manipulation," *Nature*, vol. 405, pp. 769–772, 2000.
24. N. R. Franklin, Q. Wang, T. W. Tombler, A. Javey, M. Shim, and H. J. Dai, "Integration of suspended carbon nanotube arrays into electronic devices and electromechanical systems," *Applied Physics Letters*, vol. 81, pp. 913–915, 2002.
25. J. Kong, B. J. LeRoy, S. G. Lemay, and C. Dekker, "Integration of a gate electrode into carbon nanotube devices for scanning tunneling microscopy," *Applied Physics Letters*, vol. 86, p. 112106, 2005.
26. H. B. Peng, T. G. Ristroph, G. M. Schurmann, G. M. King, J. Yoon, V. Narayanamurti, and J. A. Golovchenko, "Patterned growth of single-walled carbon nanotube arrays from a vapor-deposited Fe catalyst," *Applied Physics Letters*, vol. 83, pp. 4238–4240, 2003.
27. B. J. LeRoy, S. G. Lemay, J. Kong, and C. Dekker, "Electrical generation and absorption of phonons in carbon nanotubes," *Nature*, vol. 432, pp. 371–374, 2004.
28. S. D. Li, Z. Yu, C. Rutherglen, and P. J. Burke, "Electrical properties of 0.4 cm long single-walled carbon nanotubes," *Nano Letters*, vol. 4, pp. 2003–2007, 2004.
29. D. Kondo, S. Sato, and Y. Awano, "Low-temperature synthesis of single-walled carbon nanotubes with a narrow diameter distribution using size-classified catalyst nanoparticles," *Chemical Physics Letters*, vol. 422, pp. 481–487, 2006.
30. Y. S. Min, E. J. Bae, B. S. Oh, D. Kang, and W. Park, "Low-temperature growth of single-walled carbon nanotubes by water plasma chemical vapor deposition," *Journal of the American Chemical Society*, vol. 127, pp. 12498–12499, 2005.
31. Y. M. Li, D. Mann, M. Rolandi, W. Kim, A. Ural, S. Hung, A. Javey, J. Cao, D. W. Wang, E. Yenilmez, Q. Wang, J. F. Gibbons, Y. Nishi, and H. J. Dai, "Preferential growth of semiconducting single-walled carbon nanotubes by a plasma enhanced CVD method," *Nano Letters*, vol. 4, pp. 317–321, 2004.
32. E. O. Sunden, T. L. Wright, J. Lee, W. P. King, and S. Graham, "Room-temperature chemical vapor deposition and mass detection on a heated atomic force microscope cantilever," *Applied Physics Letters*, vol. 88, p. 033107, 2006.
33. O. Englander, D. Christensen, and L. W. Lin, "Local synthesis of silicon nanowires and carbon nanotubes on microbridges," *Applied Physics Letters*, vol. 82, pp. 4797–4799, 2003.
34. S. Dittmer, O. A. Nerushev, and E. E. B. Campbell, "Low ambient temperature CVD growth of carbon nanotubes," *Applied Physics a-Materials Science & Processing*, vol. 84, pp. 243–246, 2006.
35. D. Mann, A. Javey, J. Kong, Q. Wang, and H. J. Dai, "Ballistic transport in metallic nanotubes with reliable Pd ohmic contacts," *Nano Letters*, vol. 3, pp. 1541–1544, 2003.
36. M. Ishida, H. Hongo, F. Nihey, and Y. Ochiai, "Diameter-controlled carbon nanotubes grown from lithographically defined nanoparticles," *Japanese Journal of Applied Physics Part 2-Letters & Express Letters*, vol. 43, pp. L1356–L1358, 2004.
37. A. Reina, M. Hofmann, D. Zhu, and J. Kong, "Growth mechanism of long and horizontally aligned carbon nanotubes by chemical vapor deposition," *Journal of Physical Chemistry C*, vol. 111, pp. 7292–7297, 2007.
38. S. M. Huang, M. Woodson, R. Smalley, and J. Liu, "Growth mechanism of oriented long single walled carbon nanotubes using 'fast-heating' chemical vapor deposition process," *Nano Letters*, vol. 4, pp. 1025–1028, 2004.

39. M. Hofmann, D. Nezich, A. Reina, and J. Kong, "In-Situ sample rotation as a tool to understand chemical vapor deposition growth of long aligned carbon nanotubes," *Nano Letters*, vol. 8(12), pp. 4122–4127, 2008.
40. A. Ismach, L. Segev, E. Wachtel, and E. Joselevich, "Atomic-step-templated formation of single wall carbon nanotube patterns," *Angewandte Chemie - International Edition*, vol. 43, pp. 6140–6143, 2004.
41. S. Han, X. L. Liu, and C. W. Zhou, "Template-free directional growth of single-walled carbon nanotubes on a- and r-plane sapphire," *Journal of the American Chemical Society*, vol. 127, pp. 5294–5295, 2005.
42. C. Kocabas, M. Shim, and J. A. Rogers, "Spatially selective guided growth of high-coverage arrays and random networks of single-walled carbon nanotubes and their integration into electronic devices," *Journal of the American Chemical Society*, vol. 128, pp. 4540–4541, 2006.
43. J. W. Mintmire and C. T. White, "Universal density of states for carbon nanotubes," *Physical Review Letters*, vol. 81, pp. 2506–2509, 1998.
44. Y. Zhang, Y. Li, W. Kim, D. Wang, and H. Dai, "Imaging as-grown single-walled carbon nanotubes originated from isolated catalytic nanoparticles," *Applied physics A - Materials science & processing*, vol. 74, pp. 325–328, 2002.
45. Y. Li, J. Liu, Y. Q. Wang, and Z. L. Wang, "Preparation of monodispersed Fe-Mo nanoparticles as the catalyst for CVD synthesis of carbon nanotubes," *Chemistry of Materials*, vol. 13, pp. 1008–1014, 2001.
46. J. Q. Lu, D. A. Rider, E. Onyegam, H. Wang, M. A. Winnik, I. Manners, Q. Cheng, Q. Fu, and J. Liu, "Carbon nanotubes with small and tunable diameters from poly(ferrocenylsilane)-block-polysiloxane diblock copolymers," *Langmuir*, vol. 22, pp. 5174–5179, 2006.
47. S. Bhaviripudi, A. Reina, J. F. Qi, J. Kong, and A. M. Belcher, "Block-copolymer assisted synthesis of arrays of metal nanoparticles and their catalytic activities for the growth of SWNTs," *Nanotechnology*, vol. 17, pp. 5080–5086, 2006.
48. S. N. Lu, Z. Y. Guo, W. Q. Ding, D. A. Dikin, J. Lee, and R. S. Ruoff, "In situ mechanical testing of templated carbon nanotubes," *Review of Scientific Instruments*, vol. 77, , 2006.
49. L. An, J. M. Owens, L. E. McNeil, and J. Liu, "Synthesis of nearly uniform single-walled carbon nanotubes using identical metal-containing molecular nanoclusters as catalysts," *Journal of the American Chemical Society*, vol. 124, pp. 13688–13689, 2002.
50. W. Kim, H. C. Choi, M. Shim, Y. M. Li, D. W. Wang, and H. J. Dai, "Synthesis of ultralong and high percentage of semiconducting single-walled carbon nanotubes," *Nano Letters*, vol. 2, pp. 703–708, 2002.
51. A. Jorio, R. Saito, J. H. Hafner, C. M. Lieber, M. Hunter, T. McClure, G. Dresselhaus, and M. S. Dresselhaus, "Structural (n, m) determination of isolated single-wall carbon nanotubes by resonant Raman scattering," *Physical Review Letters*, vol. 86, pp. 1118–1121, 2001.
52. C. Fantini, A. Jorio, M. Souza, M. S. Strano, M. S. Dresselhaus, and M. A. Pimenta, "Optical transition energies for carbon nanotubes from resonant Raman spectroscopy: Environment and temperature effects," *Physical Review Letters*, vol. 93, p. 147406, 2004.
53. S. M. Bachilo, L. Balzano, J. E. Herrera, F. Pompeo, D. E. Resasco, and R. B. Weisman, "Narrow (n,m)-distribution of single-walled carbon nanotubes grown using a solid supported catalyst," *Journal of the American Chemical Society*, vol. 125, pp. 11186–11187, 2003.
54. S. M. Bachilo, M. S. Strano, C. Kittrell, R. H. Hauge, R. E. Smalley, and R. B. Weisman, "Structure-assigned optical spectra of single-walled carbon nanotubes," *Science*, vol. 298, pp. 2361–2366, 2002.
55. A. Jorio, A. P. Santos, H. B. Ribeiro, C. Fantini, M. Souza, J. P. M. Vieira, C. A. Furtado, J. Jiang, R. Saito, L. Balzano, D. E. Resasco, and M. A. Pimenta, "Quantifying carbon-nanotube species with resonance Raman scattering," *Physical Review B*, vol. 72, 2005.
56. Y. H. Miyauchi, S. H. Chiashi, Y. Murakami, Y. Hayashida, and S. Maruyama, "Fluorescence spectroscopy of single-walled carbon nanotubes synthesized from alcohol," *Chemical Physics Letters*, vol. 387, pp. 198–203, 2004.

57. G. Lolli, L. A. Zhang, L. Balzano, N. Sakulchaicharoen, Y. Q. Tan, and D. E. Resasco, "Tailoring (n,m) structure of single-walled carbon nanotubes by modifying reaction conditions and the nature of the support of CoMo catalysts," *Journal of Physical Chemistry B*, vol. 110, pp. 2108–2115, 2006.
58. P. C. Collins, M. S. Arnold, and P. Avouris, "Engineering carbon nanotubes and nanotube circuits using electrical breakdown," *Science*, vol. 292, pp. 706–709, 2001.
59. Z. Yao, C. L. Kane, and C. Dekker, "High-field electrical transport in single-wall carbon nanotubes," *Physical Review Letters*, vol. 84, pp. 2941–2944, 2000.
60. L. An, Q. A. Fu, C. G. Lu, and J. Liu, "A simple chemical route to selectively eliminate metallic carbon nanotubes in nanotube network devices," *Journal of the American Chemical Society*, vol. 126, pp. 10520–10521, 2004.
61. G. Y. Zhang, P. F. Qi, X. R. Wang, Y. R. Lu, D. Mann, X. L. Li, and H. J. Dai, "Hydrogenation and hydrocarbonation and etching of single-walled carbon nanotubes," *Journal of the American Chemical Society*, vol. 128, pp. 6026–6027, 2006.
62. G. Drabner, A. Poppe, and H. Budzikiewicz, "The Composition of the Ch4 Plasma," *International Journal of Mass Spectrometry and Ion Processes*, vol. 97, pp. 1–33, 1990.
63. W. Zhou, Y. H. Ooi, R. Russo, P. Papanek, D. E. Luzzi, J. E. Fischer, M. J. Bronikowski, P. A. Willis, and R. E. Smalley, "Structural characterization and diameter-dependent oxidative stability of single wall carbon nanotubes synthesized by the catalytic decomposition of CO," *Chemical Physics Letters*, vol. 350, pp. 6–14, 2001.
64. Y. M. Li, S. Peng, D. Mann, J. Cao, R. Tu, K. J. Cho, and H. J. Dai, "On the origin of preferential growth of semiconducting single-walled carbon nanotubes," *Journal of Physical Chemistry B*, vol. 109, pp. 6968–6971, 2005.
65. H. Park, J. J. Zhao, and J. P. Lu, "Distinct properties of single-wall carbon nanotubes with monovalent sidewall additions," *Nanotechnology*, vol. 16, pp. 635–638, 2005.
66. R. J. Luyken and F. Hofmann, "Concepts for hybrid CMOS-molecular non-volatile memories," *Nanotechnology*, vol. 14, pp. 273–276, 2003.
67. Y. C. Tseng, P. Q. Xuan, A. Javey, R. Malloy, Q. Wang, J. Bokor, and H. J. Dai, "Monolithic integration of carbon nanotube devices with silicon MOS technology," *Nano Letters*, vol. 4, pp. 123–127, 2004.
68. J. Y. Huang, S. Chen, Z. Q. Wang, K. Kempa, Y. M. Wang, S. H. Jo, G. Chen, M. S. Dresselhaus, and Z. F. Ren, "Superplastic carbon nanotubes - Conditions have been discovered that allow extensive deformation of rigid single-walled nanotubes.," *Nature*, vol. 439, pp. 281–281, 2006.
69. M. Y. Sfeir, F. Wang, L. M. Huang, C. C. Chuang, J. Hone, S. P. O'Brien, T. F. Heinz, and L. E. Brus, "Probing electronic transitions in individual carbon nanotubes by Rayleigh scattering," *Science*, vol. 306, pp. 1540–1543, 2004.
70. H. B. Son, A. Reina, M. S. Dresselhaus, and J. Kong, "Characterizing the chirality distribution of single-walled carbon nanotube materials with tunable Raman spectroscopy," *Physica Status Solidi B - Basic Solid State Physics*, vol. 243, pp. 3161–3165, 2006.

Chapter 3
Carbon Nanotube Field-Effect Transistors

Ali Javey

3.1 Introduction

In the past few decades, the electronics field has witnessed a dramatic miniaturization of transistor elements with the number of transistors on an integrated circuit doubling approximately every 2 years [1–4]. Innovation and integration of new materials, such as high-κ gate dielectrics, various metals, silicides and nitrides, has been the key for this evolutionary path of CMOS device scaling [1–4]. Notably, the active channel material has predominantly remained the same, mainly owing to the scalability and manufacturability of the Si technology. As the device dimensions, such as the channel lengths approach the sub-10 nm regime, direct tunneling between source (S) and drain (D), and severe short channel effects present a fundamental challenge in continued scaling of Si devices. As a result, tremendous research efforts have recently been undertaken by various academic and industrial research groups for integrating new semiconductors as the channel material to enable (i) more efficient transport of carriers (i.e., higher mobility) and (ii) improved electrostatics at nanoscale (i.e., non-planar channel materials) [5–7]. In most approaches, a hybrid technology is envisioned, where Si still remains the handling substrate for fabrication processing, heat transport, and mechanical support purposes, with a new semiconductor integrated on the top for enhanced device operations or added new functionalities. One such material system is carbon nanotubes. The unique electron transport properties and band structure of nanotubes, as discussed in Chapter 1, and their quasi 1-D geometries make semiconducting SWNTs ideal channel materials for high-speed and low-power electronics [8–27]. In this chapter, we summarize some of the recent experimental advancements in the field of carbon nanotube transistors and discuss the device physics of 1-D channel materials. In Section 3.2, we discuss the nanotube–metal interface properties and the ability to attain Schottky barrier free contacts by utilizing an appropriate metal material due to the lack of Fermi-level pinning in 1-D junctions. In Section 3.3, we discuss the high-κ gate dielectric integration followed by a presentation on quantum

---

A. Javey (✉)
Department of Electrical Engineering and Computer Sciences, University of California, Berkeley, CA 94720, USA

A. Javey, J. Kong (eds.), *Carbon Nanotube Electronics*, Series on Integrated Circuits and Systems, DOI 10.1007/978-0-387-69285-2_3,

capacitance associated with 1-D materials in Section 3.4. In Sections 3.5 and 3.6, we discuss the role of the molecular species absorbed on the surface of nanotubes in the chemical doping and device hysteresis. Finally, various nanotube transistor structures, including Schottky FETs, MOSFETs, and band-to-band tunneling FETs, are discussed in Sections 3.7–3.9.

3.2 Schottky Barrier Heights of Metal S/D Contacts

SWNT-FETs were first demonstrated in 1998 by Dekker et al. at Delft University [28], and soon after by groups at IBM [29] and Stanford University [30]. Since the first experimental demonstration, significant progress has been made in understanding 1D electron transport and device physics as well as the nanotube materials properties [8–19], which eventually enabled researchers to demonstrate SWNT-FETs with DC characteristics near the ideal performance limits [23–24]. In this path, one of the significant challenges to overcome was the basic understanding of the metal–nanotube junction properties, and removing the energy barriers at such interfaces that form the Source (S) and Drain (D) contacts. The observation of an energy barrier for injection of electrons and holes at the metal–SWNT interfaces was first reported by Zhou, C. et al. for nickel contacts [31], and the work by IBM and others led to the detailed understanding of the Schottky barriers that arise at certain metal interfaces (mostly for titanium) [32–41]. In 2003, the first experimental demonstration of the highly desirable metal ohmic contacts to the valence band of semiconductor SWNTs was reported, free of any barriers for holes by the use of high work function palladium (Pd) which also has a strong binding interaction with carbon nanotubes [22]. Since then, various experiments have shed light on the diameter dependence of the metal contact properties and the nature of various barriers at the contacts [42–45].

The ability to reproducibly form ohmic S/D contacts, without any tunneling or Schottky barriers (SBs), is critical for (i) understanding the intrinsic transport properties of nanotubes and (ii) exploring the performance limits of nanotube FETs. The most commonly explored device configuration for a SWNT-FET is the metal S/D contacted structure which resembles the conventional SB-MOSFETs, although heavily doped SWNT contacts have also been demonstrated, resembling the conventional MOSFETs (Fig. 3.1) [46–49]. The metal-contacted configuration is ideal for nanoscale devices since metals have significantly lower parasitic resistance as compared to heavily doped semiconductors; however, the metal–semiconductor junction resistance and leakage currents are often obstacles. In an ideal device, the SB height can be tuned to any value, including ≤ 0 eV by choosing a high- or low work function metal for holes and electrons, respectively. However, in practice, achieving zero SBs has been quite challenge for intrinsic or lightly doped planar Si structures due to the Fermi-level pinning at the metal interfaces. It is widely believed that the metal-induced gap states (MIGS), which result in planar dipoles at the metal–semiconductor interfaces pin the metal Fermi-level deep into the band gap

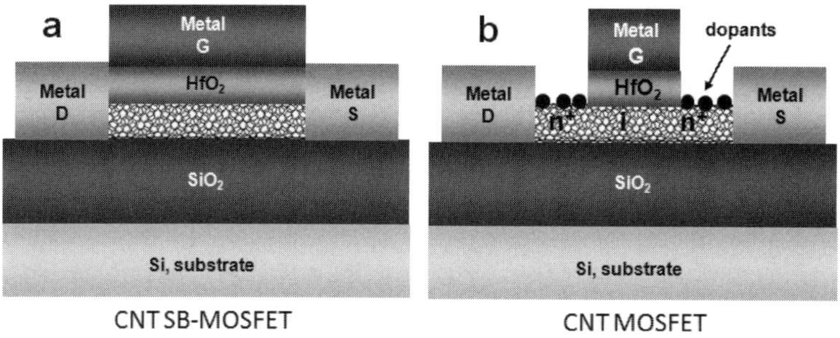

Fig. 3.1 Nanotube device structures. (**a**) SB-MOSFET with metal contacts and (**b**) MOSFET structure with heavily doped S/D contacts. Reproduced with permission from [46]. © 2005 American Chemical Society

of the semiconductor [33]. As a result, the SBs are nearly independent of the metal work function. For a 1-D semiconductor, however, the MIGS result in molecular-scale (nearly 0-D) dipoles at the interfaces with different electrostatics than the planar dipoles as first proposed by J. Tersoff et al. [34]. Unlike a planar dipole in which the potential is constant outside of the dipole layer, for a point dipole, the potential exponentially decays to zero a few nm away from the dipole interface (Fig. 3.2). Therefore, the MIGS do not introduce a strong Fermi-level pinning for 1-D channel materials. This is highly desirable and beneficial as the interfacial barrier heights and contact transparencies of SWNT-FETs can be readily tuned by the work function of the metal S/D contacts. This concept was experimentally demonstrated for Pd-contacted SWNT-FETs, where the work function of Pd S/D metal contacts was

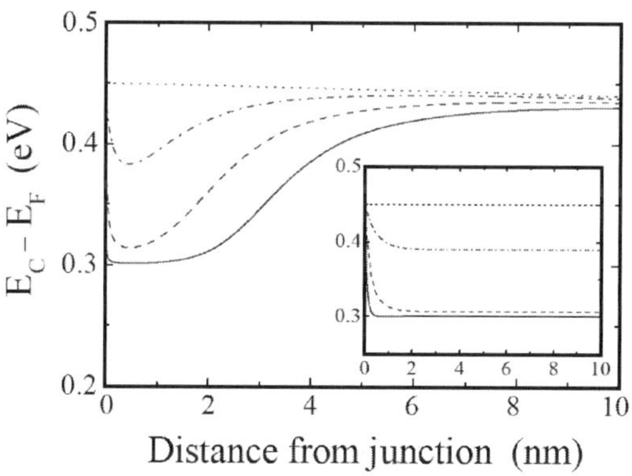

Fig. 3.2 Local conduction band edge of a carbon nanotube with a metal contact work function of $\Phi_m = 4.5$ eV as a function of distance from the metal contact. *Dotted*, *dashed-dotted*, *dashed*, and *solid lines* correspond to a MIGS density of $D_0 = 0$, 0.01, 0.1, and state/(atom-eV), respectively [34]. The inset shows the conduction band edge of a planar semiconductor. The results show that the contact properties (i.e., carrier transmission probability) of 1-D nanotubes are nearly unaffected by Fermi-level pinning due to their unique electrostatics. This is in distinct contrast to the planar case (inset). Reproduced with permission from [34]. © 2000 American Physical Society

tuned in situ by introducing molecular hydrogen gas while measuring the electron transport properties [22]. A similar observation was also made for Ti [35] and Mo [36] contacted nanotube devices where the absorption and thermal desorption of O_2 was found to modulate the metal work functions, and therefore the Schottky barrier heights at the contacts.

Pd is a noble metal with a high work function of Φ_m~5.2 eV in vacuum, and is found to enable SB free metal contacts to the valence band of SWNTs with diameters $d>\sim 1.6$ nm ($E_g<\sim 0.6$ eV) [42]. Such Pd-contacted SWNT-FETs are capable of delivering high ON currents (~25 μA) at relatively low S/D bias voltages (<1V), showing the potential of SWNTs as an ideal channel material for high-speed and low-power electronics. Notably, the temperature-dependent measurements of the ON-state conductance of the Pd-contacted SWNT-FETs exhibits metal-like behavior with the conductance of the devices monotonically increasing with the decrease in temperature [22]. The low temperature enhancement of the conductance is due to the suppression of acoustic phonon scattering, with the conductance approaching the quantum conductance limit of $\sim 4e^2/h = 155$ μS (due to 4-fold degeneracy, refer to Chapter 1) at 4 K (Fig. 3.3). This temperature-dependent result depicts the lack of SBs at the Pd-nanotube interfaces since the injection of carriers for SB junctions is primarily governed by thermionic emission and thermally assisted tunneling processes, both of which diminish at low temperatures. This is in clear contrast with the SWNT-FETs contacted with low work function metals, such as Ni, Cr, or Ti, which result in near mid gap SBs and become nearly insulating at low temperatures [31, 32]. The results can also be represented in terms of resistance. The total resistance for a carbon nanotube device is given as:

$$R = \frac{h}{4e^2} + R_C + R_{channel},$$

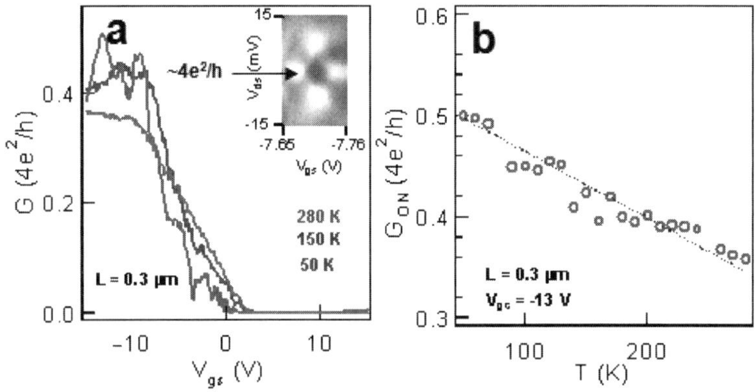

Fig. 3.3 Temperature dependent measurements of a Pd-contacted SWNT-FET with channel length L~0.3 μm using a back-gate configuration and t_{ox}~500 nm SiO_2 [22]. Adopted with permission from [22]

where R is the total resistance, R_C is the contact resistance (i.e., Schottky or tunneling barriers at the interface), and $R_{channel}$ is the resistance of the nanotube channel arising from various carrier scattering events mainly by defects, impurities, and phonons. For a device with ohmic contacts and ballistic nanotube channel (i.e., no carrier scattering), the total device resistance R approaches the quantum resistance of $h/4e^2 \sim 6.5\ k\Omega$.

A unique and well-established property of Pd is that its work function can be reversibly reduced upon surface exposure to hydrogen gas. This surface phenomenon was exploited for Pd-contacted SWNT-FETs, enabling researchers to study the FET transfer characteristics as the metal work function was reduced in situ, shedding light on the nanotube–metal interface properties [22]. Upon exposure to hydrogen, the p-channel conductance was decreased with a simultaneous enhancement of the n-channel conductance, eventually leading to ambipolar (exhibiting near symmetric n-channel and p-channel conductance) and n-type characteristics as depicted in Fig. 3.4. In contrast, SWNT-FETs with other metal contacts (such as Au) do not show any noticeable response to H_2 exposure, therefore, excluding the chemical interaction of nanotubes and hydrogen as the cause of the observed effect. The p- to n-FET transition by hydrogen can only be attributed to the lowering of the Pd metal work function. This result provides a direct experimental evidence for the lack of Fermi-level pinning for these miniaturized, 1D structures.

The ability to control the polarity of the nanotube FETs by simply applying an appropriate metal contact while maintaining the nanotubes chemically intrinsic provides a unique path for novel device engineering. In fact, Ti-contacted SWNT-FETs have been shown to exhibit ambipolar transfer characteristics [32, 43] with near mid gap SBs while Al-contacted devices are intrinsically n-type [44] with small, non-zero SBs to the conduction band (Fig. 3.5(a)). In future, a SWNT CMOS circuitry may be envisioned that involves the use of asymmetric metal contacts

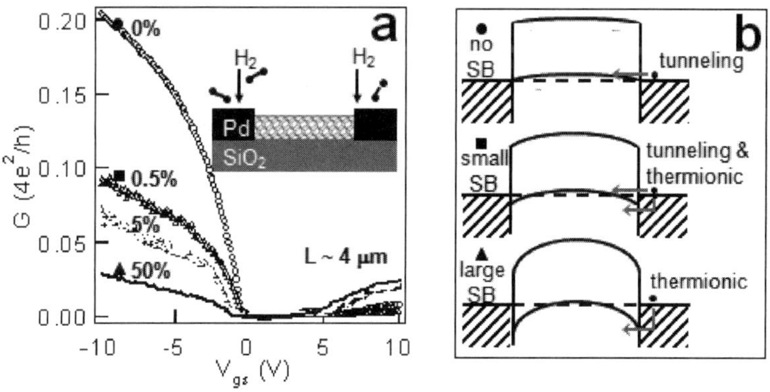

Fig. 3.4 Hydrogen response of a Pd-contacted SWNT-FET. Hydrogen reduces the work function of Pd metal contacts, therefore, increasing the SB height to the valence band while reducing the SB height to the conduction band [22]. Adopted with permission from [22]

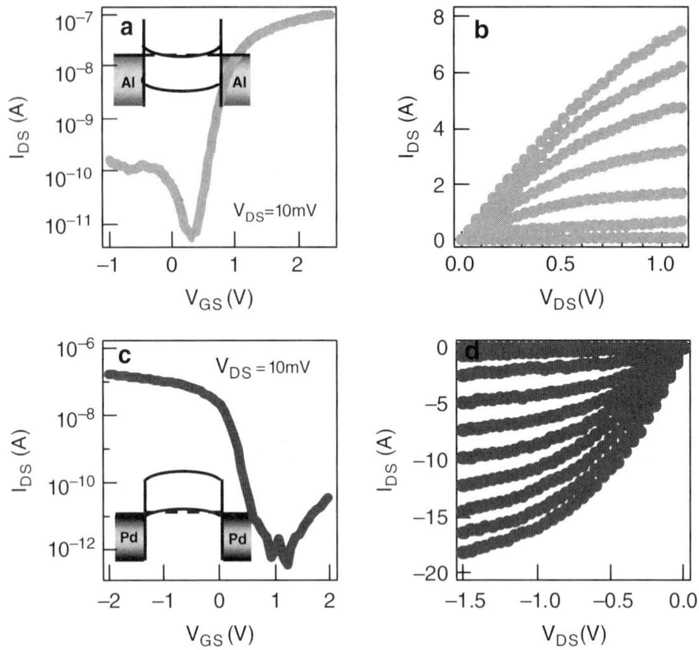

Fig. 3.5 Controlling the FET polarity by the metal contact material. (**a**) Al-contacted (S/D) nanotubes exhibit n-FET characteristics while (**d**) Pd contacts result in p-FETs. Nanotubes are chemically intrinsic. For both devices, a back-configuration with $t_{ox} \sim 10$ nm was used. Reproduced with permission from [44].

without channel doping, which is highly desirable as at such small scales, dopant fluctuation presents a major hurdle in attaining uniform device arrays. An important note is that the metal work functions highly depend on the environment, and the reported vacuum values cannot be used to accurately predict the SB heights at the metal–nanotube interfaces. For instance, researchers at IBM demonstrated that O_2 gas species absorbed on the surface of certain metals, such as Ti, results in the enhancement of the metal work function [35]. It was shown that by annealing the devices under vacuum and measuring them in situ, n-channel conduction can be significantly enhanced due to the molecular desorption of oxygen and the lowering of the SB heights to the conduction band. A more in depth analysis of the SB heights on nanotube band gap and metal work function is discussed in Chapter 8.

Beside Schottky barriers, pure tunneling barriers, independent of gate voltage, may also develop at the metal–nanotube interfaces, limiting the ON current drive and degrading the performance of the SWNT-FETs [42]. These pure tunneling barriers develop when the metal–semiconductor interactions are weak with low binding energies and long bond lengths. Metal–nanotube interactions were first examined by Y. Zhang et al. [50, 51] by TEM imaging of nanotubes coated with various metal thin films ($<$5 nm) as shown in Fig. 3.6. Ti, Rh, and Pd were found to interact strongly with nanotubes, resulting in a uniform surface coating while Au and Pt showed weak interactions with poor wetting on the nanotube surfaces, resulting in

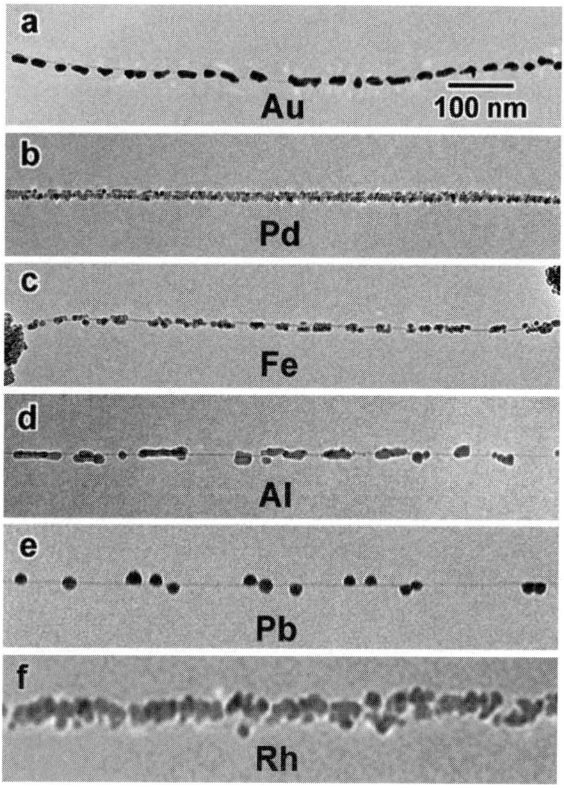

Fig. 3.6 TEM images of nanotubes coated with a thin film (<5 nm) of various metals. Reproduced with permission from [51]. © 2000 American Institute of Physics

aggregate formation on the surfaces. Interestingly, Pd and Rh are found to enable more transparent contacts to SWNTs than other high work function metals, such as Au and Pt [42, 45]. This may indeed be due to the weak interaction of Au and Pt with nanotubes that results in a vacuum layer formation at the interface, and therefore preventing an efficient injection of the carriers. Theoretical works have also highlighted the importance of the chemical binding interactions on the interface properties, with the Pd bond lengths predicted to be smaller than that of Au and Pt. In future, a more in depth analysis and characterization of the metal junctions is needed in order to better understand the injection of carriers and the various barriers that may develop at the interfaces. Furthermore, a low work function metal material for barrier free contact to the conduction band of SWNTs still needs to be developed to enable high-performance n-FETs, complementary to the Pd-contacted p-FETs.

In addition to the contact material, the diameter of nanotubes also plays a key role in determining the interface energy barriers for carrier injection [42, 45]. The diameter effect is twofold. First, the band gap of semiconducting tubes is inversely proportional to the diameter. Second, the chemical reactivity and surface properties of nanotubes are known to strongly depend on the nanotube diameter. While the former is directly correlated to the SB heights at the metal interfaces, the latter affects the junction quality and, therefore, the nature of the pure tunneling barriers.

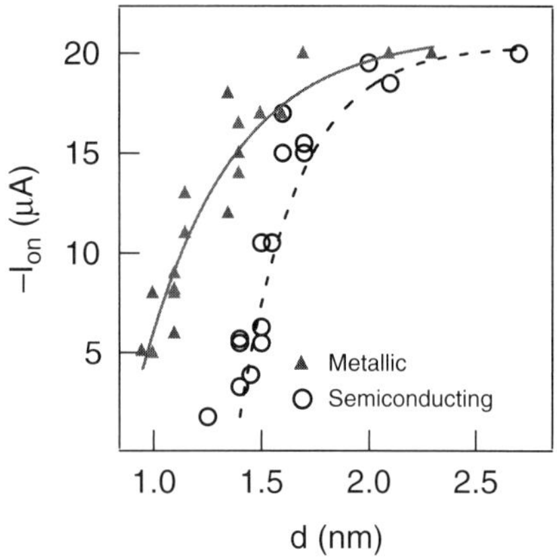

Fig. 3.7 Diameter dependence of the saturation current for semiconducting and metallic nanotubes contacted by Pd. Reproduced with permission from [42]. © 2005 American Institute of Physics

For metallic nanotubes, without a band gap, only pure tunneling barriers may form at the metal contacts. The diameter-dependent measurements of metallic nanotubes contacted by Pd and Rh have revealed the existence of pure tunneling barriers for ultra-small nanotubes ($d<1$ nm), severely limiting the injection of carriers at the contacts. This observation may be attributed to the diameter dependence of the metal–nanotube binding energy, with the binding energy decreasing for smaller diameters. On the other hand, SBs are observed for semiconducting nanotubes with diameters $d<1.6$ nm and Pd or Rh metal contacts (Fig. 3.7), arising from the enhancement of the band gap as the diameter is reduced [42]. This study reveals that still significant progress in contact technology is needed in order to achieve ohmic contacts to smaller diameter nanotubes. Such nanotubes are highly attractive for FET applications, since their larger band gaps enable lower leakage currents and higher I_{ON}/I_{OFF}.

3.3 High-κ Gate Dielectric Integration

The integration of high-κ gate dielectrics is a necessity for all future nanoscale FETs. As the device dimensions approach the 10-nm regime, severe short channel effects due to the strong electrostatic coupling of S/D electrodes to the channel presents a fundamental challenge, requiring gate oxides with ultrathin effective thicknesses. Since the physical thickness of the gate dielectric layer is already approaching the tunneling limit (~1 nm), the only feasible path for enhancing the gate control of the channel is by using high-κ gate dielectrics. Various high-κ materials, such as HfO_2 and ZrO_2 (dielectric constant, $\varepsilon = 15$–25) have been proposed as potential candidates for replacing SiO_2 ($\varepsilon = 3.9$) as the gate dielectric of nanoscale devices. Intel Corporation has already made the transition to HfO_2, starting from the 45-nm technology node. Such high-κ gate dielectrics are also necessary to attain scalable and high-performance nanotube FETs, capable of delivering high ON currents at low voltages.

High-κ gate dielectric integration with planar Si and other conventional semiconductors, including compounds semiconductors, has been quite challenging, mainly due to the interface problems and materials interactions, including phonon coupling, which result in lowering the channel mobility and degrading the subthreshold slope. Uniquely, the lack of surface-dangling bonds in carbon nanotubes enables for easy integration of high-κ gate dielectrics, without any significant perturbation to the electron transport. The integration of high-κ by using atomic layer deposition (ALD) for SWNT-FETs was first demonstrated by A. Javey et al, with the enabled nanotube devices exhibiting high ON currents (10–20 μA), high hole mobility (4,000–10,000 cm^2/Vs), and near ideal subthreshold swings (~70 mV/decade) [21, 23, 47]. Because of the lack of dangling bonds, chemically active sites are not available on the nanotube surfaces. As a result, nucleation of thin film growth by ALD cannot be initiated directly on the surface of nanotubes. Instead, the high-κ nucleation and growth takes place on the surrounding SiO_2 support substrate which results in the eventual drowning of nanotubes by high-κ as the film thickness increases beyond the nanotube diameter. The nanotube/high-κ interactions are only through weak van der Waals forces which are non-invasive to the carrier transport, without inducing surface and interface scattering [23]. This proposed nucleation and deposition mechanism is confirmed by TEM inspection of freely suspended nanotubes after the ALD process (Fig. 3.8) [23]. For such nanotubes without a supporting substrate, conformal surface coverage of high-κ is not observed, proving the lack of chemical reactivity of nanotubes during ALD.

Since ALD process involves drowning of nanotubes, gate leakage currents are often observed for HfO_2 films less than ~5 nm due to the lack of conformal coating on the surface of nanotubes. To address this problem, D. Farmer et al. presented an approach involving the physisorption of NO_2 gas species on the nanotube surfaces prior to ALD, followed by ALD and an annealing step for desorption of NO_2 molecules from the surface [53]. NO_2 molecules serve as nucleation sites, therefore, allowing for conformal high-κ coverage. In another approach, DNA wrapping

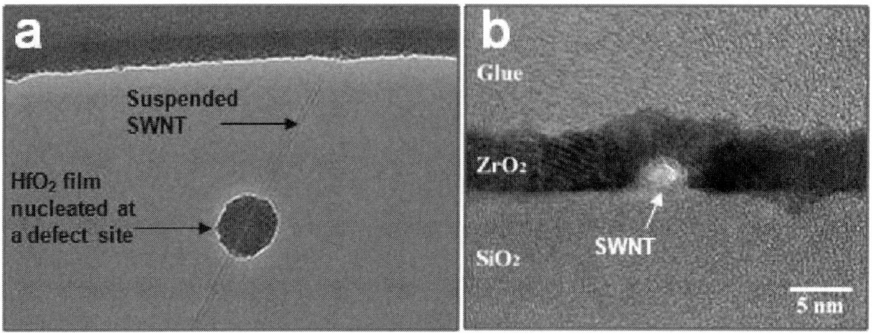

Fig. 3.8 TEM images of nanotubes after atomic layer deposition of high-κ for (**a**) suspended nanotube and (**b**) a nanotube on a SiO_2 substrate [23, 21]. Part (**a**) reproduced with permission from [23]. © 2004 American Chemical Society

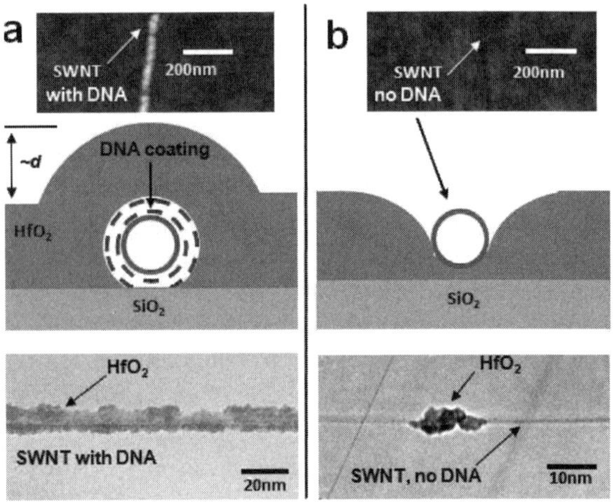

Fig. 3.9 Ultrathin, conformal ALD of HfO_2 on nanotubes enabled by DNA-induced nucleation (**a**) [52]. When DNA wrapping of nanotubes is not applied, a conformal coverage of HfO_2 is not achieved for ultrathin depositions as illustrated by the AFM image (**b**). Reproduced with permission from [52]. © 2006 American Chemical Society

of nanotubes was utilized for initiating the direct nucleation of HfO_2 on nanotube surfaces (Fig. 3.9) [52]. Both approaches enable for scaled integration of high-κ gate dielectrics down to ~2 nm, without any significant gate leakage currents while still preserving the intrinsic mobility of nanotubes.

3.4 Quantum Capacitance

In a bulk transistor structure, the gate capacitance value is simply governed by the geometric structure of the device. For a 1-D nanotube FET, however, the total capacitance (C_{tot}) depends both on the geometry (C_{g}, geometric capacitance), and also the density of states (C_{q}, quantum capacitance) [21, 54–58]. The two capacitors can be modeled in series with the total gate capacitance given as the sum of the inverses, $C_{\text{tot}}^{-1} = C_{\text{g}}^{-1} + C_{\text{q}}^{-1}$. As a result, the gate-dependent Fermi-level modulation can be expressed as $\dfrac{\partial E_{\text{F}}}{\partial V_{\text{g}}} = \dfrac{qC_{\text{g}}}{C_{\text{g}} + C_{\text{q}}}$, where $E_{\text{F}} = qV_{\text{a}}$ is the Fermi energy and V_a is the local electrostatic potential. This is assuming a S/D parasitic capacitance $C_{\text{S,D}} << C_{\text{g}} + C_{\text{q}}$.

The geometric or electrostatic component of the gate capacitance has a logarithmic dependence on the thickness of the gate dielectric layer, and can be expressed as $C_{\text{g}} = 2\pi\varepsilon/\ln(4t_{\text{ox}}/d)$, where t_{ox} is the gate dielectric thickness and ϵ is the dielectric constant of the gate insulator. This simple expression arrives from the cylinder on a plane capacitor model, and was shown to roughly match the experimentally attained C_{g} values (within a factor of ~2) from both quantum dot measurements at low temperatures (refer to Chapter 1) as well as the direct C–V measurements using the capacitance bridge set up [57].

Because of the small density of states in nanotubes, the capacitance values associated with C_q are relatively small (~4 pFcm^{-1}), and by the use of ultrathin high-κ gate dielectrics, one can readily approach an interesting regime where $C_g >> C_q$ [21, 58]. In such a quantum capacitance regime, instead of holding the charge constant, the gate holds the channel potential constant to the gate value. Furthermore, the conductance vs. gate voltage is expected to show step-wise features, with each step corresponding to $\sim T_{eff} \times 4e^2/h$ (where T_{eff} is the effective transmission probability for the carriers in the nanotube and at the contact interfaces) and arising from the contribution of an additional subband [54]. This is in contrast to the $C_g < C_q$ case, where the Fermi-level is pinned below the first subband, and therefore, significantly, higher V_g values are needed to reach the higher subbands. Also, in the quantum capacitance regime, the transconductance value is expected to be equal to the channel conductance as the channel potential shows a direct correlation to the gate voltage with $\dfrac{\partial E_F}{\partial V_g} = q$.

From the nanotube density of states and the Fermi function, the total charge density Q for a given channel potential can be derived. The quantum capacitance is then given as

$$C_q = \frac{\partial Q(V_a)}{\partial V_a} = \frac{q^2}{kTh} \sum_{i=1}^{2} \sqrt{\frac{m_i^*}{2}} \int_{E_{ii}/2}^{\infty} \frac{1}{\sqrt{E - E_{ii}/2}} \left[\sec h^2 \left(\frac{E - eV_a}{2kT} \right) + \sec h^2 \left(\frac{E + eV_a}{2kT} \right) \right] dE$$

where h is the Plank's constant, m_i^* and E_{ii} are the effective mass and the band gap of the ith sub-band, respectively.

The first experimental measurement of the quantum capacitance in a nanotube FET was demonstrated by S. Ilani et al., in which a novel set up, involving a capacitance bridge, was used to directly measure and isolate the small capacitances associated with a single nanotube device [57].

3.5 Chemical Doping

P–N junctions constitute the building blocks for virtually all conventional microelectronic device structures, and the precise control of the chemical doping profiling of semiconductor materials presents the key for device performance optimization. For planar Si structures, chemical doping is achieved by replacing the lattice atoms by impurities with three (acceptor) or five (donor) valence electrons. In carbon nanotubes, however, replacement of the carbon lattice atoms results in a severe degradation of the carrier mobility due to the destruction of the C–C sp^2 network and localization of the carriers. The doping in these surface structures is instead attained through the charge transfer from non-covalently bonded electron donating or withdrawing molecular precursors [59–65]. This is quite unique to a

1-D system where all atoms are near the surface (in the case of SWNTs, all atoms are at the surface), and therefore, efficient doping can be attained through surface engineering and charge injection. Charge injection provides a major advantage over lattice displacement as defects are not introduced in the lattice, and therefore, reducing the defect scattering of the carriers due to the dopant species.

Chemical doping of carbon nanotubes by charge injection was first demonstrated by J. Kong et al. The researchers observed the doping effect of NO_2 and NH_3 gas species on nanotube devices, therefore, allowing the detection of these molecules by SWNT-FETs [62]. Upon absorption of NO_2 on the nanotube surfaces, a dramatic *p*-doping effect was observed owing to the well-established electron withdrawing nature of NO_2. On the other hand, NH_3 absorption resulted in electron injection into the nanotubes, therefore, effectively *n*-doping the tubes. The same researchers also demonstrated the potassium (K) *n*-doping of nanotubes and doping profiling for enabling various P–N junction structures [59–61]. As compared to NH_3, K shows a significantly stronger electron donating behavior ($\sim$1 electron per K atom), there fore, enabling heavy *n*-doping of nanotubes. IBM researchers utilized this strategy for uniform n-doping of nanotubes by K to achieve metal-contacted n-FETs (K was evaporated on top of nanotube FETs in vacuum) [65], leading the way to the demonstration of the first complementary inverter logic gate on a single nanotube [66]. Later, A. Javey et al. demonstrated the first MOSFET-like nanotube structure with chemically doped (*n*+) contacts and chemically intrinsic channels by patterned doping of nanotubes with potassium [46]. A disadvantage of K is that it is not air stable, and requires the dopant coverage and measurement to be done under a vacuum environment. Beside gas molecules and alkali metals, air-stable polymers, such as polyethyleneimine [64], have also been shown to successfully dope the nanotubes. While polymers exhibit better air stability, so far, they have enabled only low to moderate doping concentrations, therefore, limiting their application for contacts. This is due to their weaker electron donating characteristics as compared to K. Still significant work needs to be done in designing appropriate dopant molecules for heavy *n*- and *p*-doping of nanotubes with long term stability in ambient air. Better theoretical understanding of the interactions between various molecular species and nanotubes is also needed in guiding the design of the ideal dopant structures.

3.6 Hysteresis and Device Passivation

A common feature of the back-gated, un-passivated SWNT-FETs is the large hysteresis in the I_{DS}–V_{GS} characteristics that are often observed when the back gate voltage is swept in different directions (Fig. 3.10(b)) [67]. This hysteresis which depends on the sweeping speed is highly undesirable for transistor applications as it induces instability in the threshold voltage and the current of the device. There are various common sources of hysteresis in a FET, including trapped charges at the semiconductor–insulator interface and the in the insulator (i.e., gate dielectric) itself. For a pristine carbon nanotube, clean of any metal catalyst or amorphous carbon residues, the lack of the surface-dangling bonds results in a clean interface with the dielectric layer, such as SiO_2, therefore, preventing trapped charges at the interface. Furthermore, today's processing technology in a clean-room environment enables the formation of high-quality, thermally grown SiO_2 layers, which can be used as the dielectric for a back-gated device configuration, free of any significant contamination and trapped charges. As a result, for a well-fabricated SWNT-FET with a back-gate configuration one would not expect a large hysteresis. However, a hysteresis as large as ~50% of the applied back gate voltage can be observed [67]. The source of this hysteresis is attributed to the absorption of polar molecules, such as water, on the surface of the SiO_2 substrate, proximal to the nanotubes. These polar molecules, with a built-in dipole, can serve as charge traps as they line up with the induced electric fields from the gate. This hypothesis was confirmed by W. Kim et al. as they demonstrated a significant reduction in the hysteresis by gentle annealing of the devices in a vacuum environment [67]. Upon exposure to ambient air, the devices once again showed large hysteresis as the water molecules re-absorbed on the device surfaces. The large sensitivity to the environment is expected for carbon nanotubes due to their large surface area to volume ratio, with all atoms exposed to the surface. This unique property makes them ideal for chemical and biological sensors, as shown in Chapter 8; however, careful consideration needs to be applied for the design and passivation of the FETs to prevent environmental response and enable device stability.

A simple approach for passivating SWNT-FETs is to spin-coat a thin film of poly(methyl methacrylate) (PMMA, ~100 nm), followed by baking at 180°C for 12 hrs during which water molecules are desorbed from the surface [67]. PMMA

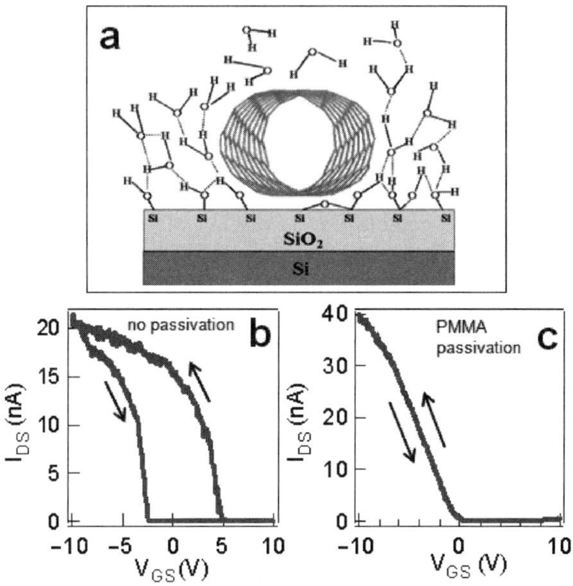

Fig. 3.10 PMMA passivation of nanotube devices for removal of hysteresis. (**a**) Schematic of a nanotube on substrate with the surrounding water molecules. Electrical characteristics of a back-gated (t_{ox}~500 nm SiO_2) nanotube FET (**b**) before and (**c**) after PMMA passivation. Reproduced with permission from [67]. © 2003 American Chemical Society

is a highly hydrophobic polymer that results in effective passivation of the devices with low permeability to water molecules. The PMMA passivation method has been shown to dramatically lower the hysteresis magnitude to near non-existence with long stability in ambient air (Fig. 3.10). However, a disadvantage of PMMA is that it is readily dissolved by a number of organic solvents, including acetone, which makes the post-passivation processing, quite challenging. In future, other passivation materials which are more robust and resistive to various chemicals need to be explored.

3.7 Near Ideal, Metal-Contacted MOSFETs

As compared to planar MOSFETs, carbon nanotubes enable better gate modulation of the semiconductor channel due to their 1-D structures [68]. Furthermore, their high mobility, compatibility with high-κ gate dielectrics, and ability to form ohmic metal junctions present an ideal path for attaining high current devices with low scattering and power loss in the channel [23, 47]. Additionally, their unique band structure with symmetric conduction and valence bands (therefore, symmetric electron and hole transport properties), along with diameter-dependent band gap in the range of ~0.2–2 eV make them ideal materials for complementary circuit design integration.

Since the first demonstration of nanotube FET in 1998 [28], significant progress has been made in addressing various fundamental and technological aspects of SWNT-FETs. Notably, in 2004, A. Javey et al. demonstrated a near-deal SWNT-FET with ballistic carrier transport, scaled channel and gate lengths (<50 nm), high-κ gate dielectrics, Pd metal ohmic contacts, and a self-aligned fabrication strategy (S/D electrodes were self-aligned in respect to the gate as shown in Fig. 3.11) [23]. Such SWNT-FETs delivered ~25 μA of ON current at only a fraction of a volt ($V_{DS} = 0.4$ V), corresponding to a current density of ~15 mA/μm as normalized to the diameter of the nanotube (d~1.7 nm). The maximum linear conductance was ~$0.5\times(4e^2/h)$ at room temperature, which is close to the ballistic limit, with a peak transconductance of ~30 μS (Fig. 3.12) [23]. The near ballistic transport, without any scattering of carriers in the channel, is quite attractive and is attributed to the long scattering mean free paths (mfp) in SWNTs. Various studies have revealed defect mfp of ~1 μm with acoustic and optical phonon mfp of >300 nm and ~15 nm, respectively [13, 69–74]. While the optical phonon mfp is quite small, they do not induce electron scattering at low V_{DS} (<~0.2 V) due to the high energy associated with these phonons [23, 75]. At room temperature, charge carriers can only emit optical phonons as the thermal energy is approximately an order of magnitude below the optical phonon energy, hence limiting the absorption of optical phonons. As a result, an optical phonon scattering can only occur at high V_{DS} where the charge carriers can gain enough energy to emit optical phonons. Furthermore, it was predicted by J. Guo et al. and proven by experimental results that the optical phonon scattering at high V_{DS} does not significantly affect

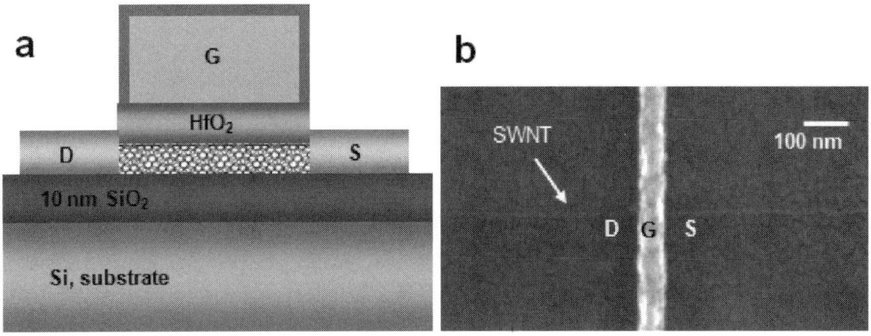

Fig. 3.11 Self-aligned nanotube FETs. (**a**) side-view schematic and (**b**) top-view SEM image of a nanotube FET with $L \sim 50$ nm, Pd metal S/D contacts, and high-κ metal gate stack. Reproduced with permission from [23]. © 2004 American Chemical Society

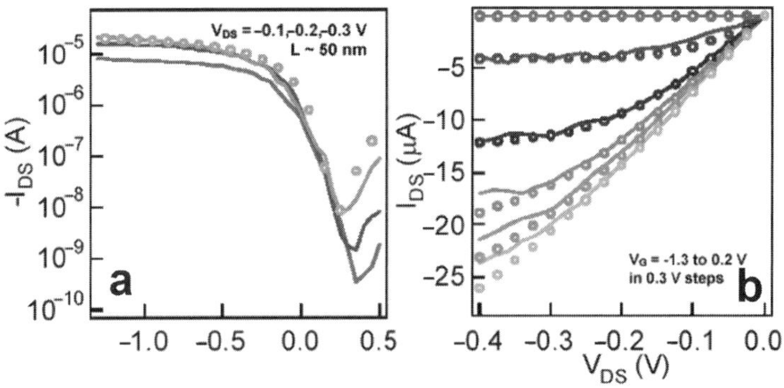

Fig. 3.12 *I–V* characteristics of the nanotube transistor shown in Fig. 3.11(b). Simulation data is shown as *circles* while experimental data is shown as *solid lines*. Reproduced with permission from [23]. © 2005 American Chemical Society

the DC characteristics of a SWNT-FET [75, 76]. This is because the backward-scattered (by optical phonons) carriers lack enough energy to overcome the barrier at the source. Therefore, if a carrier loses its energy in the channel due to optical phonons, it eventually makes it through the channel and is collected at the drain (Fig. 3.13) [75, 76]. However, optical phonon scattering at high V_{DS} is expected to affect the AC characteristics (i.e., high-frequency operation) of a device.

Beside the experimental advancements over the past decade, the field also witnessed a major progress in the quantum simulation and modeling of SWNT-FETs as discussed in Chapter 5. In particular, J. Guo et al. conducted quantum simulation of ballistic nanotube FETs with similar device input parameters as those used in the experimental work discussed here [76]. By doing so, the theoretical performance of a perfectly ballistic nanotube FET, without any channel scattering and with zero Schottky barrier heights at the contacts, was obtained and compared with the experimental results. It was found that the experimental *I–V* characteristics matched

Fig. 3.13 Valence band edge profile of an ohmically contacted CNTFET, depicting the roll of optical phonon (OP) emission on the scattering of a hole [75, 76]. Reproduced with permission from [76]. © 2004 IEEE

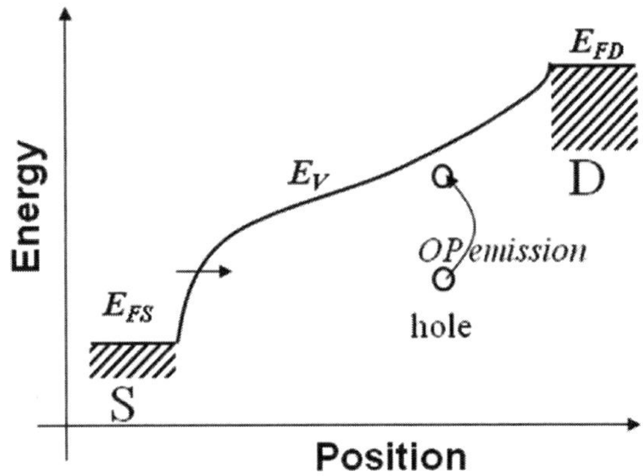

the simulation results well, hinting that the experimental SWNT-FETs with channel length $L \sim 50$ nm and high-κ gate dielectrics are indeed operating near the ballistic limit at room temperature [23, 75]. In comparison, a planar Si MOSFET with a similar channel length and a SiO_2 gate dielectric is ~40% ballistic [77]. Further scaling of SWNT-FETs was reported by Infineon with channel lengths approaching ~20 nm, demonstrating the scalability of SWNT-FETs [24].

The direct comparison of the performance characteristics of 1-D SWNT-FETs with conventional planar MOSFETs is quite challenging since their different dimensions present an obstacle in normalizing the device performance parameters. Ideally, high-frequency measurements of a device would directly reveal its intrinsic delay and switching speed. However, in practice, performing meaningful high-frequency measurements of SWNT-FETs has proven to be quite challenging due to their miniaturized scales as later discussed in the next chapter. Until the direct RF measurements are conducted, perhaps ON current densities can be used as a meaningful metric for comparison of the device performances. ON current density is particularly important as it is inversely proportional to the switching delay. Since it is speculated that parallel arrays of nanotubes are needed to enable high ON currents for logic operations, $2d$ (where d is the diameter of SWNT) is often used for the unit width normalization of SWNT-FETs. This is in part because a minimum pitch of $2d$ is needed to prevent electrostatic cross-linking of nanotubes. The ON current density of a ballistic SWNT-FET, normalized by $2d$, is shown in Fig. 3.14 as a function of I_{ON}/I_{OFF} for an operating bias $V_{DD} = 0.4$ V [76]. Also, on the same plot, the current density of the state-of-the-art Si MOSFET with a gate length of ~50 nm (90-nm technology node) is shown, clearly demonstrating the advantage of SWNTs over Si in terms of ON current density, and therefore, potentially speed. Important to note is that the I_{ON}/I_{OFF} of the shown nanotube device is only limited to 100 which may not be ideal for certain circuit designs. In future, the OFF current can be significantly improved (orders of magnitude) by using smaller diameter nanotubes with larger band gap [78] and also by incorporating a MOSFET-like structure with heavily doped contacts (instead of metal contacts) to reduce the leakage currents [45–49].

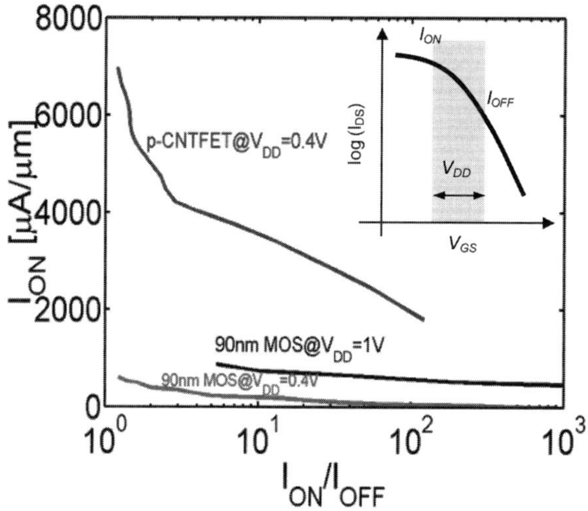

Fig. 3.14 ON current density as a function of I_{ON}/I_{OFF} for different operating bias (V_{DD}) for the SWNT-FET shown in Fig. 3.12 and a 90-nm technology node MOSFET. In terms of current density, it is clearly evident that SWNT-FETs outperform MOSFETs. Reproduced with permission from [76].

However, if the selected diameter is too small, contact degradation results in a lower I_{ON} as well [78]. Therefore, when selecting the appropriate nanotube diameter for a desired application, one must pay close attention to both I_{ON} and I_{OFF}.

3.8 SWNT MOSFETs

MOSFET-like structures with heavily doped S/D contacts (Fig. 3.1(b)) present a twofold advantage over SB-MOSFETs with metal contacts (Fig. 3.1(a)) [46–49]. First, the heavily doped contacts enable a more efficient injection of carriers from S/D into the channel than the metal-contacted devices, even when SB heights are zero. In fact, J. Guo et al. have proposed that negative SBs are needed in order to obtain injection efficiencies similar to that of the MOSFETs [76]. This is because in the MOSFET geometry, the contact Fermi-level can be deep into the conduction band (or valence band for p^+ contacts), therefore, increasing the number of carriers that can go over the potential barrier at the source for a given gate voltage. Obtaining negative SB heights, however, is not experimentally easy due to the limited metal materials available for contacts. So far, there has been no direct evidence of the existence of the negative SBs for the experimental SWNT devices.

The second advantage of MOSFET structures is that they enable lower OFF currents due to the decreased leakage at the drain [46–49]. For a metal-contacted p-FET, for instance with Pd contacts, the SB heights to the valence and conduction bands are $\sim$0 eV and $\sim E_g$, respectively, and the leakage current is predominately governed by the injection of electrons from the drain into the conduction band of the nanotube. The SB width for a SWNT scales linearly with the gate dielectric thickness with a near one to one correlation. For instance, a gate dielectric thickness of $\sim$2 nm results in a SB width of $\sim$2 nm. This is in clear contrast to the planar structures

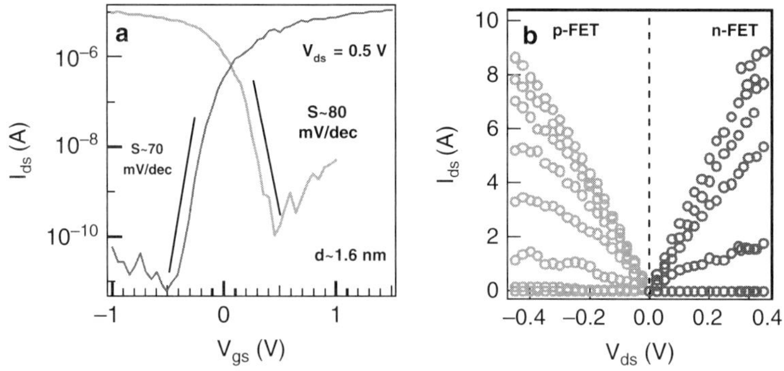

Fig. 3.15 *I–V* characteristics of nanotube n- and p-MOSFETs with "doped" contacts. Reproduced with permission from [46]. © 2005 American Chemical Society

where the SB widths scale by $\sqrt{t_{ox}}$. Since the tunneling probably increases exponentially with decreasing barrier widths, for ultrathin gate dielectrics, high leakage currents with ambipolar transfer characteristics can be readily observed for SWNT SB-MOSFETs. As a result, SB-MOSFET structures may not be very scalable, especially for applications that desire low OFF currents, and careful considerations need to be applied when designing such structures. However, a clear advantage of SB-MOSFETs over MOSFETs is the reduced parasitic resistances of the S/D contacts.

In one approach to achieve a MOSFET-like structure, patterned chemical doping of the contacts were obtained by using top-gate electrodes that under-lapped S/D as a mask [46, 48]. Both p-MOSFETs and n-MOSFETs were demonstrated with near identical characteristics, which is highly desirable for "CMOS" integration. The nanotube MOSFETs with chemically doped (with K) contacts exhibit superb characteristics with $I_{ON}/I_{OFF}{\sim}10^6$, subthreshold slope of ~70 mV/decade, and ON currents >10 μA ($d{\sim}1.5$ nm) at $V_{DS} = 0.4$ V (Fig. 3.15) [46]. In a MOSFET geometry,

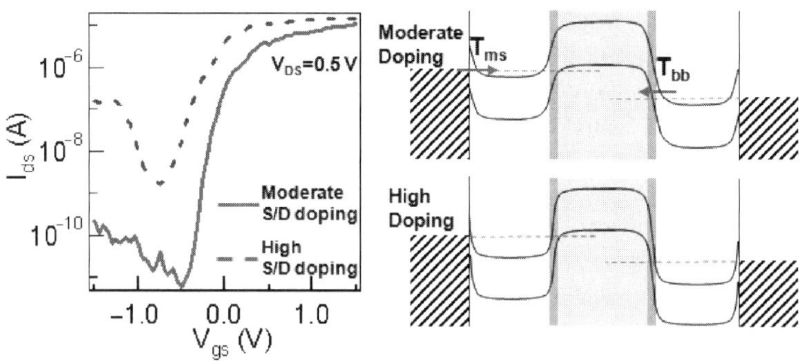

Fig. 3.16 Electrical characteristics of a nanotube n-MOSFET as a function of the contact doping density. Higher doping densities result in thinner BTBT barriers, and therefore higher leakage currents. Reproduced with permission from [46]. © 2005 American Chemical Society

the OFF-state leakage current is dominated by the band-to-band tunneling (BTBT) near the drain as illustrated in Fig. 3.16(b). The width of the BTBT barrier strongly depends on the doping densities of the contacts and the channel, as well as the gate dielectric thickness, and the applied potential (Fig. 3.16) [46]. The BTBT currents are lower than the Schottky leakage currents with similar barrier thicknesses, partly because of the lack of thermally assisted tunneling for the BTBT process as there are no allowed energy states in the band gap.

3.9 SWNT BTBT-FETs

An interesting device structure that was first demonstrated with nanotubes by J. Appenzeller et al. is the band-to-band tunneling transistor [78]. Rather than relying on the modulation of the charge density of the semiconductor channel (MOSFET) or the injection of the carriers at the metal interfaces (SB-MOSFETs), the BTBT-FET relies on the Zener or BTBT of the carriers at the contacts [78–81]. A typical BTBT-FET consists of a *p+/i/n+* configuration [82] with the p+ and n+ regions serving as the contacts [78–80]. This device structure is highly desirable for low-energy applications with small subthreshold swings (i.e., SS<60 mV/decade). For a MOSFET or a SB-MOSFET, the ideal subthreshold swing at room temperature is $SS = (kT/q)\ln(10) \sim 60$ mV/decade, arising from the exponential decay of the charge carrier distribution as a function of energy above the conduction band (for electrons) or below the valence band (for holes). Smaller SS values, and therefore sharp current drops as a function of gate bias, are highly desirable as they enable high I_{ON}/I_{OFF} at lower voltages, therefore, reducing the power consumption. This thermal energy distribution of carriers, however, does not affect BTBT-FETs as the onset of tunneling (i.e., threshold voltage of the device) occurs when the conduction and valence bands of a junction overlap. There are no available energy states below the conduction band or above the valence band (i.e., in the band gap). Therefore, below the threshold voltage, only a limited, gate-independent current can be expected due to direct band to band thermal activation of carriers. In practice, however, one has to take into consideration a number of non-idealities, including the thermal vibration of the lattice which effectively induces a "vibration" or "blurring" of the conduction and valence band edges, therefore, resulting in a finite current below the threshold voltage and increasing the SS. As proposed by M. Lundstrom and colleagues, phonon-assisted tunneling may also degrade the SS, especially at high V_{DS} [81]. However, despite these various factors, J. Appenzeller et al. and H. Dai et al. have already reported sub-60 mV/decade subthreshold swings (SS$\sim$25 mV/decade), proving the viability of the BTBT-FETs for low-power electronics [78–80].

A disadvantage of the BTBT-FETs as compared to the conventional MOSFETs is the ON current limitations. In a BTBT-FET, the ON current is governed by the tunneling rate of the carriers at the P–N interfaces. So far, the experimentally reported ON current values for tunnel transistors are below the desired values for digital logic applications. However, in future, the ON current may be dramatically enhanced through device optimization and thinning of the tunneling barriers by using ultra-thin gate dielectrics and higher doping densities. Interestingly, the Si community has also adopted the BTBT-FET structure as a major research arena with various groups exploring different device configurations and geometries. While planar Si structures may provide a near term manufacturability advantage, 1-D nanotubes are probably more optimal for tunnel transistors as thinner tunneling barriers can be more readily attained due to the electrostatics associated with 1-D structures, as previously discussed. It is yet to be seen how far the tunnel transistor performances can be improved and the effect that it may have on the future of integrated electronics.

3.10 Conclusion

Carbon nanotubes in many ways are the optimal channel material with the feasibility to be readily configured into various device structures, such as MOSFETs, SB-MOSFETs and BTBT-FETs. Their unique band structure, surface chemistry, electrical and physical properties, and 1-D geometric nature provide a number of major advantages over planar Si channel materials in terms of both device performance and scalability. Recently, a number of key experimental advancements have been made in understanding the fundamental device physics and exploring the performance limits of these 1-D materials. Better understanding of various fundamental properties, such as the metal contacts or high-κ interfaces, however is still needed, and furthermore a new processing technology needs to be developed to address their controlled assembly and device fabrication. Controlled assembly and processing of ultra-pure nanotubes will be the key for their potential large-scale integration for various digital applications. Nevertheless, these miniaturized structures have already served as the model system for exploring the device physics and performance limits of 1-D structures, and have shed light on various interesting and novel physical phenomena at nanoscale.

Acknowledgments We would like to acknowledge valuable discussions, advice, suggestions, inputs, encouragements, and unconditional support from Professor Hongjie Dai. Many of the work presented here was done in collaboration with other theoretical and experimental groups. The simulation was done by Professors Mark Lundstrom and Jing Guo while HfO_2 high-κ gate dielectrics were deposited by Damon Farmer and Professor Roy Gordon, and the ZrO_2 gate dielectrics were deposited by Hyoungsub Kim and Professor Paul McIntyre.

References

1. Peercy, P. S. The drive to miniaturization. Nature 406, 1023–1026 (2000).
2. Lundstrom, M. Moore's law forever? Science 299, 210–211 (2003).
3. Wong, P. H.-S. Beyond the conventional transistor. Solid-State Electron 49, 755–762 (2005).
4. Chau, R. et al. Silicon nano-transistors and breaking the 10 nm physical gate length barrier. Device Research Conf. 2003, 123–126 (2003).

5. Datta, S. et al. 85 nm Gate Length Enhancement and Depletion Mode InSb Quantum Well Transistors for Ultra High Speed and Very Low Power Digital Logic Applications. International Electron Devices Meeting Technical Digest, December 5–7 (2005).
6. Kim, D.-H. and J. A. del Alamo, Logic Performance of 40 nm InAs HEMTs. International Electron Devices Meeting Technical Digest, December 10–12 (2007).
7. Huang, X. Sub 50-nm FinFET: PMOS. International Electron Devices Meeting Technical Digest, p. 67. December 5–8 (1999).
8. Dekker, C. Carbon nanotubes as molecular quantum wires. Phys. Today 52, 22–28 (1999).
9. Ando, T. and Nakanishi, T. Impurity scattering in carbon nanotubes- absence of backscattering. J. Phys. Soc. Jpn. 67, 1104–1113 (1998).
10. White, C. T. and Todorov, T. N. Carbon nanotubes as long ballistic conductors. Nature 393, 240–242 (1998).
11. Zhou, X., et al. Band structure, phonon scattering, and the performance limit of single-walled carbon nanotube transistors, Phys. Rev. Lett. 95, 146805 (2005).
12. Li, S., Yu, Z., Rutherglen, C., and Burke, P. J. Electrical properties of 0.4 cm long single-walled carbon nanotubes. Nano Lett. 4, 2003–2007 (2004).
13. Purewal, M., Hong, B. H., Ravi, A., Chandra, B., Hone, J., and Kim, P. Scaling of resistance and electron mean free path of single-walled carbon nanotubes. Phys. Rev. Lett. 98, 186808 (2007).
14. McEuen, P. L., Fuhrer, M. S., and Park, H. K. Single-walled carbon nanotube electronics. IEEE Trans. Nanotechnol. 1, 78–85 (2003).
15. Dai, H. Carbon nanotubes: opportunities and challenges. Surf. Sci. 500, 218–241 (2002).
16. Avouris, P. Carbon nanotube electronics. Proc. IEEE 91, 1772–1784 (2003).
17. Lefebvre, J. et al. Single-wall carbon nanotube based devices. Carbon 38, 1745–1749 (2000).
18. Fischer, J. E. and Johnson, A. T. Electronic properties of carbon nanotubes, Curr. Opinion Solid State Mater. Sci. 1999, 28–33 (1999).
19. Dai, H., Javey, A., Pop, E., Mann, D., and Lu, Y. Electrical transport properties and field-effect transistors of carbon nanotubes. NANO, 1, 1–4 (2006).
20. McEuen, P. L. and Park, J.-Y. Electron transport in single-walled carbon nanotubes. MRS Bulletin 29, 272 (2004).
21. Javey, A. et al. High-k dielectrics for advanced carbon-nanotube transistors and logic gates. Nat. Mater. 1, 241–246 (2002).
22. Javey, A., Guo, J., Wang, Q., Lundstrom, M., and Dai, H. J. Ballistic carbon nanotube field-effect transistors. Nature 424, 654–657 (2003).
23. Javey, A. et al. Self-aligned ballistic molecular transistors and electrically parallel nanotube arrays. Nano Lett. 4, 1319–1322 (2004).
24. Seidel, R. V. et al. Sub-20 nm short channel carbon nanotube transistors. Nano Lett. 5, 147–150 (2005).
25. Kim, B. M. et al. High-performance carbon nanotube transistors on $SrTiO_3$/Si substrates. Appl. Phys. Lett. 84, 1946 (2004).
26. Durkop, T., Getty, S. A., Cobas, E., and Fuhrer, M. S. Extraordinary mobility in semiconducting carbon nanotubes. Nano Lett. 4, 35–39 (2004).
27. Lin, Y. M. et al. High-performance dual-gate carbon nanotube FETs with 40-nm gate length. IEEE Elec. Dev. Lett. 26, 823–825 (2005).
28. Tans, S., Verschueren, A., and Dekker, C. Room-temperature transistor based on a single carbon nanotube. Nature 393, 49–52 (1998).
29. Martel, R., Schmidt, T., Shea, H. R., Hertel, T., and Avouris, P. Single- and multi-wall carbon nanotube field-effect transistors. Appl. Phys. Lett. 73, 2447–2449 (1998).
30. Soh, H. et al. Integrated nanotube circuits: controlled growth and ohmic contacting of single-walled carbon nanotubes. Appl. Phys. Lett. 75, 627–629 (1999).
31. Zhou, C., Kong, J., and Dai, H. Electrical measurements of individual semiconducting single-walled nanotubes of various diameters. Appl. Phys. Lett. 76, 1597 (1999).

32. Heinze, S. et al. Carbon nanotubes as Schottky barrier transistors. Phys. Rev. Lett. 89, 6801 (2002).
33. Tersoff, J. Schottky barrier heights and the continuum of gap states. Phys. Rev. Lett. 52, 465–468 (1984).
34. Leonard, F. and Tersoff, J. Role of Fermi-level pinning in nanotube Schottky diodes. Phys. Rev. Lett. 84, 4693–4696 (2000).
35. Derycke, V., Martel, R., Appenzeller, J., and Avouris, P. Controlling doping and carrier injection in Carbon Nanotube transistors, Appl. Phys. Lett. 80, 2773 (2002).
36. Javey, A., Wang, Q., Ural, A., Li, Y., and Dai, H. Carbon nanotube transistor arrays for multistage complementary logic and ring oscillators. Nano Lett. 2, 929–932 (2002).
37. Martel, R. et al. Ambipolar electrical transport in semiconducting single-wall carbon nanotubes, Phys. Rev. Lett. 87 256805 (2001).
38. Appenzeller, J., Knoch, J., Radosavljević, M., and Avouris, P. Multimode transport in Schottky-barrier carbon-nanotube field-effect transistors. Phys. Rev. Lett. 92, 226802 (2004).
39. Appenzeller, J. et al. Field-modulated carrier transport in carbon nanotube transistors. Phys. Rev. Lett. 89, 126801 (2002).
40. Appenzeller, J., Radosavljevi, M., Knoch, J., and Avouris, P. Tunneling versus thermionic emission in one-dimensional semiconductors. Phys. Rev. Lett. 92, 048301 (2004).
41. Radosavljevic, M., Heinze, S., Tersoff, J., and Avouris, P. Drain voltage scaling in carbon nanotube transistors. Appl. Phys. Lett. 83, 2435–2437 (2003).
42. Kim, W. et al. Electrical contacts to carbon nanotubes down to 1 nm in diameter. Appl. Phys. Lett. 87, 1–3 (2005).
43. Javey, A., Shim, M., and Dai, H. Electrical properties and devices of large-diameter single-walled carbon nanotubes. Appl. Phys. Lett. 80(6), 1064–1066 (2002).
44. Javey, A., Wang, Q., Kim, W., and Dai, H. Advancements in complementary carbon nanotube field-effect transistors. IEDM Technical Digest (2003).
45. Chen, Z., Appenzeller, J. Knoch, J., Lin, Y.-M., and Avouris, P. The role of metal-nanotube contact in the performance of carbon nanotube field-effect transistors. Nano Lett. 5, 1497–1502 (2005).
46. Javey, A. et al. High performance n-type carbon nanotube field-effect transistors with chemically doped contacts. Nano Lett. 5, 345–348 (2005).
47. Javey, A. et al. Carbon nanotube field-effect transistors with integrated ohmic contacts and high-k gate dielectrics. Nano Lett. 4(3), 447–450 (2004).
48. Chen, J., Klinke, C., Afzali, A., and Avouris, P. Self-aligned carbon nanotube transistors with charge transfer doping. Appl. Phys. Lett. 86, 123108 (2005).
49. Klinke, C., Chen, J., Afzali, A., and Avouris, P. Charge transfer induced polarity switching in carbon nanotube transistors. Nano Lett. 5, 555–558 (2006).
50. Zhang, Y., Franklin, N., Chen, R., and Dai, H. A metal coating study of suspended carbon nanotubes and its implications to metal-tube interactions. Chem. Phys. Lett. 331, 35–41 (2000).
51. Zhang, Y. and Dai, H. Formation of metal nanowires on suspended single-walled carbon nanotubes. Appl. Phys. Lett. 77, 3015 (2000).
52. Lu, Y. et al. DNA functionalization of carbon nanotubes for ultrathin atomic layer deposition of high kappa dielectrics for nanotube transistors with 60 mV/decade switching. J. Am. Chem. Soc. 128, 3518–3519 (2006).
53. Farmer, D. B. and Gordon, R. G. ALD of high-kapp dielectrics on suspended functionalized SWNTs. Electrochem. Solid-State Lett. 8, G89–G91 (2005).
54. Guo, J. et al. Assessment of silicon MOS and carbon nanotube FET performance using a general theory of ballistic transistors. IEDM Technical Digest (2002).
55. Guo, J., Goasguen, S., Lundstrom, M., and Datta, S. Metal–insulator–semiconductor electrostatics of carbon nanotubes. Appl. Phys. Lett. 81, 1486–1488 (2002).
56. John, D. L., Castro, L. C., and Pulfrey, D. L. Quantum capacitance in nanoscale device modeling. J. Appl. Phys. 96, 5180–5184 (2004).

57. Ilani, S., Donev, L. A. K., Kindermann, M., and McEuen, P. L. Measurement of the quantum capacitance of interacting electrons in carbon nanotubes. Nat. Phys. 2, 687–691 (2006).
58. Rosenblatt, S. et al. High performance electrolyte gated carbon nanotube transistors. Nano Lett. 2, 869 (2002).
59. Zhou, C., Kong, J., Yenilmez, E., and Dai, H. Modulated chemical doping of individual carbon nanotubes. Science 290, 1552 (2000).
60. Kong, J., Cao, J., Anderson, E., and Dai, H. Chemical profiling of single nanotubes: intramolecular pnp junctions and on-tube single electron transistors. Appl. Phys. Lett. 80, 73–75 (2002).
61. Kong, J., Zhou, C., Yenilmez, E., and Dai, H. Alkaline metal doped n-type nanotubes as quantum dots. Appl. Phys. Lett. 77, 3977 (2000).
62. Kong, J., Franklin, N., Chou, C., Pan, S., Cho, K. J., and Dai, H. Nanotube molecular wires as chemical sensors. Science, 287, 622 (2000).
63. Kong, J. and Dai, H. Full and partial chemical gating of nanotubes by organic amine compounds. J. Phys. Chem. 105, 2890–2893 (2001).
64. Shim, M., Javey, A., Kam, N., and Dai, H. Polymer functionalization for air-stable n-type carbon nanotube field effect transistors. J. Am. Chem. Soc. 123, 11512–11513 (2001).
65. Radosavljevic, M., Appenzeller, J., Avouris, P., and Knoch, J., High performance of potassium n-doped carbon nanotube field-effect transistors, Appl. Phys. Lett. 84, 3693–3695 (2004).
66. Derycke, V., Martel, R., Appenzeller, J., and Avouris, Ph. Carbon nanotube inter- and intramolecular logic gates. Nano Lett. 1, 453–456 (2001).
67. Kim, W., Javey, A., Vermesh, O., Wang, Q., Li, Y., and Dai, H. Hysteresis caused by water molecules in carbon nanotube field-effect transistors. Nano Lett. 3, 193–198 (2003).
68. Guo, J., Wang, J., Polizzi, E., Datta, S., and Lundstrom, M. Electrostatics of nanowire transistors. IEEE Trans. Nanotech. 2, 329–334 (2003).
69. Yao, Z., Kane, C. L., and Dekker, C. High-field electrical transport in single-wall carbon nanotubes. Phys. Rev. Lett. 84, 2941–2944 (2000).
70. Javey, A. et al. High-field, quasi-ballistic transport in short carbon nanotubes. Phys. Rev. Lett. 92, 106804 (2004).
71. Park, J.-Y. et al. Electron–phonon scattering in metallic single-walled carbon nanotubes. Nano Lett. 4, 517 (2004).
72. Javey, A., Qi, P., Wang, Q., and Dai, H. 10- to 50-nm-long quasi-ballistic carbon nanotube devices obtained without complex lithography. Proc. Nat. Acad. Sci. 101, 13408 (2004).
73. Perebeinos, V., Tersoff, J., and Avouris, P. Electron–phonon interaction and transport in semiconducting carbon nanotubes. Phys. Rev. Lett. 94, 086802 (2005).
74. Zhou, X., Park, J -Y, Huang, S., Liu, J., and McEuen, P. L. Band structure, phonon scattering, and the performance limit of single-walled carbon nanotube transistors. Phys. Rev. Lett. 95, 146805 (2005).
75. Guo, J. and Lundstrom, M. Role of phonon scattering in carbon nanotube field-effect transistors. Appl. Phys. Lett. 86, 193103-05 (2005).
76. Guo, J., Javey, A., Dai, H., and Lundstrom, M. Performance analysis and design optimization of near ballistic carbon nanotube field-effect transistors. International Electron Devices Meeting, 703–6 (2004).
77. Lochtefeld, A. and Antoniadis, D. On experimental determination of carrier velocity in deeply scaled NMOS: how close to the thermal limit? IEEE Electron Device Lett. 22, 95, (2001).
78. Javey, A., Famer, D., Gordon, R., and Dai, H. Self-aligned 40 nm channel carbon nanotube field-effect transistors with subthreshold swings down to 70 mV/decade. Proceedings of SPIE – The International Society for Optical Engineering (Quantum Sensing and Nanophotonic Devices II, M. Razeghi, G.J. Brown, eds.), 5732, 14–18 (2005).
79. Appenzeller, J., Lin, Y.-M., Knoch, J., and Avouris, P. Band-to-band tunneling in carbon nanotube field-effect transistors. Phys. Rev. Lett. 93, (2004).

80. Appenzeller, J., Lin, Y. M., Knoch, J., Chen, Z. H., and Avouris, P. Comparing carbon nanotube transistors – the ideal choice: a novel tunneling device design. IEEE Trans. Elec. Dev. 52, 2568–2576 (2005).
81. Zhang, G., Wang, X., Li, X., Lu, Y., Javey, A., and Dai, H. Carbon nanotubes: from growth, placement, and assembly control to 60 mV/decade and Sub-60 mV/decade tunnel transistors. IEEE IEDM Technical Digest (2006).
82. Koswatta, S. O. et al. Simulation of phonon-assisted band-to-band tunneling in carbon nanotube field-effect transistors. Appl. Phys. Lett. 87, 253107 (2005).

Chapter 4
Measuring the AC Response of SWNT-FETs

Islamshah Amlani

4.1 Introduction

SWNT-FETs are considered promising candidates for high-frequency applications with a predicted frequency response in the terahertz regime [1–5]. The main reason for this anticipation is the ballistic transport in the channel over several hundred nanometers at room temperature leading to higher transconductance and mobility values compared to any other material. Significant progress has been made in understanding the DC properties of SWNT-FETs. Despite tremendous interest in the AC properties as well, a full RF characterization of SWNT-FETs have proved challenging to date.

The typical approach for RF and microwave characterization of any two-port system (including SWNT-FETs in common source (CS) or common gate (CG) configuration) requires measurement of the scattering parameters commonly referred to as S parameters. The 2×2 S matrix includes the reflection and transmission parameters at the input port (S_{11} and S_{12}) and the output port (S_{22} and S_{21}). Some of the most commonly quoted figure of merit for an active device such as a FET in CS configuration include cutoff frequency (f_T), Mason's Unilateral Gain (U), maximum available power gain (G_{max}), and maximum frequency of oscillation (f_{max}) [6]. These figures of merit can be easily calculated if S parameters of a device are available. The f_T is defined as the short circuit unity current gain frequency and is obtained by determining the frequency at which the forward small-signal current gain decays to unity. For a FET in CS configuration, the small-signal current gain is the ratio of amplitude of the small-signal drain current to the small-signal gate current. For this calculation, S parameters are first converted to hybrid or H parameters which relate input AC voltages and currents to output AC voltages and currents. The f_T is then the frequency at which the short-circuit current gain of the device, h_{21}, falls to 0 dB. Maximum port-to-port power gain, G_{max}, within a device is realized when the device is stable and both the input and output ports are conjugately matched to the impedance of the device, respectively. This is an important figure of merit as

I. Amlani (✉)
Motorola, Tempe, AZ 85284, USA

A. Javey, J. Kong (eds.), *Carbon Nanotube Electronics*, Series on Integrated Circuits and Systems, DOI 10.1007/978-0-387-69285-2_4,

it provides a fundamental limit on how much power gain can be achieved from a device. The third important figure of merit for a FET is the maximum frequency of oscillation or f_{max} and is defined as the highest frequency at which power gain can be obtained from the device. Above f_{max}, the device becomes passive and is unable to amplify an input signal. If the f_{max} of a device is above the measurement capability, it should be estimated by extrapolating the frequency at which the unilateral power gain (U) falls to unity. It should be noted that the maximum available gain should not be used for this extrapolation since beyond the "kink" frequency, the gain rolls off at a non-constant slope. For this reason, the unilateral gain is a more reliable indicator of f_{max}. It should also be noted that all complex S parameters must be known in order to determine f_{T}, f_{max}, and G_{max} of a device. In particular, the calculation of U requires accurate measurements since the unilateral gain is sensitive to the device loss. Apparently accurate S parameter measurements can lead to very noisy U data, a good indication that the measurements of device loss are not sufficiently accurate to predict f_{max}.

The above described S parameter measurement is usually performed with vector network analyzers (VNAs) which are almost ubiquitously designed for a 50-Ω reference impedance. While the width of the conventional transistors can be appropriately scaled to achieve the 50 Ω impedance match, the same is not possible for SWNT-FETs. Since SWNT is a macromolecule, its theoretical DC resistance is determined by quantum conductance ($h/e^2 \sim 25$ kΩ) even in the case of ballistic transport. In practice, the resistance is typically higher than the theoretical value varying anywhere from tens of kΩ to several MΩ for micrometer length devices. Although current density through a nanotube can be quite high, the actual current is only in the microampere range. This means that even though SWNT-FETs can generate sufficient current to drive other similar SWNT-FETs such as in the recently demonstrated ring oscillator [7], they are incapable of driving the 50 Ω load of conventional high-frequency measurement systems. Additionally, in all devices reported to date, pad capacitance of the test fixture dominates the intrinsic capacitance of SWNT which makes the de-embedding process non-trivial.

Consequently, various groups have resorted to nonstandard measurement techniques to assess the frequency response of SWNT-FETs [8–15]. Attempts have been made on both top-gated and back-gated SWNT devices configured as transistors, rectifiers, and resistors. In 2004, Frank and Appenzeller used an indirect measurement approach to show for the first time that there was no degradation in AC response for frequencies up to 500 MHz [8, 9]. In their approach, a DC shift in the drain current was measured in response to an AC signal of varying peak amplitude applied simultaneously to the gate and the source electrodes. Subsequently, Li et al. measured the microwave reflection coefficient, S_{11}, of a back-gated SWNT-FET configured as a resistor and embedded in a resonant circuit with a resonant frequency of 2.6 GHz [10]. The response of the SWNT-FET could only be assessed at the resonance frequency in their configuration. Measurement approaches that span across a wide frequency range are clearly needed.

The rest of the chapter is organized as follows. Section 4.2 provides a brief summary of three measurement approaches reported in the literature for assessing frequency response of top-gated SWNT-FETs. One common feature of these techniques is that they all leverage indirect measurement approaches to assess the frequency response and cannot provide quantitative information on gain. Section 4.3 describes in detail a recently presented approach by the author and collaborators to directly measure AC gain from a SWNT-FET. Section 4.4 summarizes this chapter and briefly discusses future trends.

4.2 Assessing the AC Response of Top-Gated SWNT-FETs

4.2.1 Power Measurement Using a Spectrum Analyzer

In 2005, Singh et al. reported a technique based on power measurement to assess frequency response of a SWNT-FET [12]. Their approach consisted of exciting the input of the device with a large sinusoidal signal using a signal generator and measuring scalar output power using a spectrum analyzer. When measuring single frequencies, spectrum analyzers typically have higher sensitivity than VNAs and a low output signal such as that from a SWNT-FET can be successfully resolved. Their top-gated device was fabricated on a quartz substrate and the layout was such that it could be operated either in a CS or a CG configuration. Figure 4.1 shows a schematic of the electrodes and an approximate equivalent circuit of the device in a CS configuration.

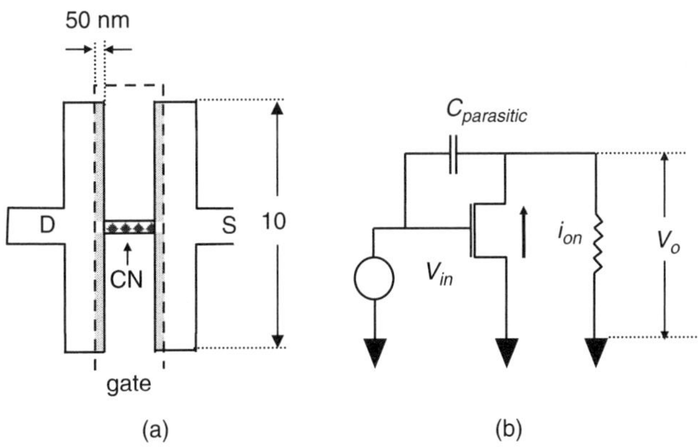

Fig. 4.1 (**a**) Schematic illustration of electrodes of a top-gated SWNT-FET. The bulk of parasitic capacitance arises due to overlap of gate and S/D electrodes shown by shaded portion. (**b**) Equivalent AC signal path for CS configuration illustrating role of parasitic gate–drain coupling capacitance on crosstalk. The model neglects the fraction ($< 1\%$) of SWNT-FET on-current that gets shunted through $C_{\text{parasitic}}$. Reproduced with permission from [12]. Copyright 2005 IEEE

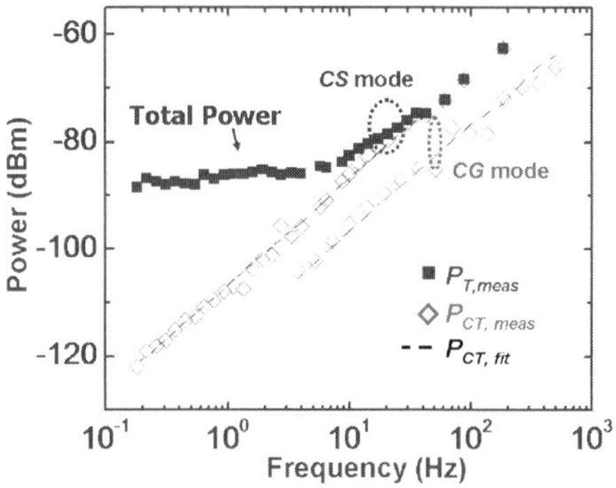

Fig. 4.2 Measured values of crosstalk power, P_{CT}, and total power, P_T, for SWNT-FETs in CS configuration. Also shown is P_{CT} for CG configuration indicating a decrease in parasitic coupling capacitance. Reproduced with permission from [12]. Copyright 2005 IEEE

In order to assess the precise frequency response of the nanotube, it is important to decouple the effect of crosstalk capacitance that depends on the pad geometry and the circuit layout. Figure 4.2 shows the total output power, P_T, for CG and CS configurations, as well as capacitive crosstalk power, P_{CT}. The crosstalk power was measured while the device was biased in the off state. The P_T is constant at ~ -90 dBm at low frequencies and can be clearly distinguished from the P_{CT} at low frequencies. Crosstalk power increases linearly with frequency and becomes dominant above 200 MHz. Hence, a conclusion can be drawn that their device operated at least up to 200 MHz. This limit is set by the parasitic crosstalk capacitance which is significantly greater than the intrinsic capacitance of the SWNT. The authors estimate that this capacitance must be reduced by several orders of magnitude, a major challenge from fabrication standpoint, in order to measure frequency response of SWNT-FETs in the hundreds of gigahertz range.

4.2.2 Homodyne Detection Using SWNT-FETs

Rosenblatt et al. have exploited gate-induced nonlinear current–voltage characteristics of SWNT-FETs to demonstrate homodyne mixing or more appropriately homodyne detection up to GHz frequencies [14]. Their device is fabricated on a high-resistivity Si substrate with a 1 μm thick thermal SiO_2 layer. Figure 4.3 shows a micrograph of the device and the experimental setup.

In their experiment, an RF signal was applied to the source while the gate of the transistor was DC biased at the maximum transconductance resulting in a rectified DC output current. This current, referred to as I_{mix}, is proportional to the transconductance of the device as well as square of the AC signal amplitude, $V_S{}^{ac}$. Figure 4.4 shows the amplitude of the mixing current as a function of $V_S{}^{ac}$ on a log–log scale for frequencies ranging from 10 MHz to 50 GHz. At frequencies above 2 GHz, the amplitude of the mixing current appears to roll off with frequency. The straight lines indicate a power law relationship with the exponent of $V_S{}^{ac}$ in the range 1.9–2.2 matching closely to the theoretical prediction. Subsequently, the authors

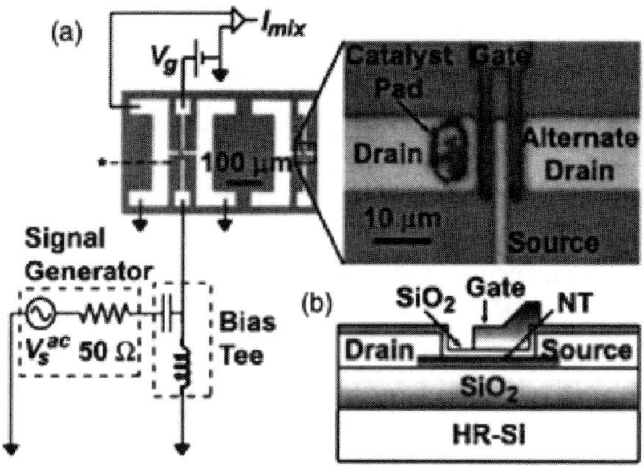

Fig. 4.3 (**a**) Optical micrograph of nanotube device along with circuit schematic. A high-frequency probe delivers the AC signal $V_S{}^{ac}$ to the source electrode while simultaneously grounding the drain electrodes. The mixing current I_{mix} was detected as a function of frequency and the gate voltage V_g. (**b**) Schematic cross section of device, layers not to scale. Reprinted with permission from reference [14]. Copyright 2005, American Institute of Physics

Fig. 4.4 Mixing current I_{mix} vs $V_S{}^{ac,\,rms}$ for a gate voltage near the peak of I_{mix}. Amplitude at 10 MHz and 1 GHz approximately overlap. The ideal power-law line of 2 is shown for reference. Reprinted with permission from reference [14]. Copyright 2005, American Institute of Physics

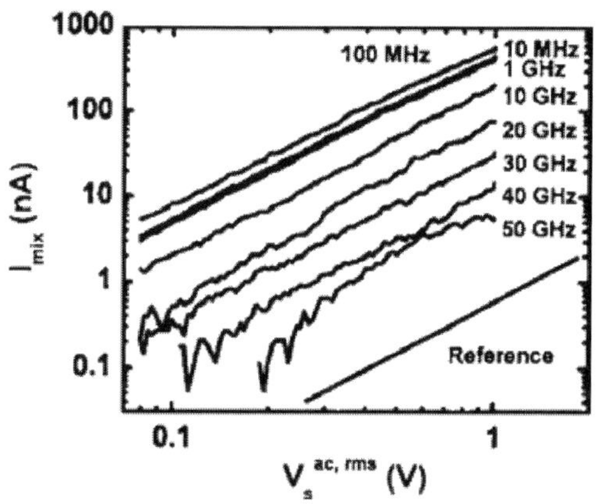

described the damping of mixing current as a function of frequency using a first-order low-pass filter model. According to their distributed mixer model, the predicted minimum cutoff frequency is limited by the RC time constant associated with the resistance of the source contact and the gate capacitance of the SWNT.

4.2.3 RF Characterization Using a Two-Tone Measurement

In 2006, Pesetski et al. reported another variant of the mixer approach to demonstrate SWNT-FET operation at microwave frequencies [15]. The authors adopted a two tone measurement technique and measured intermodulation product at the output of a SWNT-FET configured as a common source amplifier. Their approach provides immunity against the crosstalk capacitance which limits the frequency response as discussed in Section 4.2.1. Their top-gated device was fabricated on a low loss substrate and the gate significantly underlaped source and drain to reduce C_{GS} and C_{GD} thereby allowing RF measurements at microwave frequencies. Prior to the experiment, the authors first characterized their device by quantifying the expected nonlinear output while the device was biased in compression. They applied two different low frequency signals with the same amplitude to the input of the nanotube amplifier and verified that the output voltages do show peaks at second harmonic and sum and difference frequencies with amplitudes predicted from the DC I_D–V_D characteristics.

After establishing the validity of their approach, the authors generated two microwave signals with frequencies 10 kHz apart. The inset of Fig. 4.5 shows a schematic of the measurement setup. The two signals were combined using a resistive power combiner and applied to the input (gate) of the device. The non-linearity of the device produces an output signal at the difference of the two input frequencies. In their experiment, both input frequencies were varied simultaneously while maintaining a 10 KHz difference and the output signal was recorded at 10 KHz which represents the intermodulation product at the difference frequency. Figure 4.5 shows the results of the measurement. The magnitude of the output signal matches the

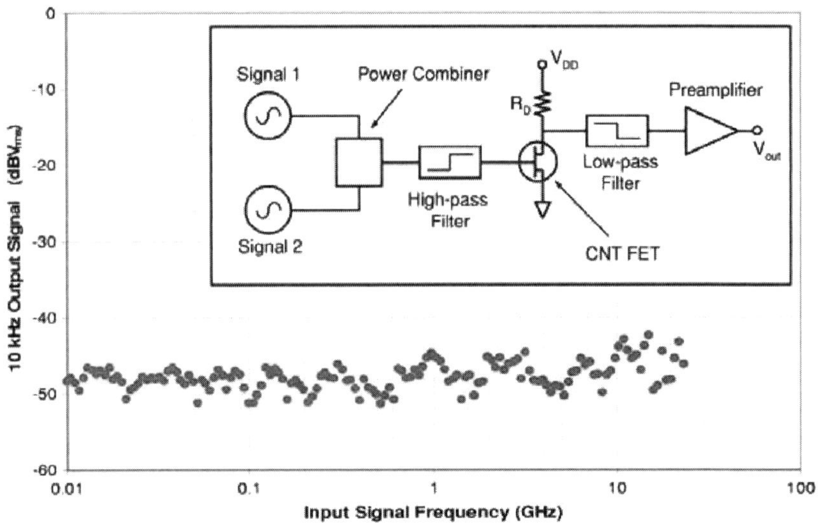

Fig. 4.5 Experimental verification of SWNT-FET operation at frequencies up to 23 GHz. The SWNT-FET amplifier is operated with an input signal containing two tones, at frequencies f and f+10 kHz. The FET acts like a mixer, producing an intermodulation product at 10 kHz. The amplitude of the 10 kHz output signal is plotted as a function of input frequency. Its value is consistent with the DC measurements for the entire frequency range. No evidence of a roll off is seen even at 23 GHz. Reprinted with permission from reference [15]. Copyright 2006, American Institute of Physics

predicted response and appears constant from DC to 23 GHz, a limitation imposed by the measurement apparatus. Thus, it can be concluded from the data that their device operated at a constant gain at least up to 23 GHz.

4.3 AC Gain from a SWNT-FET Common Source Amplifier

4.3.1 Measurement Approach

In Section 4.1, it was mentioned that one major impediment in measuring high-frequency response using conventional measurement systems is the unavailability of systems that provide an appropriate matching impedance for SWNT-FETs. For instance, since the normalized maximum transconductance of reported SWNT-FETs is in μS range, the expected voltage gain ($A_v = -g_m Z_0$) in a 50 Ω system is close to the noise floor of a typical VNA (–60 to –100 dB) and at microwave frequencies such a low signal will be easily shrouded by parasitic crosstalk around the FET. This problem, however, can be significantly minimized if the reference impedance of the measurement system becomes comparable to the impedance of the device.

We recently reported a time-domain measurement approach using a high impedance (high-Z) active probe that yields a direct determination of AC voltage gain from a SWNT-FET configured as a common source amplifier [16, 17]. The described technical approach also yields frequency domain measurement referred to as the frequency response function (FRF) of the device. Conceptually, the FRF can be thought of as a function that contains gain and phase responses of a device under test (DUT) at all frequencies of interest. Mathematically, the FRF is defined as the inverse Fourier Transform of the Impulse function. The FRF measurement requires the excitation of the DUT with energies at all relevant frequencies. The fastest way to perform the measurement is to use a broadband excitation signal that excites all frequencies simultaneously such as a narrow impulse, a chirp, or a fast-edge of a step function. The estimation of the FRF depends upon the transformation of the time domain output to frequency domain. The Fourier transform is used for this computation. In the experiment, this computation can be performed digitally using a fast Fourier transform (FFT) algorithm. This way, the theoretical advantages of the Fourier transform can be implemented in an efficient digital computation scheme.

Our measurement approach utilizes a high bandwidth real-time digital oscilloscope in conjunction with a high impedance active probe to obtain the FRF of the SWNT-FET. The procedure consists of stimulating the DUT with a fast pulse edge where the rise time of the fast-edge determines the signal bandwidth. Digital signal processing (DSP) algorithms built into the measuring tool are then used to take the derivatives of the digitized input and output pulse edges to yield the respective impulse functions. The FRF is obtained by calculating the FFT of the impulse for both the input and the output. Frequency domain output is normalized by subtracting the input FRF from that of the output. Similarly, the input phase can be subtracted from that of the output to yield the intrinsic phase of the device. For high-frequency characterization, it is important to de-embed the measured data using an on-wafer THRU structure [18] with matched characteristic impedance. In a typical 50 Ω system, this is done by subtracting output of the THRU from that of the device output to obtain de-embedded FRF. A similar de-embedding procedure for a SWNT-FET would require a THRU standard with matching characteristic impedance. Due to unavailability of such a standard, de-embedding of the measured output to remove parasitic effects was not performed.

The above described measurement technique was first tested on several high bandwidth MOSFETs using conventional 50 Ω probes and the results were found comparable to those obtained from a network analyzer establishing validity of the approach. In Section 4.3.2, we provide a brief description of the device fabrication process. Sections 4.3.3 and 4.3.4 describe DC and AC characterization results, respectively. Section 4.3.5 discusses modeling results and predictions on the intrinsic frequency response of the device.

4.3.2 Fabrication

Our top-gated SWNT-FET was fabricated on an oxidized Si substrate in a ground-signal-ground (GSG) configuration with a 50 μm probe pitch. Figure 4.6 shows a cross-sectional layout of the SWNT-FET. An optical micrograph and a scanning electron micrograph of the device are shown in Fig. 4.7. Detailed fabrication steps are described in reference [19]. Briefly, catalyst islands consisting of ultra-thin Al/Ni (6 nm/1 nm) bilayer films were patterned on the substrate followed by selective growth of SWNTs in a thermal chemical vapor deposition process. The diameter of grown nanotube was estimated to be around 2 nm based on height measurements of several other similar devices using an atomic force microscope. Source and drain

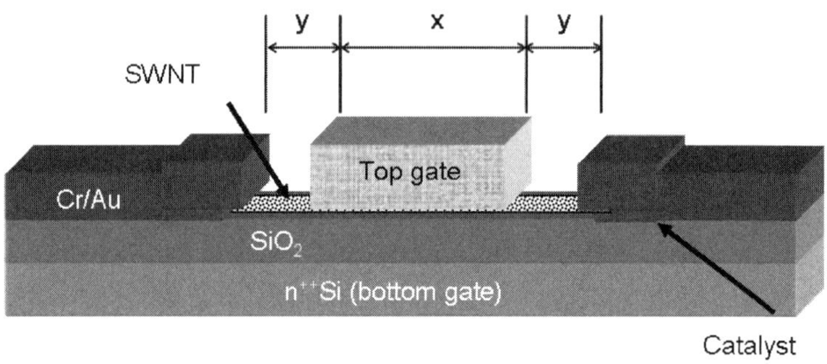

Fig. 4.6 Three-dimensional cross-sectional layout of the fabricated SWNT-FET. The channel length is defined by the top gate with length $x = 0.75$ μm. The ungated sections on either side of the channel labeled as 'y' are approximately 0.375 μm in length. Substrate is a highly doped Si, and SiO_2 is approximately 130 nm thick. A fixed negative back-gate bias is applied to the substrate during DC and AC measurements in order to electrostatically dope the ungated regions and reduce the Schottky barriers

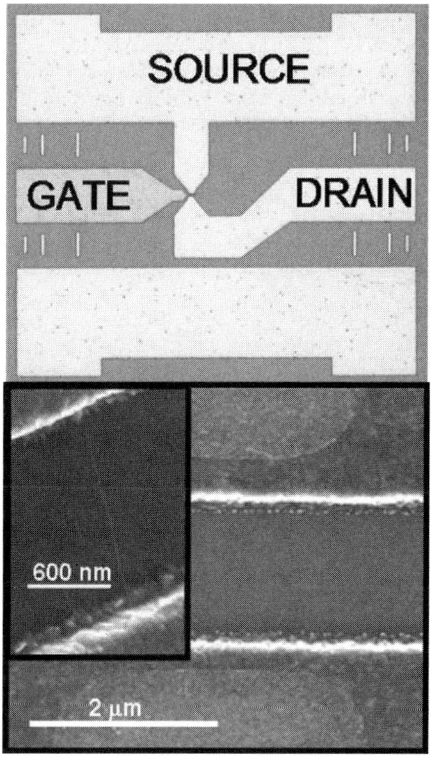

Fig. 4.7 (**a**) Optical micrograph of the SWNT-FET in a coplanar waveguide configuration. The ground-signal-ground probe pitch is 50 μm. (**b**) A representative SEM image of the device taken prior to top gate patterning step. Catalyst islands consisting of Al/Ni metal stack are patterned under the source/drain metal and appear with a different contrast in the image. The SWNT growth was carried out by chemical vapor deposition. The inset shows the SWNT bridging the gap between source and drain

electrodes were defined by patterning Cr/Au on top of the SWNT in a subsequent lithography step. The spacing between the source and drain electrode as defined by lithography was 1.5 μm. The gate dielectric (10 nm thick Al_2O_3) and gate metal (Pt/Au ~ 50 nm/200 nm) were defined in a final patterning step in which all three materials were evaporated in a single vacuum cycle. Here, Pt was used to prevent inter-diffusion between Al_2O_3 and Au during the post-fabrication anneal of nanotube devices. The lithographically defined gate length was 0.75 μm resulting in source and drain underlap regions.

4.3.3 DC Characterization

Electrical characterization of nanotube transistors at DC was carried out using a semiconductor parameter analyzer. Figure 4.8 shows DC transport characteristics of the SWNT-FET. The minimum on-resistance of this device is approximately 90 kΩ and the maximum transconductance is ~10 μS. Since source and drain electrodes of the FET underlap the gate electrode, there is an ungated region of approximately 375 nm on each sides of the channel. This effectively increases the on-resistance and

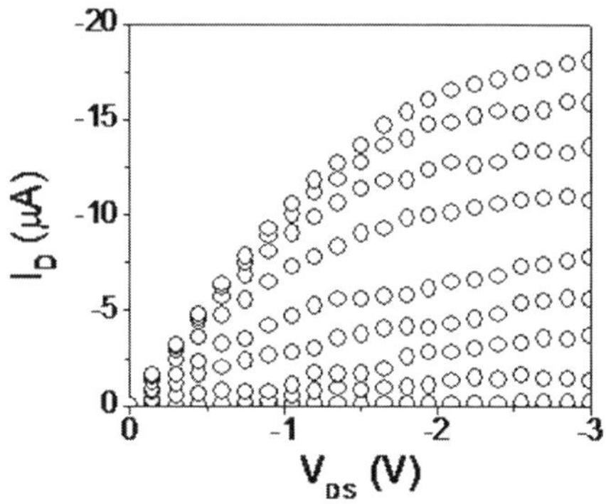

Fig. 4.8 Measured DC transport characteristics of the SWNT-FET as a function of top gate bias. A fixed back-gate bias of –5 V is applied during the measurement. V_G varies between 1.7 and –1.3 V in increments of 300 mV

reduces the current for a given gate and drain bias condition. For this reason, a fixed back-gate bias of –5 V was maintained on the back gate to reduce the resistance of the ungated regions and minimize the effect of Schottky barriers that may be present at the metal/nanotube interface. The observed high saturation current approaching 20 μA suggests that scattering along the length of the nanotube is minimal and the transport is quasi-ballistic [20].

4.3.4 AC Characterization

On-wafer AC measurements were performed on an RF probe station equipped with Cascade GSG probes having a 50 um probe pitch. The experimental setup is shown in Fig. 4.9. A function generator was used to apply a sharp edge with a short rise time to the gate of the transistor. In general, the fast edge must have a low repetition rate and the rise time must be sufficiently fast to create the desired frequency range. As a rule of thumb, the rise time is obtained by multiplying 0.35 to the inverse of

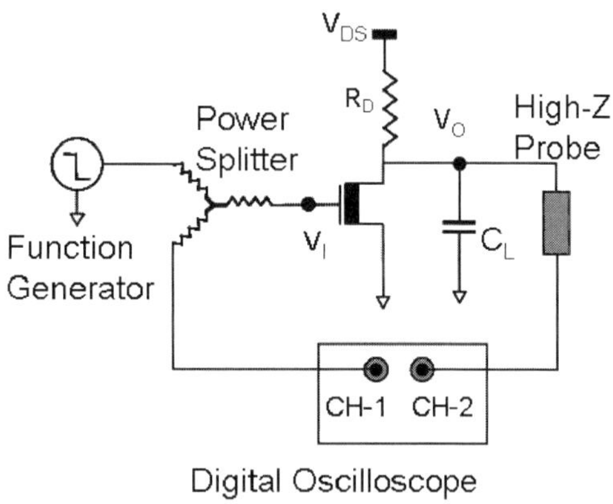

Fig. 4.9 Measurement setup showing SWNT-FET configured as a common-source amplifier. A power splitter is used to measure the input fast pulse edge, V_I, applied to the gate of the SWNT-FET. The value of drain resistor R_D is 620 kΩ. The output signal, V_O, is measured using a high-impedance (1 MΩ) active probe. C_L is the parasitic load capacitance of the measurement apparatus

the desired bandwidth [21]. For instance, if a bandwidth of 2 GHz is desired, the rise time of the edge needs to be at least 175 ps. Channel 1 of a digital oscilloscope was used to provide a trigger for the scope from the fast edge used to stimulate the device. The bandwidth of the scope was 4 GHz and all our measurements were well within that range. For best edge speed, a 50 Ω power splitter was used to direct the signal both to the scope and to the input of the device simultaneously. Output signal was monitored using an active probe with an impedance of 1 MΩ and a rated bandwidth of 1.5 GHz achieved using active circuitry. The main virtue of the described technique is that it allows the use of a high-Z probe which is critical for the measurement of the output signal of a high impedance device such as a SWNT-FET. In an active probe, the probe tip contains an active amplifier in addition to a RC network. This active amplifier drives a 50 Ω cable which is connected the 50 Ω input of the oscilloscope. The key advantage of this probe is a relatively high bandwidth, a requirement difficult to achieve with a passive probe. To obtain the maximum benefit from an active probe, it must be connected in close proximity to the signal being measured. This was not easy to achieve in our experimental setup, since a DC signal must also be applied at the drain to bias the transistor into saturation. Typically, a bias-T is used to provide isolation between DC and RF signals but this was obviously not a choice since ideally the high-Z probe must be connected directly to the drain terminal of the SWNT-FET. As a compromise, an SMA T-connector was custom designed to simultaneously connect the output of the device to both a DC power supply needed to bias the drain of the transistor and to channel 2 of the scope to sample the time-domain output. Although a reasonable approach for a proof of concept experiment, this however severely impacted the bandwidth of the device due to the increased parasitic capacitance at the drain. An off-chip drain resistor was used to isolate the DC source from the transient signal.

The gain of the transistor is determined by subtracting FRF of the input from that of the output in frequency domain. However, in order to witness a visual indication of gain in the time domain as well, it is critical to compare the output signal to the correct input signal that is applied to the device. Normally, in a 50 Ω system, this is not an issue since everything is matched to the same characteristic impedance and input applied is the same as that observed on channel 1. This is not true in the described measurement setup since the input is terminated at 50 Ω and the output is terminated at 1 MΩ. Due to the use of a power splitter which is designed for 50 Ω systems, the signal measured at channel 1 is actually lower than that applied to the device. Therefore, a correction to the channel 1 signal is required. The correction factor was experimentally obtained by directly connecting the output of the power splitter to the high-Z probe and sampling both time domain signals at channel 1 and channel 2 simultaneously. The signal at channel 1 was scaled up accordingly to accommodate for the difference and the corrected signal was used for comparison of the time domain input and output signals.

Figure 4.10 shows the time domain measurements of the transistor configured as a common source amplifier. A DC offset was applied to the gate and drain of the transistor to bias it in saturation near the maximum trans-conductance. These bias settings appear as initial states of the time domain signals at $t < 0$. A pulse

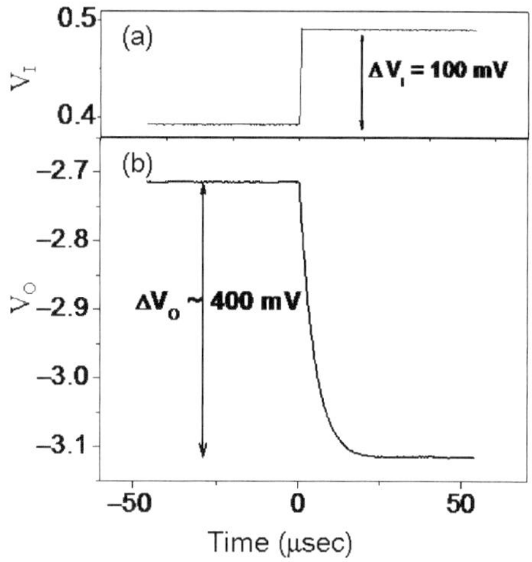

Fig. 4.10 Time domain signals (**a**) applied to the gate, V_I, and (**b**) measured at the drain, V_O, of the SWNT-FET. The output signal is amplified with a ratio of 4:1

with $\Delta V_I = 100$ mV peak-to-peak and a short rise time was applied to the gate of the transistor (Fig. 4.10(a)). Figure 4.10(b) shows the time domain output signal measured using the high-Z probe. The initial state of the output signal corresponds to the DC drain bias $V_O = -2.72$ V with a sharp change occurring at $t = 0$ as a result of the input stimulus. The amplitude of the output signal is $\Delta V_O \sim 400$ mV showing signal amplification with a ratio of 4:1. The output signal is out of phase as expected for this circuit configuration.

DSP algorithms were applied to the time domain input and output signals to display the frequency domain results simultaneously. Figure 4.11 shows the FRF of the nanotube transistor obtained by subtracting the input from the output. A gain of ∼ 12 dB is seen with a roll-off around 150 kHz and a unity voltage gain fre-

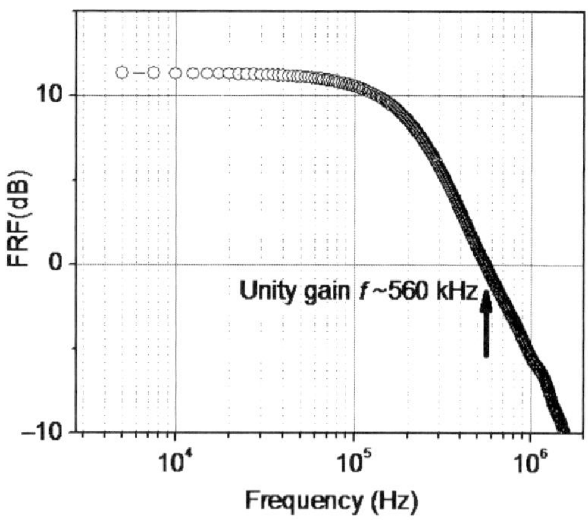

Fig. 4.11 Frequency response function (FRF) of the SWNT-FET CS amplifier. A gain of ∼12 dB with a unity voltage gain frequency of 560 kHz is observed. The device bandwidth is constrained due to the parasitic effects of the measurement apparatus. Above 560 kHz, the signal rolls off at 20 dB/decade as expected for the voltage gain of a first-order low-pass system

quency of approximately 560 kHz. Thus, the impedance of our resistor network in conjunction with the parasitic capacitance C_L forms a low pass filter with a roll off around 150 kHz.

In our test setup, parasitic capacitance consists of three parts: (1) the parasitic drain pad capacitance, (2) the shunt capacitance of the drain resistor and connecting cables, and (3) the inherent capacitance of the high-Z probe. The drain pad capacitance is estimated to be around 1 pF based on the electrode geometry. Although the specified intrinsic high-Z probe capacitance is also $\sim$ 1 pF, the connector on the high-Z probe had to be customized to accommodate an SMA connector which increased the capacitance significantly. Thus, the capacitance of the modified high-Z probe in parallel to the capacitance of the load resistor and connecting cables was the dominant capacitance of the measurement system and is estimated to be $\sim$3 pF. All parasitic capacitances must be reduced to improve the measured bandwidth. Drain pad capacitance can be reduced significantly by using improved electrode design with narrow width and a low-loss substrate such as quartz. Shunt capacitance of the bias network can also be significantly improved by fabricating a load resistor on the same substrate as the device and reducing the cables to an absolute minimum. High impedance active probes are commercially available with capacitances down to the f_F range and a bandwidth of several tens of GHz. Preferably a high-Z probe can also be designed on the same substrate as the SWNT-FET. Considering all these factors, it is speculated that improvements in the frequency response well into the GHz range can be realized.

4.3.5 Modeling

A circuit-compatible SPICE model for SWNT-FET is used to model both DC and AC characteristic of the SWNT-FET [22, 23]. The detailed description of the model is also provided in Chapter 6. Briefly, it is a physics based model that also includes device non-idealities such as the quantum confinement effects in both the circumferential and the channel length directions, the acoustic and optical phonon scattering in the channel region, the capacitance and resistance of the doped S/D SWNT region, as well as the possible Schottky Barrier (SB) resistances of S/D contacts. The current in the nanotube channel is modeled by three current sources: (1) the thermionic current contributed by the semiconduting sub-bands (I_{semi}) with the classical band theory, (2) the current contributed by the metallic sub-bands (I_{metal}), and (3) the leakage current (I_{btbt}) caused by the band to band tunneling mechanism through the semiconducting sub-bands. The detailed description and equations are provided in [22]. Real-time dynamic response is modeled accurately using a transcapacitance array as opposed to a single lumped capacitance.

The key parameter for evaluating SWNT-FET current is $\Delta\phi_B$, the channel surface potential change in response to changes in gate and source/drain bias. As shown in Fig. 4.12, there are three electrostatic coupling capacitors: the capacitance (C_{ox}) between the gate and channel, the capacitance (C_{sub}) between channel and substrate,

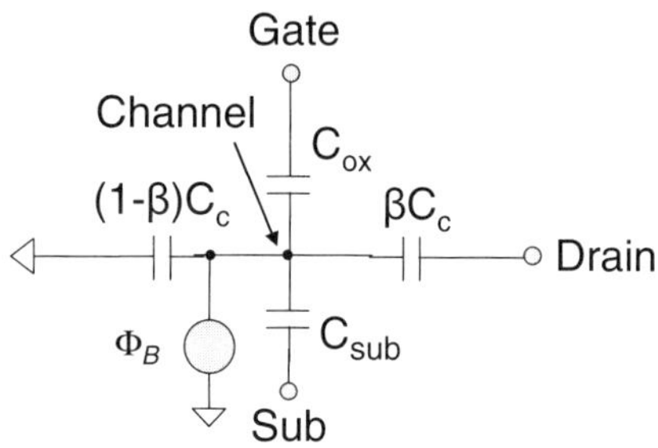

Fig. 4.12 The electrostatic capacitor model used to calculate the channel surface potential change $\Delta\phi_B$ before and after changes in gate, source, drain, or substrate bias. C_{ox}, C_{sub}, βC_c, and $(1-\beta)C_c$ are the physical coupling capacitor between the channel region and the gate, the substrate, the drain and other environment (including the source), respectively

and the capacitance (C_c) between channel and external drain (D′)/source (S′). $\Delta\phi_B$ is dynamically affected by the drain bias. The parameter βC_c is a fitting parameter that describes this effect due to two mechanisms: (1) the surface potential lowering due to the electrostatic coupling between the channel region and the external drain electrode through fringing electric field and (2) the surface potential lowering due to non-uniform channel surface potential profile caused by DIBL effect. Operationally, the parameters C_c and β are chosen to fit the sub-threshold slope and the measured short channel effect. $\Delta\phi_B$ is calculated using the charge conservation equation $Q_{cap}(\Delta\phi_B) = Q_{CNT}(\Delta\phi_B)$, where Q_{cap} is the charge induced by the electrodes, and Q_{CNT} is the total charge induced on the SWNT surface.

The small-signal SWNT-FET device model is shown in Fig. 4.13. The transcapacitance pairs are derived from the physical capacitors shown in Fig. 3.8 using Meyer capacitor model [24]. L_{KS} and L_{MS} represent the kinetic and magnetic inductance, respectively. Inductance values need to be taken into account only for frequencies above 1 THz and are thus not included in fitting the data for the measured SWNT-FET.

There are essentially two ways to model the SWNT-FET as shown in Fig. 4.14. One way is to model the top-gated section of the nanotube as the FET and the un-gated sections on either side of the top gate as resistors (Fig. 4.14a). In this approach, the effect of electrostatic doping achieved by the bias applied to the back

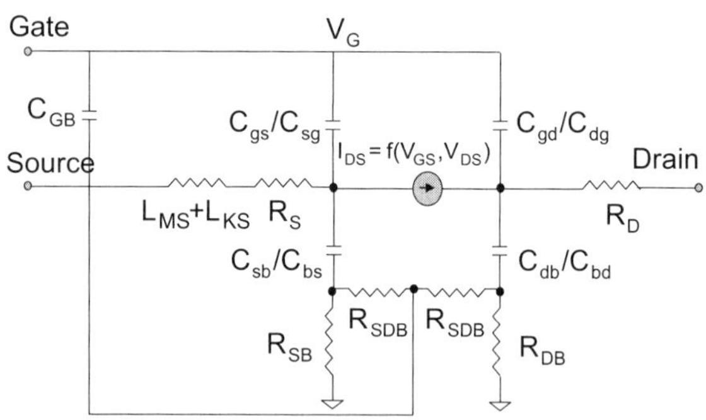

Fig. 4.13 The small-signal SWNT-FET device model. $C_{xy}=\partial Q_x/\partial V_y$. C_{gs}/C_{sg}, C_{dg}/C_{gd}, C_{sb}/C_{bs}, C_{db}/C_{bd} are transcapacitance pairs derived from the capacitors illustrated in Fig. 3.8 using charge partition methodology

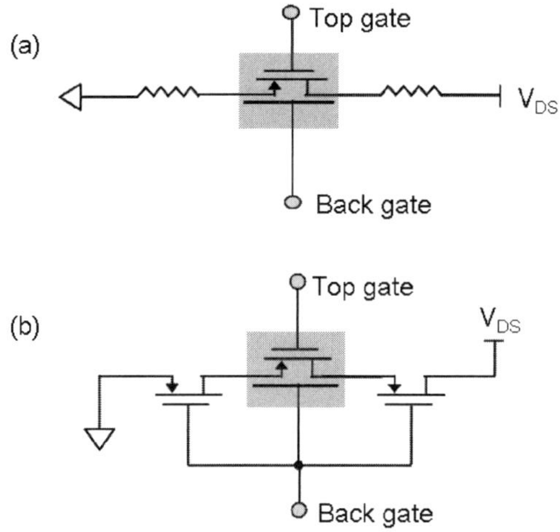

Fig. 4.14 Two approaches to model the SWNT-FET: (**a**) a FET with resistors at each end and (**b**) three FETs in series

gate is modeled by changing the doping concentration of the un-gated sections. The limitation of this approach, however, is that the source and drain resistances do not depend on the back-gate bias. The second and more accurate approach is to model the device as three separate FETs connected in series (Fig. 4.14b) with the nanotube at the two ends with only the back-gate control and the middle section with both the top and the back-gate control. Despite being computationally more intensive, the second model is used to fit the measured data due to its better accuracy.

Solid line in Fig. 4.15 shows a fit to the DC transport characteristics using the physical parameters shown in Table 4.1. The fitting parameters are mean free path in the channel and source/drain extension regions, parasitic coupling capacitance and

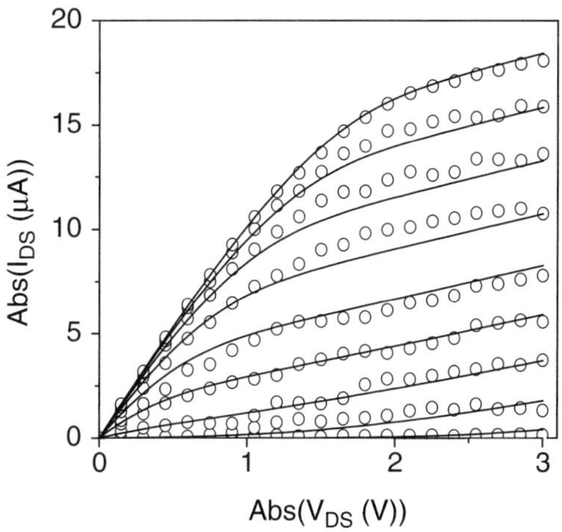

Fig. 4.15 Comparison of measured (*symbols*) and modeled (*solid line*) data. Both horizontal and vertical scales show absolute values of the drain voltage and the drain current, respectively. V_G varies between 1.7 and –1.3 V in increments of 300 mV

Table 4.1 The parameters used for modeling dc characteristics

T_{ox} (Al_2O_3)	10 nm	SWNT diameter	2 nm
K_{ox} (Al_2O_3)	9	Channel length	700 nm
T_{sub} (SiO_2)	130 nm	Source/drain SWNT length	400 nm
MFP* (channel)	100 nm	S/D work function (Φ_s)	4.6 eV
MFP* (S/D SWNT)	100 nm	SWNT work function (Φ_{SWNT})	4.5 eV
V_{sub}	−5 V	Gate capacitance (C_{ox})	70 aF/μm
β	0.13	Parasitic capacitance (C_c)	98 aF/μm

*Scattering mean free path.

β. In order to get the best fit, a coupling capacitance between the channel region and the drain is fitted as 12.7 aF/μm (this is linked to the parameter β and C_c). Due to the presence of the back gate and other parasitic couplings, the effective coupling capacitance between the top gate and the channel is only about 50% of the theoretical value if the back gate is removed. The agreement between the simulation and measurement is quite reasonable with less than 10% mismatch.

The simplified circuit for AC measurement used in the simulation is shown in the inset of Fig. 4.16 and the parameters used for modeling AC characteristic are listed in Table 4.2. The only fitting parameter is the parasitic load capacitance, C_L, which represents combined parasitic capacitance of the probe, connecting cables and the drain pad. The fitted value of 2.8 pF closely matches the estimated value. Figure 4.16 shows a comparison of the measured and simulated FRF using the parameters listed in Table 4.2. There is a 1 dB difference between the simulation and measurement.

The reasonably good agreement of modeling with both the DC and AC experimental data is obtained with the same set of physical parameters. This gives us confidence in projecting the SWNT-FET AC performance under more ideal conditions. If the load capacitance (C_L) is assumed to be zero and the probe resistance

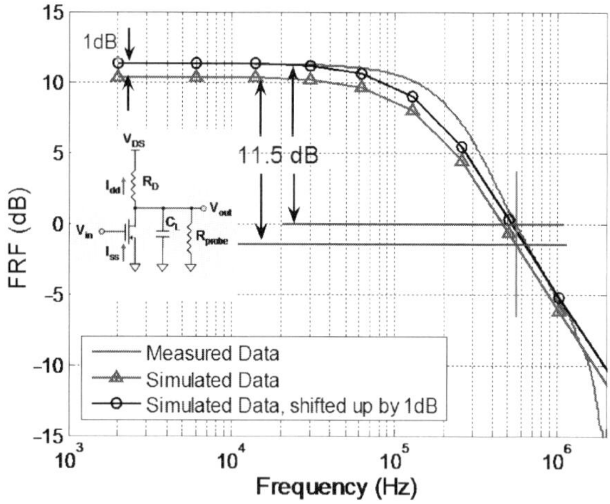

Fig. 4.16 *Solid line* without any symbol shows the measured FRF of the SWNT-FET. *Solid line* with "Δ" shows the simulation result. As can be seen, there is a 1 dB difference between the measurement and simulation. Better match can be achieved by shifting the simulation data up by 1 dB as shown by the *solid line* with "o"

Table 4.2 The parameters used for AC modeling

V_{DS}	-10 V	R_D	620 kΩ
I_{dd} (DC)	11.75 μA	R_{probe}	1 MΩ
I_{ss} (DC)	9.027 μA	C_L	2.8 pF
V_{in} (DC)	-0.3 V	V_{out} (DC)	-2.718 V

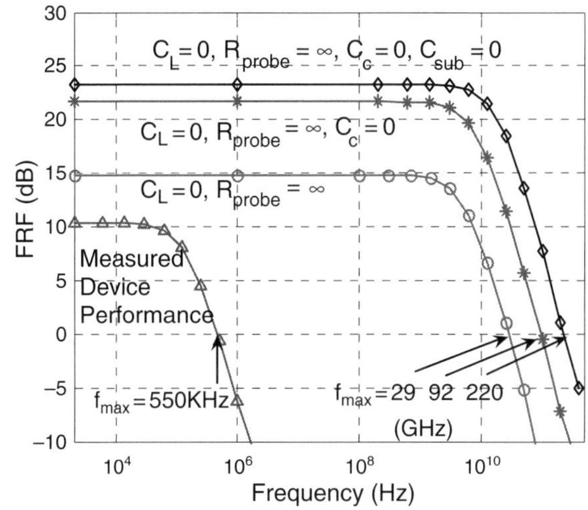

Fig. 4.17 The predicted FRF of the SWNT-FET circuit as a function of frequency under more ideal conditions. Symbol "Δ" shows the well-fitted voltage gain for the measurement circuit (with a 620 kΩ load resistor). Symbol "o" predicts the actual circuit performance without the interference from the measurement apparatus, i.e., the load capacitance is zero and the probe resistance is infinite. The curve with symbol "∗" predicts the circuit performance assuming there is no parasitic coupling capacitance C_c. Symbol "⋄" predicts the circuit performance assuming the function of the back gate is substituted with chemical doping. With a 620 kΩ load resistor, the voltage gain increases to 23 dB and the unity voltage gain frequency approaches 220 GHz

is assumed to be infinite (ideal probe), then the intrinsic low frequency gain of the circuit is about 15 dB and the voltage unity-gain frequency is up to 29 GHz (Fig. 4.17). In our device configuration, the device transconductance is degraded due to the presence of the back gate, which also controls the SWNT-FET channel potential in addition to the top gate. If back-gate coupling and the parasitic coupling are removed, the low frequency gain will be improved to 23 dB, and the voltage unity-gain frequency will increase to 220 GHz. The predicted cutoff frequency for this device based on extracted parameters is ∼50 GHz.

4.4 Conclusions

The field of AC characterization of nanoelectronics devices with intrinsically small dimensions such as a SWNT-FET is full of interesting challenges and opportunities. Here, the term intrinsic signifies the fact that the width of the material size cannot be easily scaled up using the top down lithography approach as in the case of conventional semiconductor technologies. Standard measurement methodologies are not capable of characterizing the RF and microwave properties of such devices. For rapid progress in this field, improvements are anticipated both in the fabrication and measurement technologies. If the fabrication technology sufficiently matures to allow formation of reliable ohmic contacts with resistances that approach the quantum conductance limit ($\sim$10 kΩ), then it may be possible to use numerical or on-wafer impedance transformation techniques [25] to characterize these devices with reasonable accuracy. Another brute force alternative to this is to realize a FET with several SWNT in parallel [26–28]. If multiple SWNTs of controlled chiralities can be grown in an aligned fashion and reliable quasi-ohmic contacts can be made to these nanotubes while minimizing parasitic capacitance of source and drain contacts, then one can envision a parallel nanotube based structure with impedance that is suitable for characterization using standard and well-developed approaches. The control of chiralities is a non-trivial problem but is critical to ensure reproducible ON/OFF ratios and threshold voltage characteristics. A few recently reported attempts demonstrate the feasibility of these concepts. Although the control of chiralities has not yet been achieved, the ability to grow multiple SWNTs in an aligned fashion on quartz and sapphire has been shown [29]. Similarly, another group has shown that a matte of SWNT can be used to build devices with sufficient signal for characterization using vector network analyzers [30]. Since research in this area is being pursued with great zeal and vigor, the author believes that we have only begun to scratch the surface and rapid breakthrough results are yet to be unveiled in the near future.

Acknowledgement The author would like to acknowledge collaborators, King Lee of Motorola, Dan Woodward of Tektronix, and Philip Wong and Jie Deng of Stanford University. The author would also like to thank Motorola for assistance in sample fabrication and characterization. The author would like to extend appreciation to Rudy Emrick and Vida Ilderem for their support of this work.

References

1. K. Alam and R. Lake, "Performance of 2 nm gate length carbon nanotube field-effect transistors with source/drain underlaps," *Applied Physics Letters*, vol. 87, p. 073104-1-3, 2005.
2. P. J. Burke, "AC performance of nanoelectronics: towards a ballistic THz nanotube transistor," *Solid-State Electronics*, vol. 48, pp. 1981–1986, 2004.
3. L. C. Castro, D. L. John, D. L. Pulfrey, M. Pourfath, A. Gehring, and H. Kosina, "Method for predicting f_T for Carbon Nanotube FETs," *IEEE Transactions on Nanotechnology*, vol. 4, pp. 699–704, 2005.
4. S. Hasan, S. Salahuddin, M. Vaidyanathan, and A. A. Alam, "High-frequency performance projections for ballistic carbon-nanotube transistors," *IEEE Transactions on Nanotechnology*, vol. 5, pp. 14–22, 2006.
5. J. Guo, S. Hasan, A. Javey, G. Bosman, and M. Lundstrom, "Assessment of high-frequency performance potential of carbon nanotube transistors," *IEEE Transactions on Nanotechnology*, vol. 4, pp. 715–721, 2005.

6. The RF and Microwave Handbook, edited by Muike Golio (CRC Press, 2000).
7. Z. Chen, J. Appenzeller, Y. Lin, J. S-Oakley, A. G. Rinzler, J. Tang, S. J. Wind, P. M. Solomon, and P. Avouris, "An integrated logic circuit assembled on a single carbon nanotube," *Science*, vol. 311, pp. 1735–1737, 2006.
8. J. Appenzeller and D. J. Frank, "Frequency dependent characterization of transport properties in carbon nanotube transistors," *Applied Physics Letters*, vol. 84, pp. 1771–1773, 2004.
9. D. J. Frank and J. Appenzeller, "High-frequency response in carbon nanotube field-effect transistors," *IEEE Electron Device Letters*, vol. 25, pp. 34–36, 2004.
10. S. D. Li, Z. Yu, S. F. Yen, W. C. Tang, and P. J. Burke, "Carbon nanotube transistor operation at 2.6 GHz," *Nano Letters*, vol. 4, pp. 753–756, 2004.
11. X. Huo, M. Zhang, P. C. H. Chan, Q. Liang, and Z. K. Tang, "High-frequency S parameters characterization of back-gate carbon nantoube field-effect transistors," *IEDM Technical Digest*, San Francisco, CA, pp. 691–694, 2004.
12. D. Singh, K. Jenkins, and J. Appenzeller, "Direct measurements of frequency response of carbon nanotube field effect transistors," *Electronics Letters*, vol. 41, pp. 280–281, 2005.
13. D. Singh, K. Jenkins, J. Appenzeller, D. Neumayer, A. Grill, and H.-S. P. Wong, "Frequency response of top-gated carbon nanotube field-effect transistors," *IEEE Transactions on Nanotechnology*, vol. 3, pp. 383–387, 2004.
14. S. Rosenblatt, H. Lin, V. Sazonova, S. Tiwari, and P. L. McEuen, "Mixing at 50 GHz using a single-walled carbon nanotube transistor," *Applied Physics Letters*, vol. 87, p. 153111, 2005.
15. Aaron A. Pesetski, J. E. Baumgardner, E. Folk, J. X. Przybysz, J. D. Adam, and H. Zhang, *Applied Physics Letters*, vol. 88, p. 113103, 2006.
16. I. Amlani, R. Zhang, J. Lewis, J. Deng, H.-S. P. Wong, and K. Lee, "First demonstration of AC gain from a nanotube based common-source amplifier," *IEDM Technical Digest*, San Francisco, CA, pp. 559–562, 2006.
17. I. Amlani, J. Lewis, R. Zg, K. Nordquist, S. Rockwell, and D. Woodward, "Approach to variable frequency measurements of carbon nanotube transistor," *Journal of Vacuum Science and Technology B*, vol. 24, pp. 3209–3212, 2006.
18. RF Measurements of Die and Packages, edited by S. A. Wartenberg (Artech house, Boston, 2002).
19. I. Amlani, R. Zhang, J. Tresek, and R. K. Tsui, "Field-effect and single-electron transistors based on single-walled carbon nanotubes catalyzed by Ni-Al thin films," *IEEE Transactions on Nanotechnology*, vol. 3, pp. 202–210, 2004.
20. A. Javey, J. Guo, Q. Wang, M. Lundstrom, and H. Dai, "Ballistic carbon nanotube field effect transistors," *Nature*, vol. 424, pp. 654–657, 2003.
21. Tektronix notes, "Fundamentals of Signal Integrity" (2005).
22. J. Deng and H.-S. P. Wong, "A compact SPICE model for carbon nanotube field effect transistors including non-idealities and its application — Part I: Model of the intrinsic channel region," Submitted to *IEEE Transactions on Electron Devices*, 2007.
23. J. Deng and H.-S. P. Wong, "A compact SPICE model for carbon nanotube field effect transistors including non-idealities and its application — Part II: Full device model and circuit performance benchmarking," Submitted to *IEEE Transactions on Electron Devices*, 2007.
24. T. A. Fjeldly, T. Ytterdal, M. S. Shur, Introduction to Device Modeling and Circuit Simulation (Wiley-Interscience, New York, 1998).
25. Jon Marten, IEEE Radio and Wireless Symposium Workshop, presentation entitled, "High Impedance S-parameter Measurements," San Diego CA, January 17–19, 2006.
26. A. Javey, J. Guo, D.B. Farmer, Q. Wang, E. Yenilmez, R. G. Gordon, M. Lundstrom, and H. Dai, "Self-aligned ballistic molecular transistors and electrically parallel nanotube arrays," *Nano Letters*, vol. 4, pp. 1319–1322, 2004.
27. J. Guo, S. Hasan+, A. Javey, G. Bosmon, and M. Lundstrom, "Assessment of high-frequency performance potential for carbon nanotube transistors," *IEEE Transactions on Nanotechnology*, vol. 4, pp. 715–721, 2005.

28. D. Akinwande, G.F. Close, and H.-S.P. Wong, "Analysis of the frequency response of carbon nanotube transistors," *IEEE Transactions on Nanotechnology*, vol. 5, pp. 599–605, 2006.
29. S. J. Kang, C. Kpcabas, T. Ozel, M. Shim, N. Pimparkar, M. A. Alam, S. V. Rotkin, and J. A. Rogers, "High-performance electronics using dense, perfectly aligned arrays of single-walled carbon nanotubes," *Nature Nanotechnology*, vol. 2, pp. 230–236, 2007.
30. A. Le Louarn, F. Kapche, J.-M. Bethoux, H. Happy, G. Dambrine, V. Derycke, P. Chenevier, N. Izard, M. F. Goffman, and J.-P. Bourgoin, "Intrinsic current gain cutoff frequency of 30 GHz with carbon nanotube transistors," *Applied Physics Letters* vol. 90, p. 233108(3), 2007.

Chapter 5
Device Simulation of SWNT-FETs

Jing Guo and Mark Lundstrom

5.1 Introduction

In recent years, significant progress in understanding the physics of single-walled carbon nanotube (SWNT) electronic devices and in identifying potential applications has occurred [1, 2]. In a SWNT [3, 4], low bias transport can be nearly ballistic across distances of several hundred nanometers. Deposition of high-κ gate insulators does not degrade the carrier mobility. The conduction and valence bands are symmetric, which is advantageous for complementary applications. The bandstructure is direct, which enables optical emission. Because of these attractive features, SWNTs are receiving much attention for potential nanoscale field-effect transistor (FET) applications. SWNT-FETs also provide a concrete context for exploring mesoscopic physics of one-dimensional nanostructures.

Device simulations of SWNT-FETs have been extensively reported in last 5 years [5–16]. Significant advances have been achieved in developing simulation methods, understanding device physics, and optimizing designs using modeling and simulation. This chapter is not intended to be an extensive review of the field. We focus on the device physics of SWNT-FETs as revealed by quantum device simulation. The simulation results are entirely taken from the work by the authors and their collaborators.

The chapter is organized as follows. Section 5.2 describes an atomistic simulation approach for SWNT-FETs using the non-equilibrium Green's function (NEGF) formalism. Section 5.3 discusses device characteristics at the ballistic limit. The effect of phonon scattering on device characteristics is discussed in Section 5.4. Section 5.5 assesses the high-frequency performance limits of SWNT-FETs at the ballistic limit and in the presence of scattering. Section 5.6 examines photoconductivity of SWNT-FETs, and Section 5.7 summarizes this chapter.

J. Guo (✉)
Department of Electrical and Computer Engineering, University of Florida, Gainesville, FL, 32611, USA

A. Javey, J. Kong (eds.), *Carbon Nanotube Electronics*, Series on Integrated Circuits and Systems, DOI 10.1007/978-0-387-69285-2_5,

5.2 SWNT-FET Simulation Using NEGF Approach

5.2.1 The NEGF Formalism

Conventional device simulation methods treat nanoscale devices using a top-down approach by inserting quantum extensions into semiclassical transport models. For molecular scale devices, the validity of the top-down approaches is questionable. The simulation methods for molecular scale devices should incorporate an understanding of atomistic structures, quantum effects, and multi-phenomena (e.g., heat dissipation, light emission). They should also be able to treat open boundaries for devices to connect to circuits, and be computationally efficient enough for routine simulations. New bottom-up simulation methods (from atomistic level to device level) are being explored to address these challenges.

The non-equilibrium Green's function (NEGF) formalism [17–19] provides an ideal approach for bottom-up device simulations due to the following reasons: (1) atomistic descriptions of devices can be readily implemented, (2) open boundaries can be rigorously treated, and (3) multi-phenomena (e.g., inelastic scattering, AC characteristics, light emission, and etc.) can be modeled.

Figure 5.1 summarizes how to apply the NEGF approach to a generic transistor. The transistor channel, which can be a piece of silicon, a SWNT, or a single molecule, is connected to the source and drain contacts. The channel conductance is modulated by the gate. The step-by-step procedure for the NEGF approach is described as follows [19]:

(1) Identify a suitable basis set and write down the Hamiltonian matrix H for the isolated channel. The self-consistent potential, which is a part of the Hamiltonian matrix, is included in the diagonal components of H. The size of the $N \times N$ Hamiltonian matrix is determined by the total number of basis functions in the channel region.

(2) Compute the self-energy matrices, Σ_1, Σ_2, and Σ_S, which describe how the channel couples to the source contact, the drain contact, and the dissipative processes (e.g., phonon scattering, electron–photon coupling), respectively. The source and drain self-energies can be computed using a recursive relation [19]. At the ballistic limit, $\Sigma_S = 0$. The dissipative processes can be treated by conceptually adding a "scattering contact" which in steady state takes carriers away from the initial states and put an equal number of carriers back to the final states. The carrier statistics of the "scattering contact" are determined by the distribution function of the channel, so that Σ_S needs to be iteratively solved with the Green's function using the self-consistent Born approximation [19].

(3) Compute the retarded Green's function,

$$G^r(E) = \left[(E + i0^+)I - H - \Sigma_1 - \Sigma_2 - \Sigma_S\right]^{-1} \tag{5.1}$$

Fig. 5.1 A generic transistor comprised of a device channel connected to source and drain contacts. The source–drain current is modulated by a third electrode, the gate. The quantities involved in the NEGF formalism are also shown

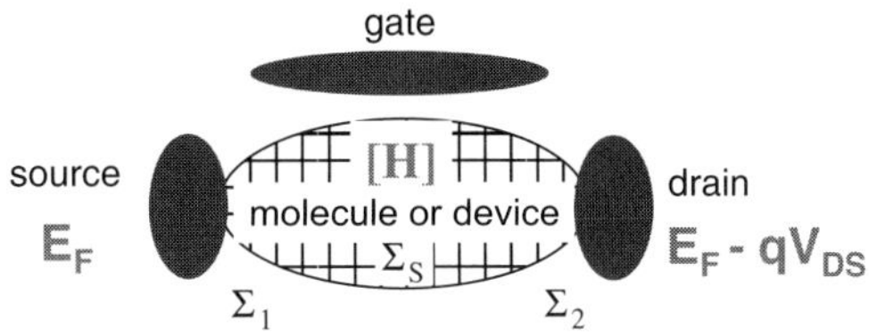

It is computationally expensive to compute the Green's function by directly inverting the matrix, because the inversion needs to be performed for a large number of energy grid points. Efficient computational techniques, such as a recursive algorithm, have been developed to reduce the computational cost by order of magnitude [20].

(4) Determine the physical quantities of interest from the Green's function matrix. For example, the electron density can be computed by integrating the diagonal entries of the following electron correlation function over energy,

$$G^n(E) = G^r \left[\Gamma_1 f_1(E) + \Gamma_2 f_2(E) + \sum_s^{in} \right] G^{r+} \tag{5.2}$$

where $f_{1,2}$ are the equilibrium Fermi functions of the two contacts, and $\Gamma_{1,2} = i(\Sigma_{1,2} - \Sigma_{1,2}^+)$ are the broadening functions of contact 1 or contact 2, and Σ_S^{in} is the in-scattering function of the dissipative processes.

(5) For a self-consistent solution, the NEGF transport equation is solved iteratively with the Poisson equation until self-consistency is achieved. The source current, for example, can then be computed as

$$I_S = (4e/h) \int_{-\infty}^{+\infty} dE \cdot \text{Trace} \left[\sum_1 f_1 A - \Gamma_1 G^n \right] \tag{5.3}$$

where $A = i(G^{\text{r}} - G^{\text{r}+})$ is the spectral function, and the factor of 4 comes from a spin degeneracy of 2 and a valley degeneracy of 2 in the carbon nanotube energy band structure.

The NEGF approach as outlined here is based on a number of simplifying assumptions such as the use of a single particle, mean field picture. See Datta for a discussion of the method [19].

5.2.2 SWNT-FET Simulation in a Real Space Basis Set

The NEGF approach as described above can be implemented in an atomistic basis set that consists of the p_Z orbitals of all the carbon atoms in the channel. There are four orbitals in the outer electron shell of a carbon atom (s, p_x, p_y, and p_z). One p_z orbital is often sufficient because the bands involving p_z orbitals are largely uncoupled from the bands involving the other orbitals, and the bands due to the s, p_x and p_y orbitals are either well below or well above the Fermi level, and therefore, unimportant for carrier transport [4, 21]. We use a tight-binding approximation to describe the interaction between carbon atoms, and only nearest neighbor coupling with a coupling parameter of $t = 3\,\text{eV}$ was considered. Notice that the so-called σ–π hybridization, which can be important for very small diameter tubes (<0.8 nm), is not treated in the model. The σ–π hybridization can be treated using a 4 orbital tight binding model [22], the extended Hückel theory [23], or an ab intio simulation [24], which is beyond the scope of this chapter.

The real space approach produces matrices in the size of the number of carbon atoms in the channel. A SWNT channel with a length of tens of nanometers consists of several thousand carbon atoms. The straightforward but computationally expensive approach is to compute G^r is by directly inverting the matrix $G_{inv} = (E+i0^+)I - H - \Sigma_1 - \Sigma_2 - \Sigma_S$. Significant computational savings can be achieved by exploiting the block tridiagonal structure of G_{inv}, which allows G^r to be computed by a recursive algorithm without inverting a large matrix [25]. If the SWNT channel consists of N_C carbon rings of a $(n, 0)$ nanotube, the computational cost of directly inverting G_{inv} goes as $O[(n^3 \times N_C^3)]$ whereas with the recursive algorithm it is only $O\,(n^3 \times N_C)$. By using the recursive algorithm, the computational cost only increases *linearly* with the tube length [21, 26].

5.2.3 SWNT-FET Simulation in a Mode Space Basis Set

A mode-space approach that significantly reduces the size of the Hamiltonian matrix when the potential around the tube is nearly invariant has also been developed [26]. The approach decouples the two-dimensional real space SWNT lattice to one-dimensional modes by performing a basis transform in the circumferential direction of the tube from the real space to the k space. The wave function in the circumferential direction of each mode is a plane wave with wave vector satisfying the periodic boundary condition. For coaxially gated SWNT-FETs, the potential is invariant around the SWNT, and the mode space approach is exact. For SWNT-FETs with planar gates, the mode space approach applies as long as the potential variation around the SWNT is smaller than the spacing between the subbands. This condition is satisfied by many SWNT-FETs demonstrated to date because the diameter of the tube is small compared to the gate oxide thickness and the spacing between the first and second subbands is as large as several hundreds of meV.

The mode space approach reduces the computational cost by orders of magnitude because (i) it reduces a two-dimensional problem to a set of one-dimensional problems and (ii) only the lowest few modes are relevant to transport and need to be treated. Routine device simulation and optimization becomes possible by using the mode space approach. In the cases when the symmetry around the tube is broken, the real space treatment becomes necessary. For example, for SWNTs with atomistic vacancies or defects, the variation around the circumferential direction of the SWNT is large, and the mode space approach no longer applies. A recent study based on a real space treatment showed that a single atomistic vacancy can decrease the drive current of a SWNT-FET by as large as 25% [27].

5.2.4 *Treatment of Metal–SWNT Contacts*

Most SWNT-FETs demonstrated to date operate as Schottky barrier transistors. Schottky barriers (SBs) exist at the metal–SWNT contacts [28]. An atomistic treatment of the metal contacts is not practical for routine device simulation, so a phenomenological treatment has been developed. In the mode space treatment, the self-energy of a mode is [19]

$$\Sigma_{\mathrm{C}}(E) = \tau g_{\mathrm{s}} \tau^{+}, \tag{5.4}$$

where τ is the coupling between the mode and the metal contact, g_{s} is the surface Green's function. The contact broadening function can be computed as,

$$\Gamma_{\mathrm{C}}(E) = i(\Sigma_{\mathrm{C}} - \Sigma_{\mathrm{C}}^{+}) = 2\pi \tau D_{\mathrm{s}} \tau^{+} \tag{5.5}$$

where D_{S} is the density-of-states (DOS) of the metal contact and satisfies $2\pi D_{\mathrm{s}} = i(g_{\mathrm{s}} - g_{\mathrm{s}}^{+})$. If one assumes that the density of states in metal D_{S} and the coupling between the mode and the contact τ are energy-independent in the energy range of interest, Eq. (5.5) can be simplified as

$$\Sigma_{\mathrm{c}} = -i\alpha t, \tag{5.6}$$

where $t \approx 3\,\mathrm{eV}$ is the C–C binding parameter, and α is a unitless parameter whose value is determined by the coupling strength between the SWNT and the metal contact and the DOS of the metal contact. It can be treated as a fitting parameter, whose value is determined by best fitting between the simulation results and the measurements [29].

5.3 Device Characteristics at the Ballistic Limit

Significant advances have been achieved in understanding the device physics of SWNT-FETs. SWNT-FETs have been made in two ways as shown in Fig. 5.2. The first kind of SWNT-FETs operate like an unconventional Schottky barrier FETs (Fig. 5.2(a)). At low gate voltages, a potential barrier is created in the channel region, the thermionic emission (TE) current over the top of the barrier is small and the transistor is turned off. At high gate voltages, the gate modulates the tunneling current by tuning the thickness of the Schottky barrier at the source end of the

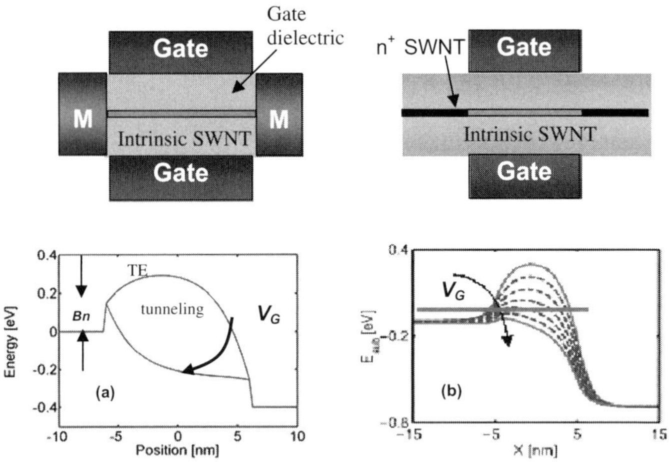

Fig. 5.2 Device schematic and conduction band profile of two kinds of SWNT-FETs. (**a**) A metal source and drain FET operating as a SBFET, and (**b**) a FET with doped source and drain extensions operating as a MOSFET. The different lines in the band diagrams correspond to different applied gate voltage values

channel. SWNT-FETs demonstrated at early stages all operate as SBFETs due to their simple fabrication process [30, 31]. SWNT-FETs with doped source and drain extensions, which operate like a conventional MOSFET, have been demonstrated recently [32–34]. As shown in Fig. 5.2(b), the gate modulate the channel conductance at both the off and on states. The doped source and drain extensions can be created by either electrostatic gating or chemical doping. The simpler electrostatic gating approach is good for proof of concept, but the chemical doping approach is technologically more relevant. We first discuss device physics and limitations of SWNT SBFETs, and then point out potential advantages of SWNT MOSFETs.

Figure 5.3 shows the (simulated) *I–V* characteristics of a mid-gap SWNT SBFET. The *I–V* characteristic is ambipolar. For $V_{GS}>V_{DS}/2$, the conduction band edge is pushed down by the gate voltage. Electrons can tunnel from the source contact into the channel region and produce electron current. For $V_{GS} < V_{DS}/2$, the valence band is lifted up by the gate voltage. Holes can tunnel from the drain contact into the valence band and produce hole current. The minimal leakage current is achieved at $V_{GS} = V_{DS}/2$, at which the electron current is equal to the hole current [35, 36]. In terms of band profile, it requires the conduction band bending at the source end of the channel to be symmetric to the valence band bending at the drain end of the channel. By considering that strong tunneling occurs in these devices because of the small effective mass and the thin barriers, the minimum leakage current of a SB SWNT-FET can be estimated as,

$$I_D = \frac{8ek_BT}{h} \langle T \rangle \, e^{-(E_G - qV_{DS})/2k_BT}, \tag{5.7}$$

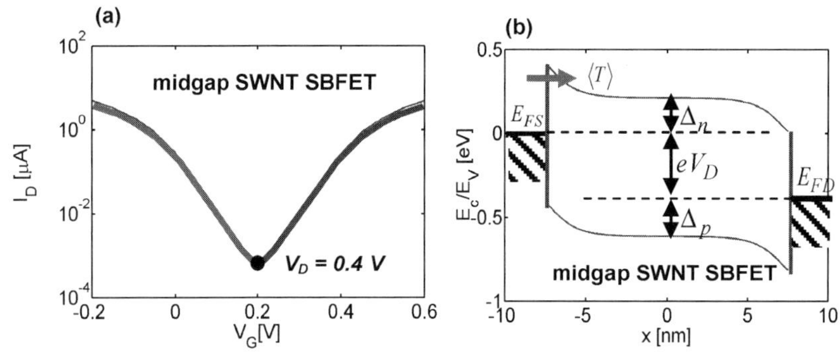

Fig. 5.3 Operation of a mid-gap SWNT SBFET. (**a**) The log (I_D) vs. V_{GS} characteristic, and (**b**) the energy band diagram at $V_{GS} = V_{DS}/2$. (Reprinted with permission from [36] © [2004] IEEE)

where <T> is the average current transmission coefficient. Equation (5.7) shows that the minimum leakage current exponentially depends on the SWNT bandgap and the source–drain voltage. We next explore the scaling properties of SWNT SBFETs in terms of the SWNT diameter and power supply voltage.

Figure 5.4 shows the I_D vs. V_G characteristics of SWNT SBFETs with three different nanotube diameters. The SWNT bandgap is approximately inversely proportional to the tube diameter, and can be expressed as $\sim 0.8\mathrm{eV}/d$ (in nm) [4]. Varying the tube diameter from 1 nm to 2 nm changes the SWNT bandgap from ~0.8 eV (a value in between the Si and Ge bandgaps) to ~0.4 eV (a value smaller than the Ge bandgap). Because the minimal leakage current exponentially increases as the tube bandgap decreases, the minimum leakage current increases dramatically as the tube diameter increases. At the same time, the on-current increases because it is easier to make good contacts to materials with smaller bandgaps, but the on–off ratio decreases significantly as the nanotube diameter increases [36].

We next examine power supply voltage scaling. Figure 5.5(a) shows the I_D vs. V_G characteristics of the SWNT SBFET with three different power supply voltages. For each power supply voltage, we define the off-current at the minimal leakage point

Fig. 5.4 Scaling of nanotube diameter. I_D vs. V_G characteristics at $V_D = 0.4$ V for the SWNT SBFET with different nanotube diameter. The solid line with circles is for (13, 0) SWNT (with d~1 nm), the solid line is for (17, 0) SWNT (with d~1.3 nm), and the dashed line is for (25, 0) SWNT (with d~2 nm). (Reprinted with permission from [36] © [2004] IEEE)

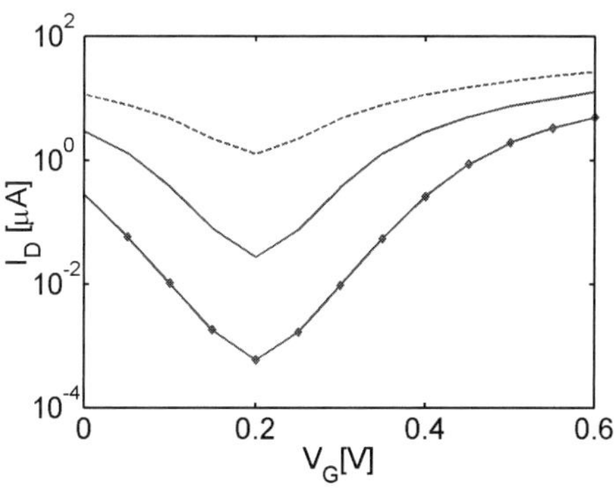

($V_{G,off} = V_D/2$ and $V_D = V_{DD}$), and the on-current is defined at $V_{G,on} = V_{G,off} + V_{DD}$ and $V_D = V_{DD}$. Figure 5.5(a) shows that the minimal leakage current increases exponentially as the power supply voltage increases, just as Eq. (5.7) predicts [35, 36]. Figure 5.5(a) also shows that the on-current increases with V_{DD}. The off-current vs. on-current for different power supply voltages is plotted in Fig. 5.5(b). The trade-off for reducing the off-current by lowering the power supply voltage is the degradation of on-current. The choice of power supply voltage will depend on the type of circuit applications.

We also simulated SWNT SBFETs with different metal-SWNT SB heights. When the gate insulator is thin, varying the SB height has a small effect on the qualitative feature of the I–V curve. The transistor is ambipolar regardless of the SB height. The reason is that the SB thickness is roughly the gate oxide thickness due to electrostatic screening length [8, 37] and the effective mass of carriers in SWNTs is small. The barrier is thin when the gate oxide is thin, and it is nearly transparent for any physical value of the SB height. Although the hole conduction current is larger than the electron conduction current for a SWNT SBFET with a zero SB height for holes, the transistor still qualitatively shows ambipolar characteristics when the oxide is thin ($<$10 nm) [5, 29].

One advantage of an ambipolar SWNT SBFET is that it can be used as either an n-type or a p-type FET in a CMOS application if the threshold voltage of the transistor is carefully designed [38], but the large leakage current increases the standby power. Acceptable leakage currents require a bandgap of at least ~0.8 eV (a nanotube diameter of less than 1 nm), but it is more difficult to make low SB contact to a SWNT with a smaller diameter. Carriers must tunnel into the channel at on-state, which lowers on-current. A zero SB to a SWNT is possible by using a right combination of the contact material and SWNT diameter [28, 39], but reducing the SB height to zero is still not enough for reaching the true ballistic performance limit at on-state, because even then a significant fraction of the current is carried by

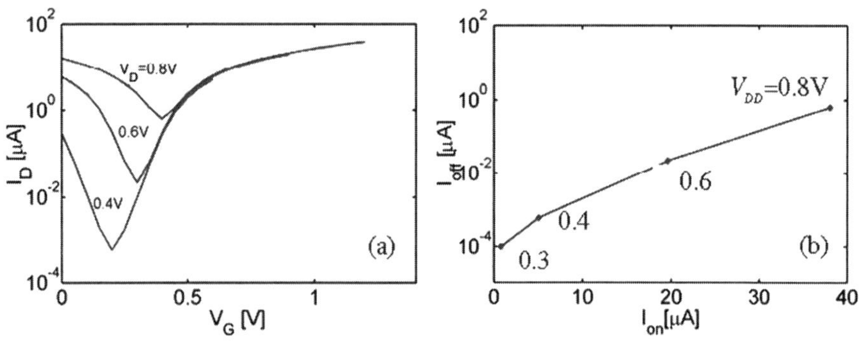

Fig. 5.5 Scaling of power supply voltage. (**a**) I_D vs. V_G characteristics under different power supply voltages for the SWNT-FET. For each power supply voltage, the drain is biased at the power supply voltage, $V_D = V_{DD}$. (**b**) The off-current vs. on-current for different power supply voltages. (Reprinted with permission from [36] © [2004] IEEE)

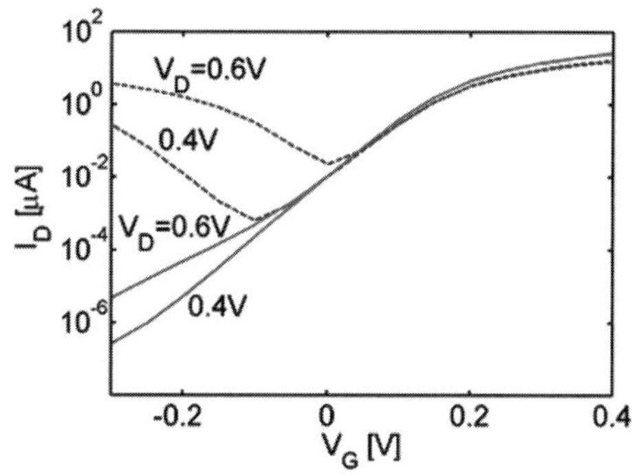

Fig. 5.6 I_D vs. V_G characteristics for the~SWNT MOSFET (*the solid lines*) and the SWNT SBFET (*the dashed lines*) at $V_D = 0.4$ V and 0.6 V

electrons below the Fermi level, which must tunnel into the semiconductor [40]. A SWNT MOSFET improves both the ballistic on-current and suppresses the minimal leakage current, as shown in Fig. 5.6. For a typical MOSFET, the Fermi level in the n^+ source and at the beginning of the channel is well above the conduction band, which is effectively a negative Schottky barrier [40]. The type of doping for the source and drain extensions defines the type of the transistor and suppresses ambipolar conduction. The n^+ doped source and drain extensions do not conduct hole current. The source–drain current, however, increases when the gate voltage decreases to sufficiently negative value, which leads to band-to-band tunneling from contacts to the channel region (Fig. 5.6).

5.4 Role of Phonon Scattering

The excellent carrier transport properties of SWNTs have been a topic of strong interest. Since the geometry of the SWNTs result in a smooth surface, surface roughness scattering can be expected to be negligible. Backscattering is suppressed by the reduction in phase space for one-dimensional conductors. The result is that mean-free-paths (mfps) of several hundred nanometers and an extraordinarily high mobility up to 20,000 cm^2/Vs are commonly observed under low bias conditions [41, 42]. Under high bias (>0.2 V), however, scattering by optical phonons (OPs) and zone boundary (ZB) phonons dominates, and the mfps decrease substantially to the order of 10 nm [43–45]. The effects of phonon scattering in metallic tubes and in semiconducting tubes for low-field transport (e.g., low-field mobility measurements and calculations) have been studied in literature. In this section, we focuses on high-field transport in the presence of phonon scattering in both SWNT SBFETs and SWNT\MOSFETs [46–49].

We first perform non-self-consistent simulations to examine the direct effect of phonon scattering on the on-current of SWNT SBFET [46]. The result shows that even for a channel several times longer than phonon scattering mfp, the direct effect of phonon scattering is small. For a SWNT-FET with an on-state conduction band profile as shown in Fig. 5.7, the ballistic on-current is computed to be 4.7 μA. By using the ballistic band profile, a non-self-consistent transport simulation in the presence of phonon scattering was performed. The simulated on-current in the presence of both acoustic phonon (AP) and OP/ZB scattering is $I_{nsc} \approx 4.4$ μA. The current is surprisingly close to the ballistic value (∼94%), although the channel length (50 nm) is several times longer than the phonon scattering mfp (∼10 nm). A more careful examination shows that the small degradation from the ballistic current is mostly due to the near elastic AP scattering instead of the OP/ZB scattering.

Figure 5.7 explains why the short mfp OP/ZB scattering has a small direct effect on the DC source–drain current. Electrons can tunnel into the channel from the source contact in the energy range close to the source Fermi level, as shown by the dashed line in Fig. 5.7. The source injected electrons accelerate as they travel along the channel and can possess enough energy to emit an OP/ZB phonon. After emitting an OP, whose energy is much larger in a SWNT than in common semiconductors, a backscattered electron encounters a much higher and thicker Schottky barrier at the source end, and can hardly return back to the source. The result is that the electrons rattle around in the channel and finally exit to the drain. The OP and ZB scattering, though occurs even in a short-channel SWNT-FET, has a small direct effect on the current under modest gate biases [47].

In order to describe the indirect effect of phonon scattering through self-consistent electrostatics [50], we next show self-consistent simulation results. The simulated device is a planar gate SWNT-FET with a 50 nm-long channel and an 8 nm-thick HfO_2 top gate insulator. Figure 5.8(a) shows that phonon scattering results in larger electron density in the channel region. The reason is that phonon scattering lowers the average carrier velocity, so the charge density in the channel must increase to maintain a similar source–drain current. Due to the self-consistent potential produced by the larger electron density, the band profile in the channel region moves up, as shown in Fig. 5.8(b). The source–drain current is reduced through self-consistent electrostatics. As shown by the current spectrums plotted in Fig. 5.8(b), compared to the non-self-consistently computed current spectrum

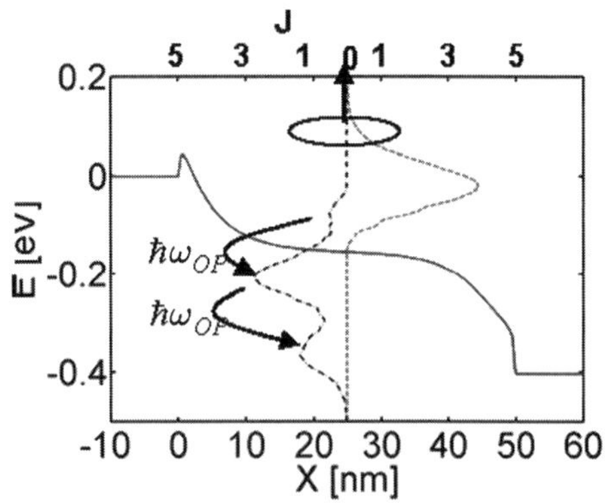

Fig. 5.7 The non-self-consistently computed source current spectrum (*the red dashed line*) and the drain one (*the black dash-dot line*) in the presence of phonon scatterings. The first subband profile of the simulated SWNT SBFET is shown by the solid line. (Reprinted with permission from [46] © [2005] AIP)

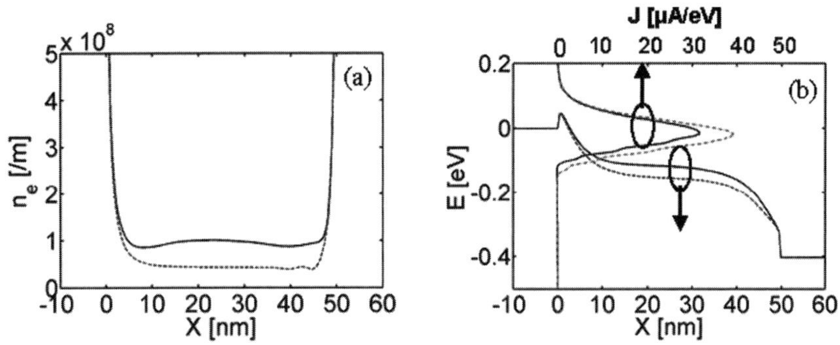

Fig. 5.8 The indirect effect of phonon scattering through self-consistent electrostatics. (**a**) The electron densities and (**b**) the first conduction subband profiles at the ballistic limit (*the blue dashed lines*) and in the presence of phonon scattering (*the black solid lines*). The top axis in (b) shows the current spectrum calculated in the presence of phonon scattering. The *solid line* shows the self-consistently computed current spectrum (using the solid band profile). For comparison, the *red dash-dot line* shows the non-self-consistently computed current spectrum (using the dashed band profile). (Reprinted with permission from [46] © [2005] AIP)

(the red dash-dot line), the self-consistently computed current spectrum (the solid line) delivers a smaller current density in a narrower energy window. The self-consistently computed on-current is $I_{sc} \approx 3.7$ μA, and is about 80% of the ballistic on-current. In summary, the direct effect of scattering reduces the ballistic current by 6%, but the indirect effect reduces it by an additional 14%.

The strength of the indirect effect strongly depends on the effectiveness of gate control. For a SWNT-FET with thin, high-κ gate insulator, the gate insulator capacitance is larger. The self-consistent potential produced by the charge in the channel region is small, and the indirect effect is also small. On the other hand, for a SWNT-FET with thick gate insulator and a small gate capacitance, the self-consistent potential produced by the charge in the channel is larger. The indirect effect through self-consistent electrostatics is also more important.

The indirect effect of phonon scattering also occurs in a SWNT MOSFET. Figure 5.9(a) and (b) sketch the subband profile and an OP emission event under a modest gate bias in a SWNT SBFET and in a SWNT MOSFET, respectively. For the SWNT SBFET, the current is controlled by quantum tunneling through the Schottky barrier at the source end of the channel. For the SWNT MOSFET, the current is controlled by thermionic emission over the top of the barrier at the beginning of the channel. For both transistors, OP/ZB phonon scattering can result in pile-up of charge in the channel, and affects the potential profile near the source through two-dimensional electrostatic effect. The indirect effect in the SWNT SBFET, however, is expected to be more severe than in the SWNT MOSFET. As shown in Fig. 5.9, the potential profile at the beginning of the channel in the SWNT SBFET varies much more rapidly than in the SWNT MOSFET. As a result, the source-injected electrons in the SWNT SBFET can gain enough energy and emit an optical phonon within a distance much shorter than that in the SWNT MOSFET. The

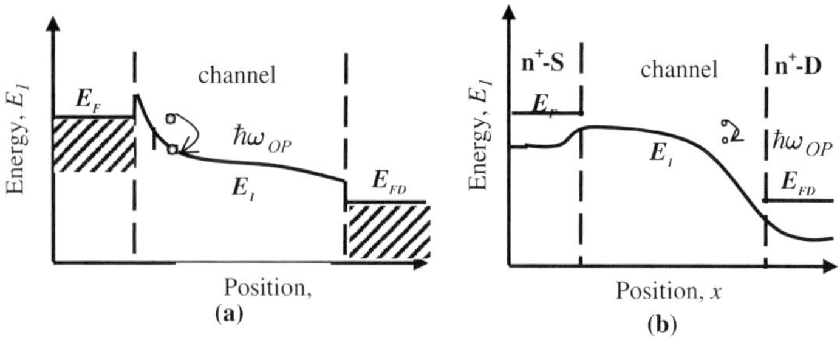

Fig. 5.9 The schematic sketch of the first subband profile and an OP emission event for (**a**) a SBFET and (**b**) a MOSFET with a similar channel length and on-current under a modest gate bias. (Reprinted with permission from [46] © [2005] AIP)

pile-up of charge in a SWNT SBFET occurs closer to the beginning of the channel, and has a larger effect on the potential profile at the beginning of the channel, which controls the source–drain current. The indirect effect, therefore, is expected to be more important in the SWNT SBFET.

It is also interesting to examine the effect of phonon scattering on the DC off-current of the FETs. We found that phonon scattering has a small effect on the minimal leakage current of the SWNT SBFET, but significantly affect the minimal leakage current of the SWNT MOSFET, as shown in Fig. 5.10. In the presence of phonon scattering, the minimal leakage current is much larger and occurs at a larger gate voltage [49]. The reason is explained in Fig. 5.11. As the gate voltage decreases, the source–drain current first exponentially decreases in the subthreshold region. As the gate voltage is sufficiently negative, the source–drain current reaches a minimum value and then exponentially increases due to the onset of band-to-band tunneling. At the ballistic limit, band-to-band tunneling occurs when the valence band in the channel aligns with the conduction band in the source extension, as shown in Fig. 5.11(a). In the presence of phonon scattering, band-to-band tunneling starts to play an important role before the valence band in the channel aligns

Fig. 5.10 The I_D vs. V_G characteristics at $V_D = 0.1$ V for the SWNT MOSFET at the ballistic limit (*the dashed line*) and in the presence of phonon scattering (*the solid line with circles*). (Reprinted with permission from [49] © [2005] AIP)

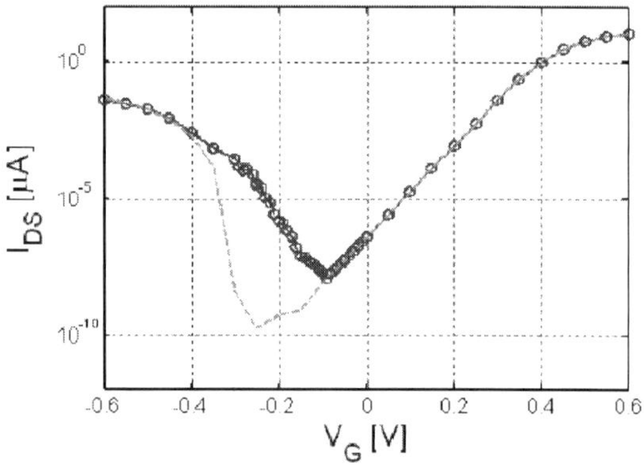

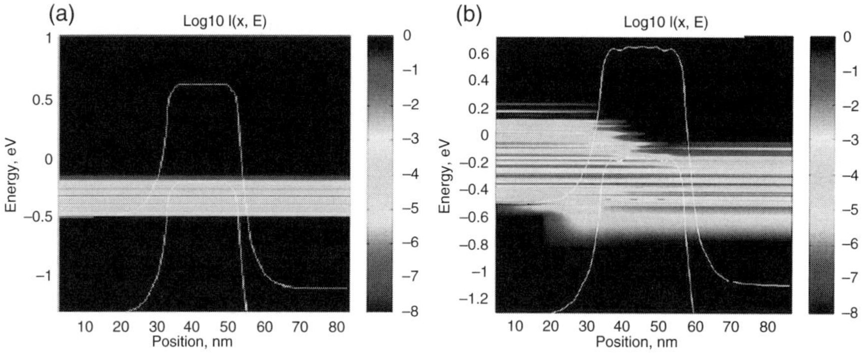

Fig. 5.11 The current spectrum at the minimal bias point (**a**) for the ballistic limit and (**b**) in the presence of phonon scattering. The conduction and valence subband profiles of the SWNT MOSFET are also shown. (Reprinted with permission from [49] © [2005] AIP)

with the conduction band in the source extension as gate voltage decreases due to a phonon-assisted tunneling process, as shown in Fig. 5.11(b). The minimal leakage current, therefore, is reached at a larger gate voltage. The results indicate that it is important to treat phonon-assisted scattering in order to accurately compute the minimal leakage current. The steep subthreshold slope in the band-to-band tunneling current promises a new type of transistor with a subthreshold swing less than 60 mV/dec [5]. Phonon-assisted tunneling again plays an important role in determining the exact value of the subthreshold slope for the band-to-band tunneling current.

5.5 High-Frequency Performance Limits

The excellent carrier transport properties of SWNTs lead to strong interest for high-speed device applications. The low-field mobility is as high as 20,000 cm^2/Vs [42, 51]. In addition, the SWNT-FET also provides a concrete context to study time-dependent quantum transport in one-dimensional nanostructures. The first experiments on AC characteristics were reported by Appenzeller and Frank in 2004, with a measurement frequency up to 580 MHz [52, 53]. The subsequent progress has been rapid. Measurements with a frequency up to 2.6 GHz [54], 10 GHz [55], and 50 GHz [56] have been reported. A five-stage ring oscillator built on a single tube has been demonstrated [38]. Theoretical works predict THz operation at the ballistic limit [9, 14, 57–59]. In this section, we discuss our work on assessing the high-frequency performance limits at the ballistic limit and in the presence of phonon scattering.

Figure 5.12 shows a small-signal equivalent circuit model for a SWNT-FET. The high-frequency performance of a SWNT-FET can be assessed using the model under quasi-static approximation. The equivalent circuit model for the intrinsic SWNT-FET is shown within the dashed rectangle in Fig. 5.12. The parameters are

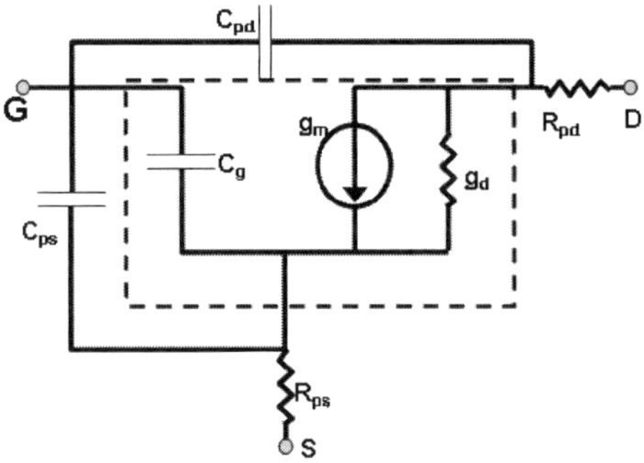

Fig. 5.12 A small-signal equivalent circuit model of a SWNT-FET for quasi-static simulations. The dashed rectangle shows the equivalent circuit of the intrinsic SWNT-FET. Parasitic capacitances and resistances are also shown. (Reprinted with permission from [59] © [2005] IEEE)

obtained by running self-consistent quantum simulations and numerically evaluating the derivatives. The intrinsic gate capacitance C_g and the transconductance g_m are

$$C_\mathrm{g} = \left.\frac{\partial Q_\mathrm{ch}}{\partial V_\mathrm{g}}\right|_{V_\mathrm{d}}, \quad g_\mathrm{m} = \left.\frac{\partial I_\mathrm{d}}{\partial V_\mathrm{g}}\right|_{V_\mathrm{d}} \tag{5.8}$$

where Q_ch is the total charge in the SWNT channel and I_d is the source–drain current. Here, we assume that gate electrostatic control is good, so that the charge in the channel is equal to that in the gate when parasitic capacitance is zero. The source–drain conductance $g_\mathrm{d} = \partial I_\mathrm{d}/\partial V_\mathrm{d}|_{V_\mathrm{g}}$ is obtained by running DC simulations with slightly different V_d. At low V_d (linear region), $1/g_\mathrm{d}$ accounts for the channel resistance, which includes the quantum resistance of a ballistic channel. The output conductance, g_d, is small at high V_d because the current saturates.

The parasitic capacitance between the gate and the source (drain) is treated as $C_\mathrm{ps}(C_\mathrm{pd})$, and the parasitic resistance of the source (drain) contact is treated as $R_\mathrm{ps}(R_\mathrm{pd})$. An estimation shows that the parasitic capacitance plays a more important role than the parasitic resistance in the state-of-the-art SWNT-FETs [14]. For a SWNT-FET with an equivalent circuit in Fig. 5.12, the cut-off frequency is

$$f_\mathrm{T} \approx \frac{1}{2\pi}\frac{g_\mathrm{m}}{C_\mathrm{g} + C_\mathrm{pS} + C_\mathrm{pD}}. \tag{5.9}$$

The parasitic capacitances lower the cut-off frequency by orders of magnitude in state-of-the-art SWNT-FETs. In addition, parasitic resistance also imposes challenges for probing the intrinsic response of the SWNT-FET. The parasitic capacitance is dominant because the effective channel width (the SWNT diameter) is often much smaller than the width of the contacts fabricated by microelectronic processes. Lowering the parasitic capacitance is the most important issue to improve the high-frequency performance of state-of-the-art SWNT-FETs. Two possible solutions to significantly reduce the effect of parasitic capacitances are (i) to fabricate a SWNT-FET with one-dimensional needle-like source and drain contacts and (ii) to fabricate a SWNT-FET with a closely packed SWNT array as the channel. To assess the performance limit, the intrinsic cut-off frequency can be computed by assuming zero parasitic capacitance,

$$f_{\mathrm{T}} \approx \frac{1}{2\pi}\frac{g_{\mathrm{m}}}{C_{\mathrm{g}}}. \tag{5.10}$$

Figure 5.13(a), which plots the intrinsic cut-off frequency as a function of the gate voltage for a SWNT SBFET, shows that the intrinsic cut-off frequency drops significantly as V_{G} increases to a large enough value. Figure 5.13(b), which plots the intrinsic gate capacitance vs. the gate voltage at $V_{\mathrm{D}} = V_{\mathrm{DD}}$, indicates that the drop of the cut-off frequency is due to the increase of the intrinsic gate capacitance at large V_{G}.

In order to understand why the gate capacitance increases at large gate overdrive in the ballistic limit, we sketch the band diagrams at a low V_{G} and a high V_{G} for a ballistic SWNT SBFET, as shown in Fig. 5.14(a and b), respectively. When V_{G} is low, only $+k$ states in the channel are occupied and contribute to the intrinsic gate capacitance because the drain injection is negligible. In contrast, at high V_{G}, the first subband edge in the channel is low, and the bottom of the $-k$ states are populated due to the drain injection and reflection of soucre-injected carriers by the barrier at the drain end of the channel. As a result, the channel capacitance significantly increases when the $-k$ states at the bottom of the subband begins to be populated, where a singularity in density-of-states (DOS) exists. Thus the cut-off frequency decreases.

We next compare the intrinsic cut-off frequency of the ballistic SWNT-FET to that of a ballistic n-type Si MOSFET. Figure 5.15, which plots the projected intrinsic cut-off frequencies vs. the transistor channel length, shows that the cut-off frequency of the SWNT-FET is about 50% higher than that of the n-type Si MOSFET. The larger cut-off frequency of the SWNT-FET at the ballistic limit is due to larger band-structure-limited velocity of the SWNT channel.

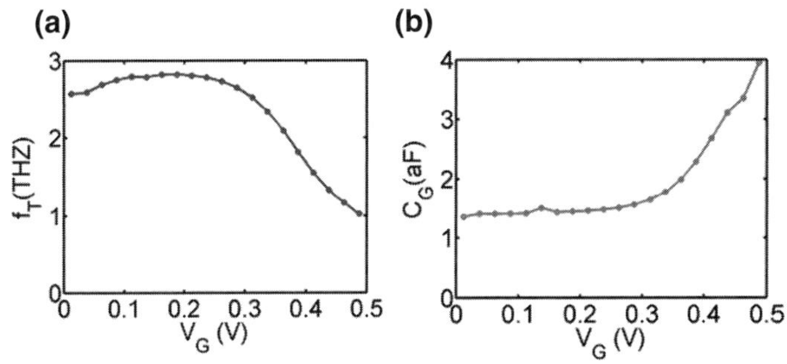

Fig. 5.13 (**a**) The intrinsic cut-off frequency and (**b**) the intrinsic gate capacitance versus the gate voltage for the ballistic SWNT SBFET. The channel length is 50 nm and the top HfO_2 gate insulator thickness is 8 nm [29]. No parasitic capacitance and resistance are included. (Reprinted with permission from [59] © [2005] IEEE)

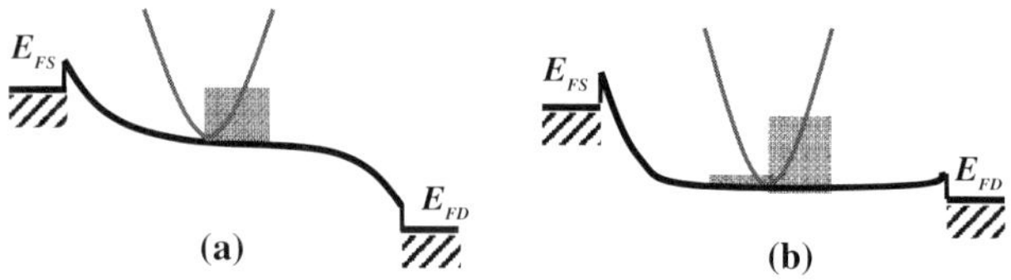

Fig. 5.14 Sketch of conduction band profile at (**a**) a low gate overdrive and (**b**) a high gate overdrive for a ballistic SWNT SBFET. The occupied states are covered by the gray windows. (Reprinted with permission from [59] © [2005] IEEE)

Phonon scattering has a large effect on the intrinsic cut-off frequency of a SWNT SBFET, though its direct effect on DC current is small [60]. We simulated a SWNT-FET with good gate electrostatic control (with a 5 nm-thick, coaxial high-κ gate insulator), in which the indirect effect through self-consistent electrostatics is also small. The transconductance and on-current in the presence of phonon scattering is above 80% of the ballistic value for a channel length up to 200 nm. In contrast, Fig. 5.16, which plots the intrinsic cut-off frequency versus the channel length at the ballistic limit (the circles) and in the presence of phonon scattering (the crosses), shows that phonon scattering significantly lowers f_T. The dashed line is a fitting of the ballistic result by $f_T = 110\,GHz \cdot \mu m/L_{ch}$, and the solid line is a fitting of the scattering result by $f_T = 40\,GHz \cdot \mu m/L_{ch}$. Although the transistor delivers a near ballistic DC on-current (>80%), the cut-off frequency in the presence of phonon scattering is only about 40% of the ballistic value for the simulated channel lengths. The reason is that phonon scattering leads to random walks of electrons and lowers the average carrier velocity, as shown in the inset of Fig. 5.16. The charge piles up in the channel and the intrinsic gate capacitance C_g increases significantly. The cut-off frequency, which is determined by g_m/C_g decreases mostly due to the increase of C_g.

The SWNT-FET examined above has a thin high-κ gate insulator and a large gate insulator capacitance C_{ins}. The SWNT-FET operates close to the so-called quantum capacitance (C_Q) limit [61]. The indirect effect of phonon scattering through self-consistent electrostatics is small. The gate capacitance, which is the series

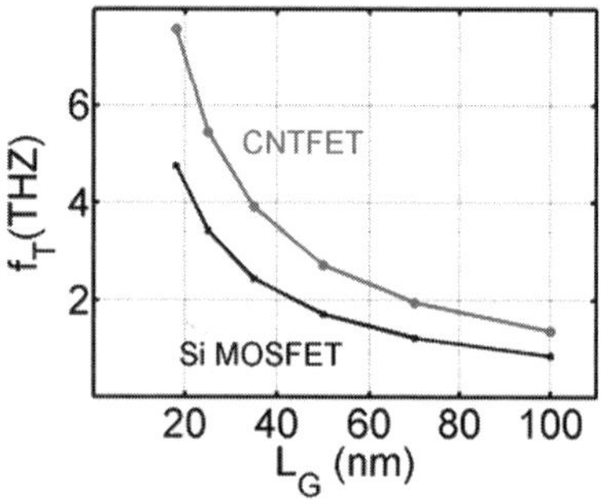

Fig. 5.15 The intrinsic cut-off frequencies for the ballistic SWNT SBFET and a ballistic double-gate Si MOSFET versus the channel length [59]. The performance limits are assessed and no parasitic capacitance and resistance is included. (Reprinted with permission from [59] © [2005] IEEE)

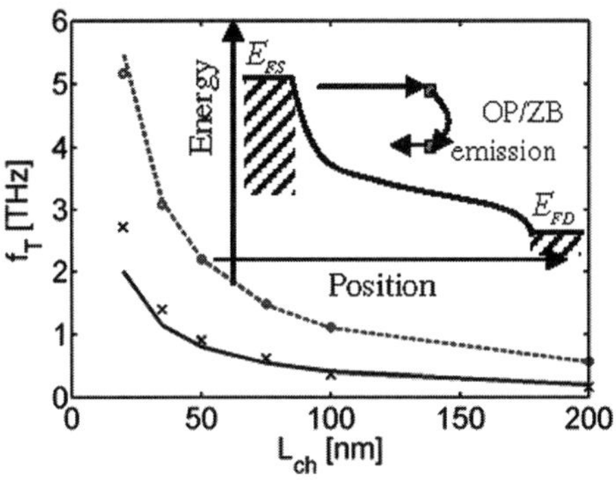

Fig. 5.16 The cut-off frequency versus the channel length at on-state ($V_D = V_G = 0.5$ V). The *circles* are numerically computed f_T at the ballistic limit and the *dashed line* is a fitting curve of $f_T = 110\,\text{GHz}\,\mu\text{m}/L_{ch}$. The *crosses* are numerically computed f_T in the presence of phonon scattering and the *solid line* is a fitting curve of $f_T = 40\,\text{GHz}\,\mu\text{m}/L_{ch}$. The inset sketches a source-injected electron, which gets backscattered by emitting an OP/ZB phonon. (Reprinted with permission from [60] © [2005] IEEE)

combination of C_{ins} and C_Q, is close to the quantum capacitance because $C_{ins}>C_Q$. We also examined the case when the gate insulator is thick and the transistor operates close to the conventional MOSFET limit [13]. Phonon scattering does not lead to an increase of the intrinsic gate capacitance because $C_{ins}<<C_Q$ thus $C_g \approx C_{ins}$. It does, however, severely degrade the DC source-–drain current and transconductance indirectly through the self-consistent electrostatic effect. The intrinsic cut-off frequency, which is determined by g_m/C_g, still decreases, but mostly due to the decrease of g_m, instead of due to the increase of C_g. In summary, how scattering affects the performance of these devices depends not only on the physics of scattering, but also on device design.

5.6 Optoelectronic Phenomena

Since the first demonstration of electroluminescence (EL) from a SWNT-FET a few years ago [62], significant progress has been achieved in SWNT optoelectronics [63–69]. The channel of the device consists of a single molecule, which makes it the smallest light emitting device demonstrated to date. Although at its early stage, the advance of SWNT optoelectronics might eventually lead to important applications. In this section, we apply the NEGF approach to simulate photoconductivity of a SWNT SBFET [70].

Figure 5.17 shows the simulated device geometry. A back-gated SWNT SBFET is illuminated. The light is polarized along the SWNT channel. We simulate the photo-current as a function of the light intensity and photon energy. A band-to-band transition picture is used. Notice that the electrostatic binding energy of the so-called exciton (an electron–hole pair) can be large in a SWNT due to its one-dimensional geometry which results in decreased electrostatic screening [71]. Excitons may play an important role in SWNT optoelectronic devices. Although the treatment of excitons in a SWNT device simulation warrants careful future studies, it is beyond the scope of this discussion.

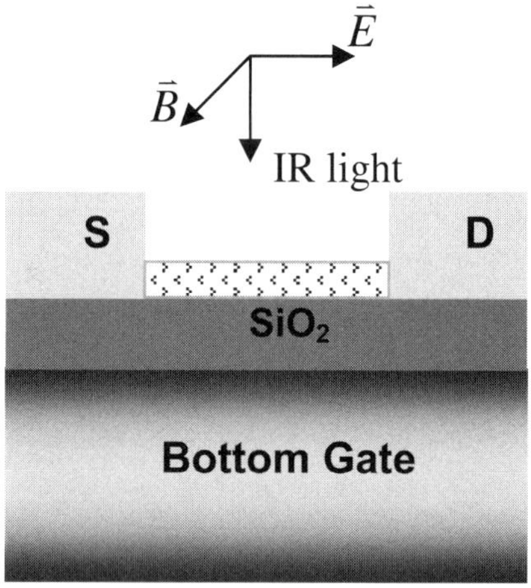

Fig. 5.17 The simulated SWNT SBFET under infrared (IR) light illumination. The electric field, , is polarized along the SWNT channel. The intrinsic channel has a length of L_{ch}=15 nm. The SiO_2 bottom oxide thickness is 8 nm. (Reprinted with permission from [70] © [2006] AIP)

Figure 5.18(a) shows the I_D–V_G characteristics of the transistor without light illumination (the solid line) and with different illumination intensities. The dark current shows a strong dependence on the gate voltage and the minimum leakage current is obtained at $V_G = V_D/2$. In contrast, the photocurrent, which is determined by the illumination intensity and the quantum efficiency, is nearly independent of the gate voltage. As the illumination intensity increases, the photocurrent becomes dominant and the dependence of I_D on V_G becomes much weaker. The source–drain current increases as the illumination intensity increases. The increases is especially obvious for bias points close to $V_G = V_D/2$, at which the dark current reaches the

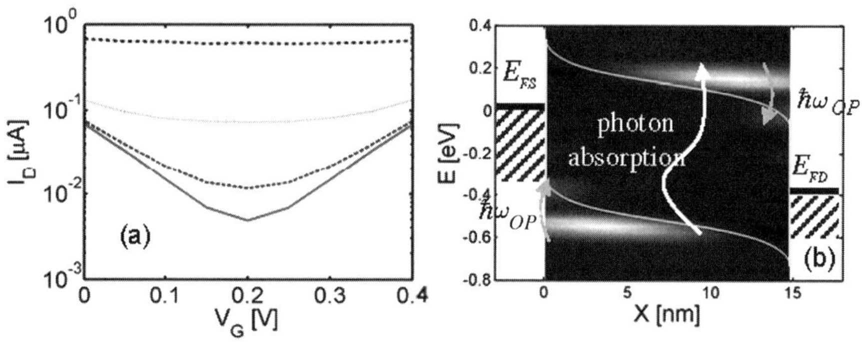

Fig. 5.18 (**a**) I_D vs. V_G characteristics without light illumination (*the solid line*) and with three different illumination power densities, 10^5 W/cm^2 (*the dashed line*), 10^6 W/cm^2(*the dotted line*), and (*the dash-dot line*). (**b**) The energy and position resolved current spectrum, *J*(*E*, *x*), on a grayscale plot at $V_G = V_D/2 = 0.2$ V. A brighter color represents a larger value. (Reprinted with permission from [70] © [2005] AIP)

minimum value. The results are in qualitative agreement with the experiment on a longer channel device [64].

Figure 5.18(b) plots the energy and position resolved current spectrum, $J(E, x)$, under a illumination intensity of 10^7 W/cm^2 and photon energy $\hbar\omega = 0.7$ eV at $V_G = V_D/2$. The conduction subband profile is symmetric to the valence subband profile. Absorption of a photon by an electron in the valence band creates an electron–hole pair. Because the conduction band is coupled more strongly to the drain than to the source due to the SB for electrons at the source end of the channel, electrons flow to the drain. On the other hand, because the valence band is coupled more strongly to the source due to the SB for holes at the drain end of the channel, holes flow to the source. The asymmetric couplings of electrons and holes to different contacts separate photo-generated electrons and holes, and the SWNT SBFET operates as a photodetector.

We next examine the effect of phonon scattering on the photocurrent spectrum. Figure 5.18(b) shows that a photon-generated electron can be accelerated by the electric field. Enough kinetic energy can be gained near the drain, and an OP/ZB phonon can be emitted. The backscattered electron, however, is unlikely to return back to the source because it encounters a much higher and thicker SB. Phonon scattering has a small *direct* effect on the photocurrent, but it significantly changes the energy-resolved photocurrent spectrum. At the ballistic limit, only one peak appears in the spectrum of the electron photo-current. In the presence of phonon scattering, additional peaks appear at energies of $n\hbar\omega_{OP}$ (where n is an integer) below the main peak due to OP/ZB phonon emission, as shown in Fig. 5.18(b).

We next investigate how the source–drain current depends on the photon energy and the SWNT diameter. Figure 5.19 plots the source–drain current versus the photon energy for three zigzag SWNTs with different diameters and subband gaps. When the photon energy is low, the dominant component of the source–drain current is the dark current, which exponentially increases as the tube diameter increases [36]. Even for the photon energies below the subband gap energy, a considerable photocurrent flows due to photon-assisted tunneling [72]. (In a quantum mechanical treatment, electrons in the valence subband can absorb a photon with an energy smaller than the subband gap, and then tunnel into the conduction band if an electric field exists.) The peak value of the source–drain current is reached when the photon energy of the light is slightly above the subband gap energies because the source/drain contact broadening and weak quantum confinement along the short SWNT channel direction shift the density-of-states (DOS) peak slightly away from the subband edges.

Multiple absorption peaks appear in the infrared, visible, and ultraviolet regions for each simulated device because the quantum confinement in the circumferential direction of the SWNT leads to discrete one-dimensional subbands. The absorption peaks are determined by the SWNT diameter. As the SWNT diameter increases, the bandgap decreases, and the absorption peaks shift to lower photon energies. The result is in agreement with the experiment [64]. The absorption wavelength, therefore, can be engineered by properly choosing the SWNT diameter, which offers another dimension for device design.

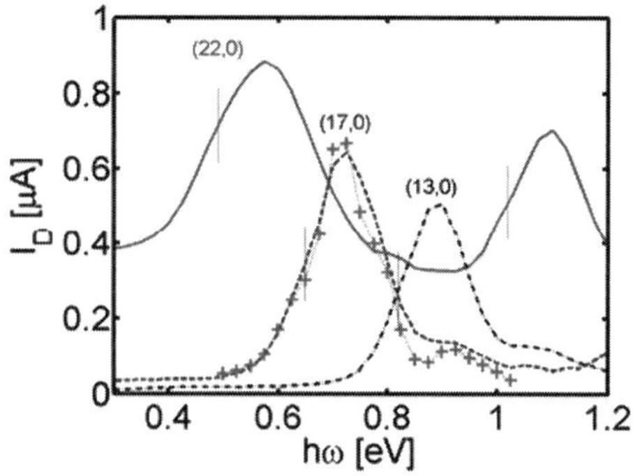

Fig. 5.19 The source–drain current versus the photon energy in for a (22, 0) SWNT with $E_g \approx 0.49$ eV (*the magenta solid line*), a (17, 0) SWNT with $E_g \approx 0.63$ eV (*the blue dashed line*), and a (13, 0) SWNT with $E_g \approx 0.82$ eV (*the black dash-dot line*) under the illumination intensity of at $V_G = V_D/2 = 0.2$ V in the presence of electron–phonon coupling. For comparison, the pink dotted line with crosses shows the current of the (17, 0) SWNT without electron–phonon coupling. The vertical bars show the subband gaps (Reprinted with permission from [70] © [2006] AIP)

The effect of electron–phonon coupling on the photocurrent is also examined in detail. The pink dotted line with symbols in Fig. 5.19 plots the photocurrent without electron–photon coupling for the (17, 0) SWNT, in order to compare to that in the presence of phonon scattering (the blue dashed line). Electron–phonon coupling (i) slightly broadens the main photocurrent peak near $\hbar\omega_{\mathrm{IR}} \approx 0.7\,\mathrm{eV}$, (ii) smoothes the oscillations on the I_{D} vs. $\hbar\omega_{\mathrm{IR}}$ curve, and (iii) increases I_{D} by about 10% at energies of about $\hbar\omega_{\mathrm{OP}}$ above the main peak. Electron–phonon coupling reduces the carrier life time and broadens the singularities of the DOS at the edge of the one-dimensional subbands. As a result, the photocurrent peak near $\hbar\omega_{\mathrm{IR}} \approx 0.7\,\mathrm{eV}$ is slightly broadened. Electron–phonon coupling breaks coherent transport in the channel and washes out quantum interference effect. The oscillations in the I_{D} vs. $\hbar\omega_{\mathrm{IR}}$ curve, therefore, are smoothened. Furthermore, the increase of the source–drain current in the presence of phonon scattering near $\hbar\omega_{\mathrm{IR}} \approx 0.7\,\mathrm{eV} + \hbar\omega_{\mathrm{OP}}$ is due to a phonon-assisted photocurrent [73].

5.7 Summary

A bottom-up simulation approach based on the NEGF formalism has been developed for SWNT-FETs. The approach has been applied to simulate DC characteristics, high-frequency performance limits, and optoelectronic characteristics of SWNT-FETs. The DC simulation has matured to such a point that the measured *I–V* characteristics can be quantitatively described by device simulation [29]. Device simulation has been playing an important role in understanding experiments and suggesting design optimizations for SWNT-FETs. Many challenges, however, still remains. Although SWNT-FETs are being extensively explored for electronics applications, the first application may be in the area that large scale integration is less important (such as sensors [74, 75]). Extending the approach to simulate new device applications beyond CMOS, such as SWNT sensors, is necessary. The Kondo effect and single electron charging [76] have been observed in SWNT-FETs. The effects are especially important at low temperatures. The current NEGF approach is limited to a single particle picture and a mean field theory. In order to capture these effects, extending the approach to treat strong electron–electron correlation is required. The high-frequency characteristics have been simulated based on quasi-static approximation. In order to capture non-quasi-static effects, a time-dependent transport formalism needs to be used. Excitons can qualitatively change the device optoelectronic characteristics. Treating excitons in device simulations imposes another challenge. The field of SWNT device simulation has benefited from the rapid progress of experiments in the last decade. We expect that close interaction between theorists and experimentalists continues to play an important role in addressing these challenges in the future.

Acknowledgements The authors would like to thank their collaborators, A. Javey, H. Dai, S. Datta, M. Alam, M. P. Anantram, S. Hasan, S. Koswatta, N. Neophytou, Y. Yoon, and Y. Ouyang, who contributed to the work described here.

References

1. P. Avouris, J. Appenzeller, R. Martel, and S. J. Wind, "Carbon nanotube electronics," *Proceedings of the IEEE*, vol. 91, pp. 1772–1784, Nov, 2003.
2. P. L. McEuen, M. S. Fuhrer, and H. K. Park, "Single-walled carbon nanotube electronics," *IEEE Transactions on Nanotechnology*, vol. 1, pp. 78–85, Mar, 2002.
3. S. Iijima and T. Ichihashi, "Single-shell carbon nanotubes of 1-nm diameter," *Nature*, vol. 363, pp. 603–605, Jun 17, 1993.
4. R. Saito, G. Dresselhaus, and M. S. Dresselhaus, *Physical properties of carbon nanotubes*. London: Imperial College Press, 1998.
5. J. Appenzeller, Y. M. Lin, J. Knoch, Z. H. Chen, and P. Avouris, "Comparing carbon nanotube transistors – The ideal choice: A novel tunneling device design," *IEEE Transactions on Electron Devices*, vol. 52, pp. 2568–2576, Dec, 2005.
6. M. P. Anantram, "Current-carrying capacity of carbon nanotubes," *Physical Review B*, vol. 62, pp. R4837–R4840, Aug 15, 2000.
7. G. Pennington and N. Goldsman, "Semiclassical transport and phonon scattering of electrons in semiconducting carbon nanotubes," *Physical Review B*, vol. 68, pp. 045426, Jul 15, 2003.
8. D. L. John, L. C. Castro, J. Clifford, and D. L. Pulfrey, "Electrostatics of coaxial Schottky-barrier nanotube field-effect transistors," *IEEE Transactions on Nanotechnology*, vol. 2, pp. 175–180, Sep, 2003.
9. K. Alam and R. Lake, "Performance of 2 nm gate length carbon nanotube field-effect transistors with source/drain underlaps," *Applied Physics Letters*, vol. 87, Aug 15, 2005.
10. G. Flori, G. Iannaccone, and G. Klimeck, "Performance of carbon nanotube field-effect transistors with doped source and drain extensions and arbitrary geometry," *IEDM Technical Digest*, vol. 2005, p. 529, 2005.
11. T. S. Xia, L. F. Register, and S. K. Banerjee, "Quantum transport in carbon nanotube transistors: Complex band structure effects," *Journal of Applied Physics*, vol. 95, pp. 1597–1599, FEB 1, 2004.

12. Y. Q. Xue and M. A. Ratner, "Scaling analysis of electron transport through metal-semiconducting carbon nanotube interfaces: Evolution from the molecular limit to the bulk limit," *Physical Review B*, vol. 70, Nov, 2004.
13. J. Guo, M. Lundstrom, and S. Datta, "Performance projections for ballistic carbon nanotube field-effect transistors," *Applied Physics Letters*, vol. 80, pp. 3192–3194, Apr 29, 2002.
14. P. J. Burke, "AC performance of nanoelectronics: towards a ballistic THz nanotube transistor," *Solid-State Electronics*, vol. 48, pp. 1981–1986, Oct–Nov, 2004.
15. A. Verma, M. Z. Kauser, and P. P. Ruden, "Effects of radial breathing mode phonons on charge transport in semiconducting zigzag carbon nanotubes," *Applied Physics Letters*, vol. 87, Sep 19, 2005.
16. V. Perebeinos, J. Tersoff, and P. Avouris, "Electron–phonon interaction and transport in semiconducting carbon nanotubes," *Physical Review Letters*, vol. 94, p. 027402, Mar 4, 2005.
17. S. Datta, *Electronic transport in mesoscopic systems*. Cambridge, UK: Cambridge University Press, 1995.
18. S. Datta, "Nanoscale device modeling: the Green's function method," *Superlattices and Microstructures*, vol. 28, pp. 253–278, Oct, 2000.
19. S. Datta, *Quantum transport: atom to transistor*. Cambridge, UK, New York: Cambridge University Press, 2005.
20. R. Lake, G. Klimeck, R. C. Bowen, and D. Jovanovic, "Single and multiband modeling of quantum electron transport through layered semiconductor devices," *Journal of Applied Physics*, vol. 81, pp. 7845–7869, Jun 15, 1997.
21. J. Guo, S. Koswatta, N. Neophytou, and M. Lundstrom, "Carbon nanotube field-effect transistors," *International Journal of High Speed Electronics and Systems*, vol. 16, pp. 897–912, 2006.
22. D. Tomanek, S. G. Louie, H. J. Mamin, D. W. Abraham, R. E. Thomson, E. Ganz, and J. Clarke, "Theory and observation of highly asymmetric atomic-structure in scanning-tunneling-microscopy images of graphite," *Physical Review B*, vol. 35, pp. 7790–7793, May 15, 1987.
23. J. Cerda and F. Soria, "Accurate and transferable extended Huckel-type tight-binding parameters," *Physical Review B*, vol. 61, pp. 7965–7971, Mar 15, 2000.
24. X. Blase, L. X. Benedict, E. L. Shirley, and S. G. Louie, "Hybridization effects and metallicity in small radius carbon nanotubes," *Physical Review Letters*, vol. 72, pp. 1878–1881, Mar 21, 1994.
25. A. Svizhenko, M. P. Anantram, T. R. Govindan, B. Biegel, and R. Venugopal, "Two-dimensional quantum mechanical modeling of nanotransistors," *Journal of Applied Physics*, vol. 91, pp. 2343–2354, Feb 15, 2002.
26. J. Guo, S. Datta, M. Lundstrom, and M. P. Anantram, "Toward multi-scale modeling of carbon nanotube transistors," *The International Journal on Multiscale Computer Engineering*, vol. 2, pp. 257–277, 2004.
27. N. Neophytou, D. Kienle, E. Polizzi, and M. P. Anantram, "Influence of defects on nanotube transistor performance," *Applied Physics Letters*, vol. 88, Jun 12, 2006.
28. Z. H. Chen, J. Appenzeller, J. Knoch, Y. M. Lin, and P. Avouris, "The role of metal-nanotube contact in the performance of carbon nanotube field-effect transistors," *Nano Letters*, vol. 5, pp. 1497–1502, Jul, 2005.
29. A. Javey, J. Guo, D. B. Farmer, Q. Wang, E. Yenilmez, R. G. Gordon, M. Lundstrom, and H. J. Dai, "Self-aligned ballistic molecular transistors and electrically parallel nanotube arrays," *Nano Letters*, vol. 4, pp. 1319–1322, Jul, 2004.
30. S. Heinze, J. Tersoff, R. Martel, V. Derycke, J. Appenzeller, and P. Avouris, "Carbon nanotubes as Schottky barrier transistors," *Physical Review Letters*, vol. 89, p. 106801, Sep 2, 2002.
31. J. Appenzeller, J. Knoch, V. Derycke, R. Martel, S. Wind, and P. Avouris, "Field-modulated carrier transport in carbon nanotube transistors," *Physical Review Letters*, vol. 89, p. 126801, Sep 16, 2002.

32. A. Javey, J. Guo, D. B. Farmer, Q. Wang, D. W. Wang, R. G. Gordon, M. Lundstrom, and H. J. Dai, "Carbon nanotube field-effect transistors with integrated ohmic contacts and high-k gate dielectrics," *Nano Letters*, vol. 4, pp. 447–450, Mar, 2004.
33. J. Chen, C. Klinke, A. Afzali, and P. Avouris, "Self-aligned carbon nanotube transistors with charge transfer doping," *Applied Physics Letters*, vol. 86, p. 123108, Mar 21, 2005.
34. Y. M. Lin, J. Appenzeller, and P. Avouris, "Novel carbon nantoube FET design with tunable polarity," *IEDM Technical Digest*, San Francisco, CA, pp. 687–690, Dec, 2004.
35. M. Radosavljevic, S. Heinze, J. Tersoff, and P. Avouris, "Drain voltage scaling in carbon nanotube transistors," *Applied Physics Letters*, vol. 83, pp. 2435–2437, Sep 22, 2003.
36. J. Guo, S. Datta, and M. Lundstrom, "A numerical study of scaling issues for Schottky-Barrier carbon nanotube transistors," *IEEE Transactions on Electron Devices*, vol. 51, pp. 172–177, Feb, 2004.
37. J. Guo, J. Wang, E. Polizzi, S. Datta, and M. Lundstrom, "Electrostatics of nanowire transistors," *IEEE Transactions on Nanotechnology*, vol. 2, pp. 329–334, Dec, 2003.
38. Z. H. Chen, J. Appenzeller, Y. M. Lin, J. Sippel-Oakley, A. G. Rinzler, J. Y. Tang, S. J. Wind, P. M. Solomon, and P. Avouris, "An integrated logic circuit assembled on a single carbon nanotube," *Science*, vol. 311, p. 1735, Mar 24, 2006.
39. A. Javey, J. Guo, Q. Wang, M. Lundstrom, and H. J. Dai, "Ballistic carbon nanotube field-effect transistors," *Nature*, vol. 424, pp. 654–657, Aug 7, 2003.
40. J. Guo and M. S. Lundstrom, "A computational study of thin-body, double-gate, Schottky barrier MOSFETs," *IEEE Transactions on Electron Devices*, vol. 49, pp. 1897–1902, Nov, 2002.
41. T. Durkop, B. M. Kim, and M. S. Fuhrer, "Properties and applications of high-mobility semiconducting nanotubes," *Journal of Physics-Condensed Matter*, vol. 16, pp. R553–R580, May 12, 2004.
42. X. J. Zhou, J. Y. Park, S. M. Huang, J. Liu, and P. L. McEuen, "Band structure, phonon scattering, and the performance limit of single-walled carbon nanotube transistors," *Physical Review Letters*, vol. 95, p. 146805, Sep 30, 2005.
43. Z. Yao, C. L. Kane, and C. Dekker, "High-field electrical transport in single-wall carbon nanotubes," *Physical Review Letters*, vol. 84, pp. 2941–2944, Mar 27, 2000.
44. J. Y. Park, S. Rosenblatt, Y. Yaish, V. Sazonova, H. Ustunel, S. Braig, T. A. Arias, P. W. Brouwer, and P. L. McEuen, "Electron–phonon scattering in metallic single-walled carbon nanotubes," *Nano Letters*, vol. 4, pp. 517–520, Mar, 2004.
45. A. Javey, J. Guo, M. Paulsson, Q. Wang, D. Mann, M. Lundstrom, and H. J. Dai, "High-field quasiballistic transport in short carbon nanotubes," *Physical Review Letters*, vol. 92, p. 106804, Mar 12, 2004.
46. J. Guo, "A quantum-mechanical treatment of phonon scattering in carbon nanotube transistors," *Journal of Applied Physics*, vol. 98, p. 063519, Sep 15, 2005.
47. J. Guo and M. Lundstrom, "Role of phonon scattering in carbon nanotube field-effect transistors," *Applied Physics Letters*, vol. 86, p. 193103, May 9, 2005.
48. S. Koswatta, S. Hasan, M. Lundstrom, M. P. Anantram, and D. E. Nikonov, "Ballisticity of nanotube FETs: role of phonon energy and gate bias," availabe at http://arxiv.org/cond-mat/0511723, 2005.
49. S. O. Koswatta, M. S. Lundstrom, M. P. Anantram, and D. E. Nikonov, "Simulation of phonon-assisted band-to-band tunneling in carbon nanotube field-effect transistors," *Applied Physics Letters*, vol. 87, Dec 19, 2005.
50. A. Svizhenko and M. P. Anantram, "Role of scattering in nanotransistors," *IEEE Transactions on Electron Devices*, vol. 50, pp. 1459–1466, Jun, 2003.
51. M. S. Fuhrer, B. M. Kim, T. Durkop, T. Brintlinger, and E. Cobas, "High mobility semiconducting nanotubes for nanoelectronics.," *Abstracts of Papers of the American Chemical Society*, vol. 227, pp. U266–U266, Mar 28, 2004.
52. J. Appenzeller and D. J. Frank, "Frequency dependent characterization of transport properties in carbon nanotube transistors," *Applied Physics Letters*, vol. 84, pp. 1771–1773, Mar 8, 2004.

53. D. J. Frank and J. Appenzeller, "High-frequency response in carbon nanotube field-effect transistors," *IEEE Electron Device Letters*, vol. 25, pp. 34–36, Jan, 2004.
54. S. D. Li, Z. Yu, S. F. Yen, W. C. Tang, and P. J. Burke, "Carbon nanotube transistor operation at 2.6 GHz," *Nano Letters*, vol. 4, pp. 753–756, Apr, 2004.
55. X. Huo, M. Zhang, P. C. H. Chan, Q. Liang, and Z. K. Tang, "High-frequency S parameters characterization of back-gate carbon nantoube field-effect transistors," *IEDM Technical Digest*, San Francisco, CA, pp. 691–694, Dec, 2004.
56. S. Rosenblatt, H. Lin, V. Sazonova, S. Tiwari, and P. L. McEuen, "Mixing at 50 GHz using a single-walled carbon nanotube transistor," *Applied Physics Letters*, vol. 87, p. 153111, 2005.
57. L. C. Castro, D. L. John, D. L. Pulfrey, M. Pourfath, A. Gehring, and K. H., "Method for predicting fT for carbon nanotube FETs," *IEEE Transactions on Nanotechnology*, vol. 4, pp. 699–704, 2005.
58. S. Hasan, S. Salahuddin, M. Vaidyanathan, and A. A. Alam, "High-frequency performance projections for ballistic carbon-nanotube transistors," *IEEE Transactions on Nanotechnology*, vol. 5, pp. 14–22, Jan, 2006.
59. J. Guo, S. Hasan, A. Javey, G. Bosman, and M. Lundstrom, "Assessment of high-frequency performance potential of carbon nanotube transistors," *IEEE Transactions on Nanotechnology*, vol. 4, pp. 715–721, Nov, 2005.
60. Y. Yoon, Y. Ouyang, and J. Guo, "Effect of phonon scattering on intrisic delay and cut-off frequency of CNTFETs," *IEEE Transactions on Electron Devices*, vol. 52, pp. 2467–2470, 2006.
61. A. Rahman, J. Guo, S. Datta, and M. S. Lundstrom, "Theory of ballistic nanotransistors," *IEEE Transactions on Electron Devices*, vol. 50, pp. 1853–1864, Sep, 2003.
62. J. A. Misewich, R. Martel, P. Avouris, J. C. Tsang, S. Heinze, and J. Tersoff, "Electrically induced optical emission from a carbon nanotube FET," *Science*, vol. 300, pp. 783–786, May 2, 2003.
63. M. Freitag, J. Chen, J. Tersoff, J. C. Tsang, Q. Fu, J. Liu, and P. Avouris, "Mobile ambipolar domain in carbon-nanotube infrared emitters," *Physical Review Letters*, vol. 93, p. 076803, Aug 13, 2004.
64. M. Freitag, Y. Martin, J. A. Misewich, R. Martel, and P. H. Avouris, "Photoconductivity of single carbon nanotubes," *Nano Letters*, vol. 3, pp. 1067–1071, Aug, 2003.
65. M. Freitag, V. Perebeinos, J. Chen, A. Stein, J. C. Tsang, J. A. Misewich, R. Martel, and P. Avouris, "Hot carrier electroluminescence from a single carbon nanotube," *Nano Letters*, vol. 4, pp. 1063–1066, Jun, 2004.
66. M. S. Arnold, J. E. Sharping, S. I. Stupp, P. Kumar, and M. C. Hersam, "Band gap photobleaching in isolated single-walled carbon nanotubes," *Nano Letters*, vol. 3, pp. 1549–1554, Nov, 2003.
67. F. Wang, G. Dukovic, L. E. Brus, and T. F. Heinz, "Time-resolved fluorescence of carbon nanotubes and its implication for radiative lifetimes," *Physical Review Letters*, vol. 92, p. 177401, Apr 30, 2004.
68. Z. C. Wu, Z. H. Chen, X. Du, J. M. Logan, J. Sippel, M. Nikolou, K. Kamaras, J. R. Reynolds, D. B. Tanner, A. F. Hebard, and A. G. Rinzler, "Transparent, conductive carbon nanotube films," *Science*, vol. 305, pp. 1273–1276, Aug 27, 2004.
69. S. Zaric, G. N. Ostojic, J. Kono, J. Shaver, V. C. Moore, M. S. Strano, R. H. Hauge, R. E. Smalley, and X. Wei, "Optical signatures of the Aharonov–Bohm phase in single-walled carbon nanotubes," *Science*, vol. 304, pp. 1129–1131, May 21, 2004.
70. J. Guo, M. A. Alam, and Y. Yoon, "Theoretical investigation on photoconductivity of single intrinsic carbon nanotubes," *Applied Physics Letters*, vol. 88, p. 133111, Mar 27, 2006.
71. V. Perebeinos, J. Tersoff, and P. Avouris, "Scaling of excitons in carbon nanotubes," *Physical Review Letters*, vol. 92, p. 257402, Jun 25, 2004.
72. D. A. Stewart and F. Leonard, "Photocurrents in nanotube junctions," *Physical Review Letters*, vol. 93, p. 107401, Sep 3, 2004.

73. V. Perebeinos, J. Tersoff, and P. Avouris, "Effect of exciton-phonon coupling in the calculated optical absorption of carbon nanotubes," *Physical Review Letters*, vol. 94, Jan 21, 2005.
74. J. Kong, N. R. Franklin, C. W. Zhou, M. G. Chapline, S. Peng, K. J. Cho, and H. J. Dai, "Nanotube molecular wires as chemical sensors," *Science*, vol. 287, pp. 622–625, Jan 28, 2000.
75. R. Chen, Y. Zhang, D. Wang, and H. Dai, "Non-covalent sidewall functionalization of single-walled carbon nanotubes for protein immobilization," *Journal of the American Chemical Society*, vol. 123, pp. 3838–3839, 2001.
76. H. W. C. Postma, T. Teepen, Z. Yao, M. Grifoni, and C. Dekker, "Carbon nanotube single-electron transistors at room temperature," *Science*, vol. 293, pp. 76–79, Jul 6, 2001.

Chapter 6
Carbon Nanotube Device Modeling and Circuit Simulation

H.-S. Philip Wong, Albert Lin, Jie Deng, Arash Hazeghi, Tejas Krishnamohan and Gordon Wan

6.1 Introduction

The development of new technology requires tools at all levels of abstraction. Modeling tools for detailed calculations of the energy band diagrams and device current–voltage characteristics [1] are essential first steps for device physics understanding. At the same time, modeling tools at higher levels of abstraction are required for device design space exploration and circuit design. As an example, for Si CMOS technology, industry-standard tools such as PISCES [2] and SPICE [3] are essential for device design and circuit simulation, respectively. Higher level abstraction tools [4] are used to describe and synthesize circuits at the system level.

In this chapter, we describe the development of a device-level model [5] that can be used as a rapid device design space exploration tool (an independent and parallel effort on SWNT-FET modeling can be found in [6]). It is simple enough to be run in a mixed-mode device/circuit simulation environment so that circuit issues can be studied at the device design level. We also describe the development of a circuit-compatible, compact device model [7] capable of large-scale circuit simulations. Using this circuit-compatible device model for SPICE, circuits consisting of a few hundred carbon nanotube transistors can be simulated.

6.2 Schottky Barrier SWNT-FET Modeling

Considerable effort has been put on modeling SWNT-FETs [8–10]. There are generally two approaches in modeling SWNT-FETs: one is the more numerically intensive NEGF "Non-Equilibrium Green's Function" approach, which is discussed in Chapter 5, and the other one is a simpler modeling methodology based on the ballistic transport assumption [8]. Even though many physical aspects of the problem are captured in the NEGF approach, due to very intensive calculations involved, it is

H.-S. Philip Wong (✉)
Center for Integrated Systems and Department of Electrical Engineering, Stanford University, Stanford, CA 94305, USA

A. Javey, J. Kong (eds.), *Carbon Nanotube Electronics*, Series on Integrated Circuits and Systems, DOI 10.1007/978-0-387-69285-2_6,

very difficult to develop simple intuitive descriptions of the device physics. It is also difficult to use NEGF modeling to rapidly explore device design spaces. Therefore, the model discussed here is based on the ballistic transport assumption as carbon nanotubes are believed to have scattering lengths in excess of a hundred nanometers [11].

The theoretical basis of this work has been reported by Natori et al. [8] using the ballistic assumption. However, some of the non-idealities that are known to exist in experimental devices were not captured by the Natori model, such as the Schottky barriers at the source/drain contacts, band-to-band tunneling current (BTBT) and ambipolar conduction. In Section 6.2.2, we provide a mesoscopic approach to model the Schottky barriers by altering the carrier distributions and using the evanescent-mode analysis [12] to model the potential profile near the Schottky regions. Our results are compared with the Natroi model to illustrate the significant impact of the Schottky barriers at the contacts. In Section 6.2.3, we apply our model to illustrate several device design issues including drive current dependence on the Schottky barrier height and ambipolar conduction. In Section 6.2.4, we illustrate the use of this simple model in a mixed-mode device/circuit simulation to capture the dynamic waveform of a Schottky barrier carbon nanotube inverter.

6.2.1 The Ballistic Model

The SWNT-FET structure modeled is depicted in Fig. 6.1(a). The nanotube potential at the surface relative to the equilibrium source Fermi level is denoted by μ, as shown in Fig. 6.1(b). Assuming the channel is ballistic between the two ideal source/drain reservoirs, the $+k_l$ states are filled according to source Fermi level, μ_s, and the $-k_l$ states are filled according to the drain Fermi level, μ_d (l denotes the direction along tube axis). Poisson's equation in the radial direction demands,

$$V_{gs} = V_{fb} + \frac{\mu}{q} + \frac{Q_{cnt}}{C_{ins}}, \tag{6.1}$$

where V_{gs} is the gate bias, C_{ins} is the insulator capacitance, V_{fb} is the flatband voltage and Q_{cnt} is the charge per unit length in the nanotube:

$$Q_{cnt} = \sum_{k_t} \sum_{k_l} q \cdot (f_s(E(k_l)) + f_d(E(k_l))) \tag{6.2}$$

And k_l and k_t are the wave vectors parallel and perpendicular to the nanotube axis. For a sufficiently long tube, (6.2) can be rewritten as [13]

$$Q_{cnt} = q \sum_{k_t} \int_{E_{i\,min}}^{E_{i\,max}} \frac{1}{2} g_{i,l}(E) \cdot (f_s(E, \mu) + f_d(E, \mu))\, dE, \tag{6.3}$$

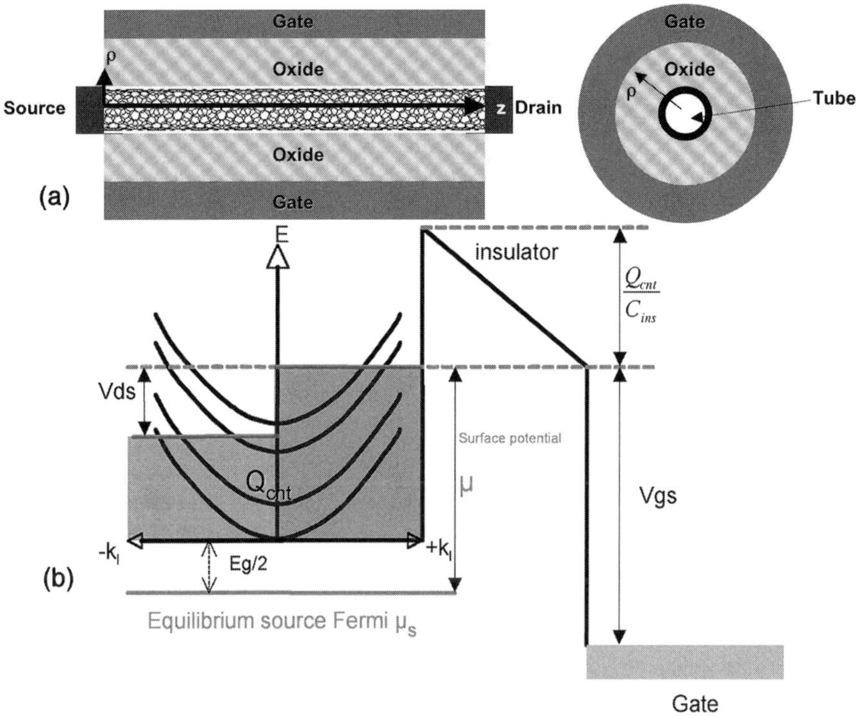

Fig. 6.1 (**a**) Device geometry cross-sections along the channel direction (z) and normal to the nanotube (ρ). Insulator thickness is denoted by t_{ox}, tube length with L and tube diameter with d. (**b**) Energy profile in the ρ direction with superimposed E-k diagram. Reproduced with permission from [13]. Copyright 2006 IEEE

where $g_{i,l}(E)$ is the 1D universal nanotube DOS [12] for the ith subband, and f_s and f_d are source/drain Fermi–Dirac distribution functions, respectively. The ½ factor is used because only one k_l state is being counted at a time; $E_{i\,\min}$ and $E_{i\,\max}$ denote the min/max energy of the ith subband and μ is the surface potential or the potential by which the source Fermi level is raised relative to its equilibrium position. Solving (6.1) to (6.3) self-consistently reveals the surface potential μ (away from the contacts). (It should be noted that (6.3) significantly depends on the band gap (diameter) of the tube.) To calculate the ballistic current (I_d), the Landauer–Büttiker expression can be used, assuming unity transmission for channel

$$I_d = \sum_{k_t} \sum_{k_l} q \cdot v_g(k_l) \cdot [f_s(E(k_l)) - f_d(E(k_l))] \cdot T(E(k_l)), \tag{6.4}$$

where $v_g(k_l)$ denotes the carrier group velocity in the axial direction. This expression is often written in the form

$$I_\mathrm{d} = \frac{2q}{h} \sum_{k_t} \int_{E_{i\,\min}}^{E_{i\,\max}} (f_\mathrm{s}(E,\mu) - f_\mathrm{d}(E,\mu))\,\mathrm{d}E, \tag{6.5}$$

where the term $2q/h$ (known as the quantum conductance) is the product of group velocity and the 1D density of states along the axis of the tube direction. This is the maximum conductance per channel for a ballistic conductor. f_s and f_d denote the S/D reservoir Fermi–Dirac distribution functions:

$$f_\mathrm{s}(E) = \frac{1}{1 + \mathrm{e}^{\frac{E-\mu}{k_\mathrm{B}T}}} \tag{6.6}$$

$$f_\mathrm{d}(E) = \frac{1}{1 + \mathrm{e}^{\frac{E-\mu+V_\mathrm{ds}}{k_\mathrm{B}T}}} \tag{6.7}$$

6.2.2 Modeling the Schottky Barriers

The ballistic model captures some important characteristics such as the quantum capacitance. It overestimates the current compared to the experimental data, because most of the fabricated devices [14] exhibit Schottky barriers (SB) at the contacts. The contact metal work function, surface preparation, and annealing conditions are known to be the main parameters that affect the SB height [11, 15–17], although the details of how process conditions affect the SB height are not yet fully understood. Nevertheless, the Schottky barriers are always present in the fabricated devices and their effect on device performance should be explored. The energy profile of the device in presence of SB is depicted in Fig. 6.2.

Due to the often small band gap as well the confined electrostatic nature of the tubes (small diameter), coupled with a tight gate control due to thin high-k gate dielectrics, it is obvious that tunneling is the dominant transport mechanism across the contacts as reported in [18]. The Schottky barriers act as scattering sites near the contacts and thus the assumption under which (6.2) is derived is no longer valid. However, the transport of carriers along the nanotube is still ballistic in-between the Schottky barriers where the potential tail from the Schottky barriers has decayed. In a simple approach, the Schottky barriers can be modeled as mesoscopic carrier scattering sites as depicted in Fig. 6.3.

Carriers which are initially at thermal equilibrium with the source contact are scattered by the SB before entering the channel, where they are no longer at thermal equilibrium. [1] However, beyond the energy relaxation length, a pseudo-distribution

[1] There will be no equilibrium within one energy relaxation length of the SB in the contact; however the energy relaxation length by which the chemical potential changes is shorter than the screen length associated with the electrostatic potential [15].

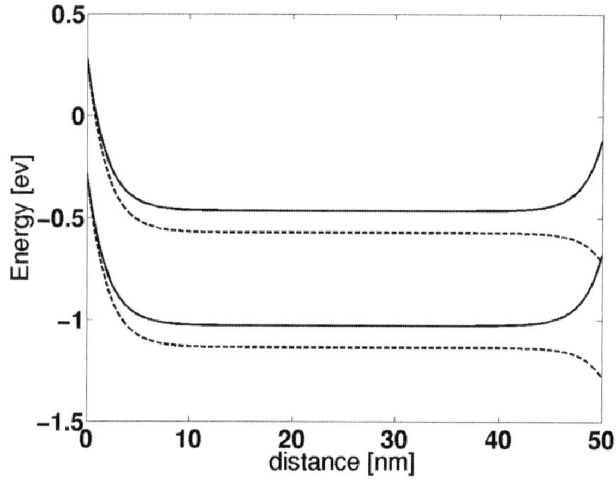

Fig. 6.2 Energy profile in the z direction for ϵ_r=20, t_{ox}=4 nm, L=50 nm, d=1.5 nm, for a (19,0) tube V_{gs}=1 V, *Solid line* is for V_{ds}=0.4 V and *dashed line* for V_{ds}=1 V, the potential at the vicinity of the contacts is calculated by evanescent-mode analysis method, [11]. Note that for V_{ds}=0.4 V, a SB exists at the drain end while for V_{ds}=1 V there is no SB going from the channel to the drain contact. Mid-gap barrier height is assumed, E_g=0.56 eV. Also note that there is a strong dependence of μ on V_{ds}. Reproduced with permission from [13]. Copyright 2006 IEEE

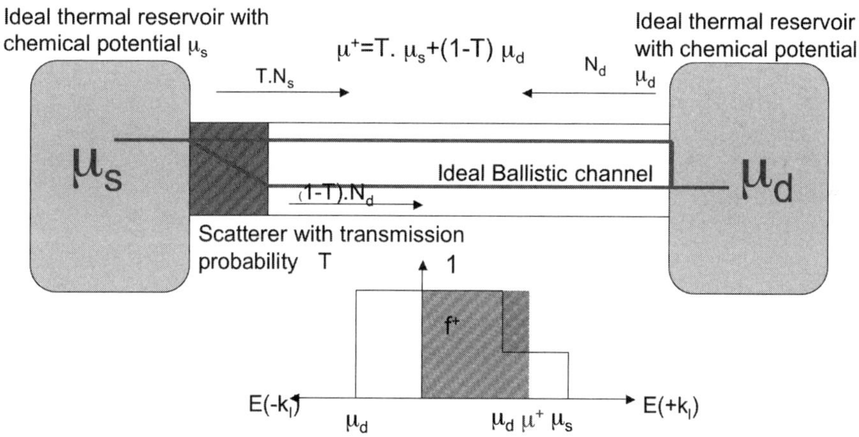

Fig. 6.3 Energy distribution of carriers and chemical potential at 0 K in the presence of a scatterer before the ballistic channel. Only the source scatterer on the source side is shown for simplicity, drain scatterer is treated analogously. Reproduced with permission from [13]. Copyright 2006 IEEE

function can be used to calculate the number of carriers assuming carrier conservation in the scatterer. [2] Let us first consider only one "scatterer" near the source end and let the transmission probability across the scatterer at energy E be $T(E)$. From N_s number of carriers at this energy in the source reservoir, $N_s T(E)$ will make it to the channel. Similarly, from N_d number of carriers injected from the drain contact, $N_d(1–T(E))$ number of carriers are reflected from the source scatterer. From the total number of carriers occupying the $+k_l$ states, a fraction of T would have the source reservoir distribution and a fraction of $(1–T)$ would have the drain reservoir distribution [19]. The total distribution would then (Fig. 6.3) be

$$f_{+k_l} = T(E) \cdot f_s(E) + (1 - T(E)) \cdot f_d(E). \tag{6.8}$$

Similarly if the drain SB is taken into account (at bias conditions when the drain SB exists, see Fig. 6.2), the following pseudo-distribution functions can be derived for the $+k_l$ and $-k_l$ components:

$$f^+ = \frac{T_s f_s + T_d f_d - T_s T_d f_d}{1 - (T_s - 1) \cdot (T_d - 1)} \tag{6.9}$$

and

$$f^- = \frac{T_d f_d + T_s f_s - T_s T_d f_s}{1 - (T_s - 1) \cdot (T_d - 1)}. \tag{6.10}$$

These two new distribution functions have to be used instead of f_s and f_d in (6.3) and (6.5), in the presence of SB. To calculate the transmission probability across each Schottky barrier, the WKB approximation is used. At each point along the channel, the total energy (relative to source Fermi level) is $E=E(k_l)-qV(z)$ where $V(z)$ is the electrostatic potential in the z direction, using approximate dispersion [20]

$$E(k_l) = \pm\sqrt{\Delta^2 + \left(\frac{3a_{c-c}V_\pi k_l}{2}\right)^2}, \tag{6.11}$$

where $\Delta=E_g/2$, a_{c-c} is the graphene lattice spacing, and V_π is the C–C bonding energy introduced in [20]. Using the WKB method, the transmission probability $T(E)$ is given by

$$\ln(T(E_l)) = -\frac{4}{3a_{c-c}V_\pi} \int_{z_i}^{z_f} \left(\Delta^2 - (E + qV(z))^2\right)^{1/2} dz, \tag{6.12}$$

[2] At 0 K a pseudo-Fermi level "chemical potential" can be calculated in the same way since the distribution functions are step functions (see Fig. 6.3).

where the integration is performed between z_i and z_f, the two classical turning points for the potential barrier, i.e., the two values of z for which $E=V(z)$. This expression takes into account both the conduction and valence bands, and therefore it is able to treat ambipolar transport and band-to-band tunneling. This would give corresponding distributions of (6.9) and (6.10) for any energy E_l. To calculate the potential profile near the contacts, we use the evanescent mode analysis similar to that used in Ultra-Thin-Body Silicon-On-Insulator (UTB-SOI) devices [12], implemented for the radial structure of SWNT-FET. The evanescent-mode analysis decouples the problem into two separate boundary problems: one with the gate boundary condition and the other with the S/D boundary conditions. In other words, it decouples the vertical electric field generated by the gate and the corresponding charge in the tube from the lateral electrical field generated by the drain for the zero charge case. Solving the Laplace equation inside the tube near the source contact, we obtain

$$V_{cnt}^{*}(\rho, z) = A \cdot J_0(\rho/\lambda_{cnt}) \cdot e^{-z/\lambda_{cnt}} \tag{6.13}$$

for the SWNT and

$$V_{ins}^{*}(\rho, z) = (A_{ins} \cdot J_0(\rho/\lambda_{cnt}) + B_{ins} \cdot Y_0(\rho/\lambda_{cnt})) \cdot e^{-z/\lambda_{cnt}} \tag{6.14}$$

inside the insulator (which is usually thicker than the tube itself), where J_0 and Y_0 denote the Bessel and Neumann functions. The potential decays in the z direction with the characteristic length (λ_{cnt}). Matching the above equations and the corresponding derivatives at the tube/insulator interface ($\rho=\rho_{\text{cnt}}$) and setting $V^{*}_{\text{ins}}=0$ at the gate/insulator interface ($\rho=\rho_{\text{cnt}}+t_{\text{ins}}$), where t_{ins} is the gate dielectric thickness, λ_{cnt} can be obtained from [12]

$$\frac{Y_0'\left(\frac{\rho_{cnt}}{\lambda_{cnt}}\right)}{J_0'\left(\frac{\rho_{cnt}}{\lambda_{cnt}}\right)} = k\frac{Y_0\left(\frac{\rho_{cnt}}{\lambda_{cnt}}\right)}{J_0\left(\frac{\rho_{cnt}}{\lambda_{cnt}}\right)} + (1-k)\cdot\frac{Y_0\left(\frac{\rho_{cnt}+t_{ins}}{\lambda_{cnt}}\right)}{J_0\left(\frac{\rho_{cnt}+t_{ins}}{\lambda_{cnt}}\right)}, \tag{6.15}$$

where $k = \frac{\varepsilon_{cnt}}{\varepsilon_{ins}}$. λ_{cnt} can be considered as an "effective" length scale of the device [12] or the characteristic length for decaying potentials from source and drain SB (see Fig. 6.2). To obtain a self-consistent solution of the charge equation and the potential profile, (6.3) is first solved using the ideal reservoir distributions f_{s} and f_{d}. Once the transmission coefficients T_{s} and T_{d} are determined, (6.3) is iterated again using f^{+} and f^{-} which were defined earlier. Self-consistency is reached after 4–5 iterations. The current is then calculated using a modified version of (6.5):

$$I_{\text{d}} = \frac{2q}{h}\sum_{k_t}\int_{E_{\min}}^{E_{\max}} (f^{+}(E,\mu) - f^{-}(E,\mu))\text{d}E \tag{6.16}$$

6.2.3 Schottky Barrier Device Characteristics

Figure 6.4 shows $I_d - V_{gs}$ and $I_d - V_{ds}$ plots for similar structures with and without the Schottky barrier. The drain current is noticeably reduced due to the Schottky barrier. The slight kink-like feature seen in the $I_d - V_{gs}$ plot is due to the fact that more subbands are occupied as the gate bias is increased. The upper limit of the current is still an order of magnitude higher than the best experimental data, which do not exceed 10–20 μA per tube. Although the gate geometry of experimental devices is not as ideal as the tight cylindrically wrapped gate used here, the low on-current hints that most of the fabricated devices may still suffer from contact resistance issues [21, 22]. The other issue is carrier scattering due to excitation of optical phonons at higher biases in experimental devices, which will degrade channel transmission and the on-current. These could be the main reasons for the low ON current observed in experimental devices, although for channel lengths shorter than 100 nm, scattering is less important. Further research is needed to fully understand the interface and contact properties of the SWNTs, in order to eliminate contact issues.

Figure 6.5 shows the effect of the SB height on the ON current. Despite the very short length scale (λ_{cnt}=1~8 nm) and the thin Schottky barrier, the barrier height still affects the ON current noticeably. It should be noted, however, that devices

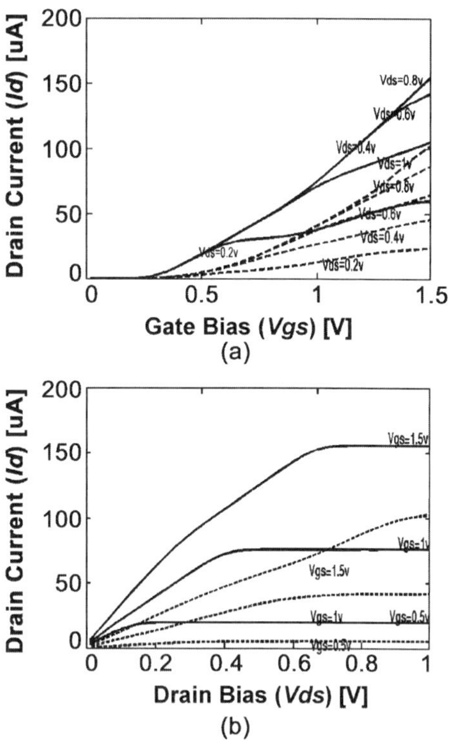

Fig. 6.4 (**a**) I_d–V_{gs} and (**b**) I_d–V_{ds} plots for a (19,0) tube, d=1.5 nm, t_{ins}=4 nm, ϵ=20, λ_{cnt}=1.9 nm, L=100 nm at T=300 K. *Solid lines* are for the no-SB case and *dashed lines* are for the mid-gap SB case. Note that the drain current saturates slower in the presence of SB. Reproduced with permission from [13]. Copyright 2006 IEEE

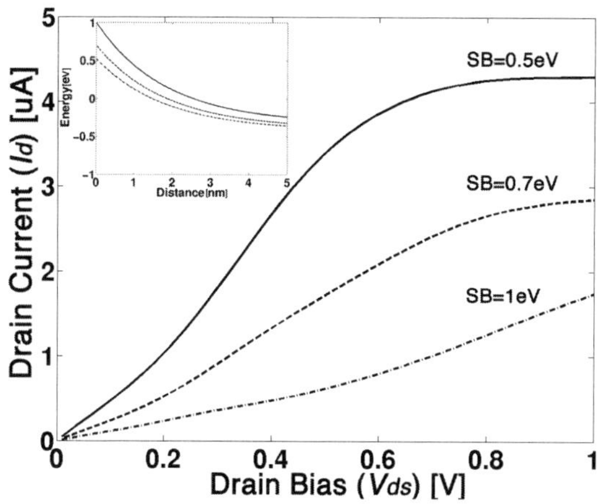

Fig. 6.5 I_d–V_{ds} plots for a (10,0) tube, d=0.8 nm, t_{ins}=4 nm, ϵ=20, λ_{cnt}=1.8 nm, E_g=1 eV, L=100 nm. V_{gs} is 1 V. *Solid line* represents mid-gap SB height ($E_g/2$). *Dashed line* is for SB height of 0.7 eV. *Dashed dotted line* is for SB height=1 eV (i.e., contacts aligned to valence band). Note that the current saturates slower for a higher SB height. Inset shows energy diagram near the source region for SB=1 eV (*solid line*), SB=0.7 eV (*dashed line*) and SB=0.5 eV (*dash-dotted line*). V_{ds}=0.5 V in the inset. Reproduced with permission from [13]. Copyright 2006 IEEE

exhibit symmetric ambipolar characteristics due to the fully symmetric band structure for electrons and holes. A higher than mid-gap SB height can be regarded as a p-type device with negative V_{gs} and V_{ds} biases. Therefore, in practice, the mid-gap SB is the worst case.

Figure 6.6 shows ambipolar conduction for the Schottky barrier SWNT-FET. The symmetry bias condition at which electron and hole currents are equal and thus total current is minimum, depends on Schottky barrier height. For mid-gap Schottky barriers, this point is always at V_{gs}=V_{ds}/2. Ambipolar conduction increases OFF state leakage current and degrades ON/OFF ratio and thus is not desirable. The component of OFF current, which is due to holes tunneling across the Schottky barriers, can be eliminated by doping the contact and making the device unipolar.

6.2.4 Mixed-Mode Simulations

To predict the performance of SWNT-FET-based logic, we consider the case of two series inverters in a ring oscillator. In this manner, the dynamics of the entire switching waveform can be appropriately accounted for (as compared to simple *CV/I* estimations). This is particularly important for SWNT-FETs where the device saturation characteristics are quite different from Si FETs [23].

Symmetrical pFETs and nFETs with no SB are compared to pFETs and nFETs with mid-gap SB in this example. The SWNT-FETs are ambipolar devices and the bias condition for minimum current (I_{OFF}) depends on the Schottky barrier height.

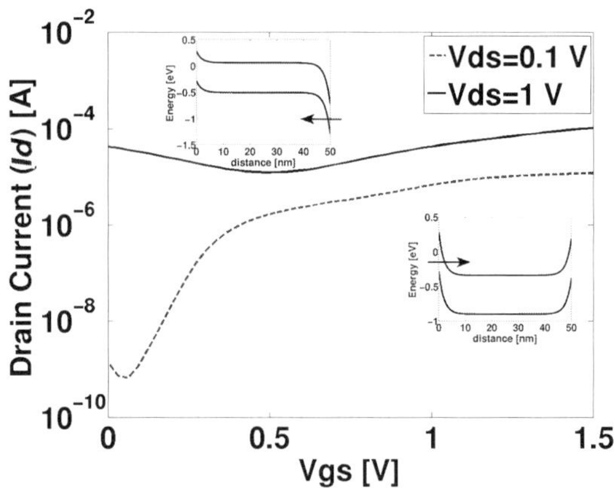

Fig. 6.6 I_d–V_{gs} plots for a (19,0) tube, d=1.5 nm, t_{ins}=4 nm, ϵ=20, λ_{cnt}=1.9 nm, L=100 nm at T=300 K, SB=E_g/2. *Solid line* is for V_{ds}=1 V and *dashed line* for V_{ds}=0.1 V. Mid-gap SB structure is symmetrical for holes and electrons, thus minimum current bias point is V_{gs}=V_{ds}/2. Insets show energy band diagrams along the device for V_{gs}=0.1 V and V_{ds}=1 V (*upper inset*) and V_{gs}=1 V and V_{ds}=0.1 V (*lower inset*). *Arrows* indicate tunneling carriers (holes for upper inset and electrons for lower inset). Reproduced with permission from [5]. Copyright 2007 IEEE

It is assumed that the gate flat band voltage can be adjusted such that OFF current is minimized for both P and N devices (possible dual-metal gate electrode). This is possible since unlike conventional CMOS, N and P SWNT-FET devices are fully symmetrical for symmetrical Schottky barriers. Lastly, since no proven scheme exists for SWNT-FET layout at this time, no load capacitances in addition to the gate capacitances are included. (Fabricated devices often use highly inefficient and very large contacts/interconnects, which will not represent the true potential performance of these devices if included.)

First, a realistic input waveform (vs a perfect step) is generated by passing a step input through a buffer (2 inverters). The output here is then used as the input to the 2 series inverters. The goal here is to examine the performance, such as the rise/fall times, of the 2 inverters for SWNT-FETs with and without SBs. Convergence of the numerical simulation is achieved quickly within 2~3 cycles. Sample waveforms are shown in Fig. 6.7. From these waveforms, it is clear that the presence of SBs in SWNT-FETs results in slower transitions and longer delays. This example illustrates the use of this Schottky Barrier SWNT-FET Model for mixed-mode simulations.

6.3 Compact Model for Circuit Simulation

In order to simulate carbon nanotube circuits with a fair complexity beyond a few transistors, a compact model, such as BSIM [24], must be used. A compact model must contain the essential device physics and yet be simple enough to allow rapid execution of the code. All the reported models [25, 26] to date used a single lumped

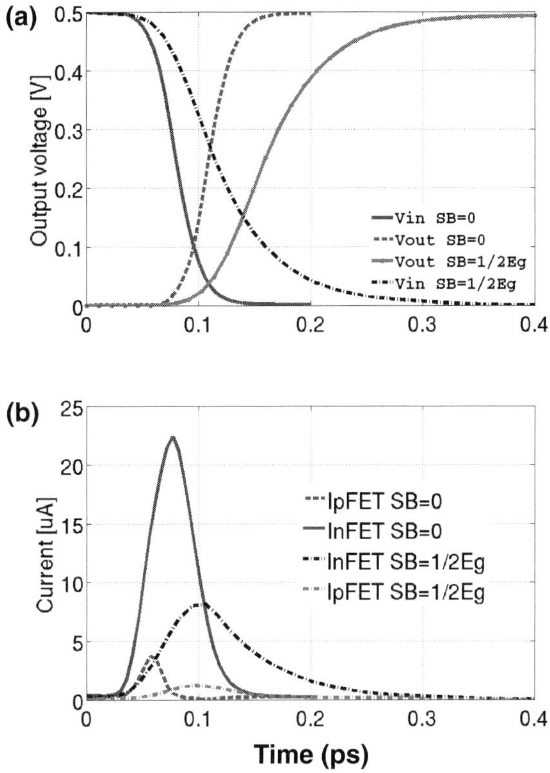

Fig. 6.7 (**a**) Input and output waveforms for two series inverters. *Solid line* is the input to the second gate for SB=0; *dashed line* is the output of the second gate for SB=0; *dotted line* is the input to the second gate for SB=$E_g/2$; and *dash-dotted line* is the output of the second gate for SB=$E_g/2$. (**b**) Current for the pFET and nFET of the second inverter as a function of time. Each FET contains one (19,0) nanotube with L=20 nm; supply voltage is 0.5 V

gate capacitance and the ideal ballistic model to evaluate the dynamic performance. These assumptions lead to inaccurate predictions of the circuit performance. To evaluate SWNT-FET circuit performance with improved accuracy, a SWNT-FET device model with a more complete circuit-compatible structure that also includes typical device non-idealities is necessary.

We start with a description of the model, implemented in HSPICE macro language (and also in VerilogA).[3] Then we illustrate the application of this model for circuit performance estimation.

6.3.1 Overview of Carbon Nanotube Transistor Compact Model

A circuit-compatible SWNT-FET device model is introduced here which accounts for some practical non-idealities for improved accuracy. It also includes a complete

[3] As a short-hand, "CNT" and "CNFET" are used instead of "SWNT" and "SWNT-FET" in model names and parameters in the actual HSPICE and VerilogA implementation.

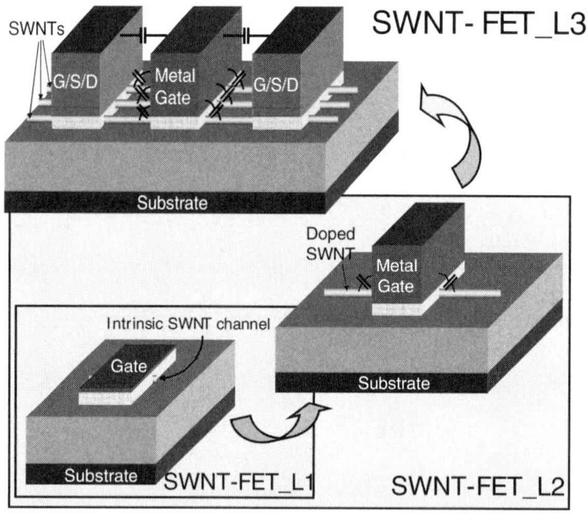

Fig. 6.8 3-Level hierarchy of the circuit-compatible compact SWNT-FET device model. The top diagram (SWNT-FET_L3) shows the device structure modeled, which allows multiple nanotubes per device and accounts for the inter-SWNT charge screening effects [28, 29]. SWNT-FET_L2 models the source/drain extension regions and contacts of the device and its parasitics [28]. SWNT-FET_L1 is the core of the model and describes the intrinsic SWNT channel region of the SWNT-FET (along with its parasitics) [27]. Reproduced with permission from [28]. Copyright 2007 IEEE

transcapacitance network in order to produce better predictions of the dynamic performance and transient response [27, 28]. Figure 6.8 (top, SWNT-FET_L3) shows the modeled SWNT-FET device structure. The SWNT-FET is a CMOS-like FET[4] with semiconducting SWNTs regions forming the channel and highly doped SWNT regions forming the source/drain extension regions.

The model is organized hierarchically in three main levels (Fig. 6.8). The first level (SWNT-FET_L1) is the core of the model and is used to describe the portion of the SWNT under the metal gate, which forms the SWNT-FET channel region. This level assumes near-ballistic transport, with acoustic and optical phonon scattering, and includes parasitic capacitances and resistance [27]. Level 2 (SWNT-FET_L2) builds upon Level 1 by including effects (e.g., parasitic capacitances and resistance) from the highly doped source/drain extension regions. It also includes Schottky Barrier resistances from the source/drain contacts to the source/drain SWNT regions [28]. Finally, Level 3 (SWNT-FET_L3) augments Level 2 and deals with multiple SWNTs per device. It includes the parasitic gate capacitance and the SWNT-to-SWNT charge screening effects for the SWNTs within the SWNT-FET device [28, 29].

[4]The CMOS-like SWNT-FET was chosen for modeling (versus the Schottky Barrier controlled SWNT-FET) due to its superior performance and fabrication feasibility [27].

6.3.2 Model of the Intrinsic SWNT Channel Region (SWNT-FET_L1)

SWNT-FET_L1 models the intrinsic SWNT channel current by considering three contributing sources: (1) thermionic current contributed by the semiconducting subbands, I_{semi}; (2) band-to-band tunneling (BTBT) current through the semiconducting subbands, I_{BTBT}; and (3) current contributed by the metallic subbands, I_{metal}. The circuit schematic of the Level 1 Model is shown in Fig. 6.9. These three current contributions are modeled as dependent current sources (the I_{metal} dependent current source is equivalently modeled as a dependent conductance G_{metal}).

For the remainder of this chapter, we present the model equations and derivations using an n-type SWNT-FET as an example. Similar equations hold for p-type SWNT-FETs. The HSPICE and VerilogA implementations of this SWNT-FET Model include both n-type and p-type SWNT-FETs.

To derive the total device current, we first look at I_{semi}. For semiconducting subbands, the hole current is usually negligible compared to the electron current in the CMOS-like SWNT-FET due to the heavily-doped n-type source/drain CNT regions. Thus, the current contributed by electrons in quantum state (m,l) is

$$J_{m,l}\left(V_{\text{XS}}, \Delta\Phi_B\right) = 2qnv_{\text{F}}, \tag{6.17}$$

where $J_{m,l}$ is a function of V_{xs}, the potential difference between node x (where x denotes either the drain or source) and the source, and $\Delta\Phi_{\text{B}}$, the change in channel surface potential due to a change in the bias. The factor of 2 is due to the electron spin degeneracy; q is the electron charge; n is the number of electrons in the quantum state (m,l) and is a function of V_{xs} and $\Delta\Phi_{\text{B}}$; and v_{F} is the Fermi velocity along the axial direction. Detailed equations and derivations for (6.17) can be found in [27].

The total I_{semi} current from the drain to the source is then the sum of all the current components flowing from the drain to the source (+k components) minus

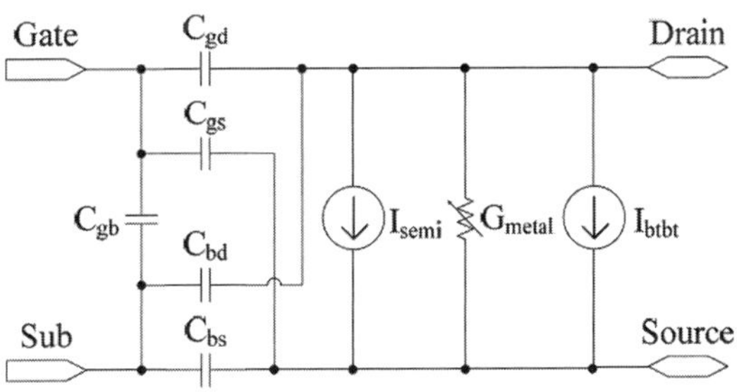

Fig. 6.9 Circuit schematic of the model for the intrinsic SWNT Channel Region (SWNT-FET_L1). Three current contributions are considered: (1) thermionic current from the semiconducting subbands (I_{semi}); (2) current from the metallic subbands (reduced to G_{metal}); and (3) band-to-band tunneling current from the semiconducting subbands (I_{btbt}). Parasitic capacitances are also included for improved transient and dynamic simulation accuracy. Reproduced with permission from [27]. Copyright 2007 IEEE

the current components flowing from the source to the drain ($-k$ components). It has a similar form as (6.4):

$$I_{\text{semi}}\left(V_{\text{ch,DS}}, V_{\text{ch,GS}}\right) = 2\sum_{\substack{k_m \\ m=1}}^{M}\sum_{\substack{k_l \\ l=1}}^{L}\left[T_{\text{LR}}\,J_{m,l}(0, \Delta\Phi_{\text{B}})|_{+k} - T_{\text{RL}}\,J_{m,l}(V_{\text{ch,DS}}, \Delta\Phi_{\text{B}})|_{-k}\right] \tag{6.18}$$

$V_{\text{ch,DS}}$ and $V_{\text{ch,GS}}$ denote the Fermi level near the source-side of the channel. The factor of 2 (in addition to the factor of 2 in $J_{m,l}$) accounts for the two times degeneracy of the subbands. M and L are the number of subbands and quantum states in that subband, respectively, so that k_m denotes the wave-number of the m th subband in the circumferential direction and k_l denotes the wave-number of the l th quantum state in the axial direction. And T_{LR} and T_{RL} are the transmission probabilities.

Perhaps the most important parameter in (6.18) is the channel surface potential change $\Delta\Phi_{\text{B}}$. Accurate computation of the channel surface potential change $\Delta\Phi_{\text{B}}$ is critical in calculating the correct current and predicting dynamic and transient performance. A brief discussion of the $\Delta\Phi_{\text{B}}$ derivation follows.

Figure 6.10 shows the electrostatic capacitance model superimposed on the energy band diagram for a SWNT-FET. $V_{\text{ch,GS}}$ is the potential difference from the gate to the source-side channel region; $V_{\text{ch,BS}}$ is the potential difference from the bulk (substrate) to the source-side channel region; $V_{\text{ch,S'S}}$ is the potential difference from the external source outside the channel region to the source-side channel region; and $V_{\text{ch,D'S}}$ is the potential difference from the external drain outside the channel region to the source-side channel region. Thus, C_{ox} is the electrostatic coupling capacitor from the gate to the channel; C_{sub} is the capacitance between the channel and substrate; and $(1-\beta)C_{\text{c}}$ and βC_{c} are the coupling capacitances from the channel to the source and drain, respectively (the total channel to source/drain capacitance is C_{c}) [27]. Note that β and C_{c} are fitting parameters in the model. For more accurate simulation results, $\Delta\Phi_{\text{B}}$ can found by applying charge conservation (instead of using the quantum capacitance approach) across the electrostatic capacitances. Conservation of charge dictates that

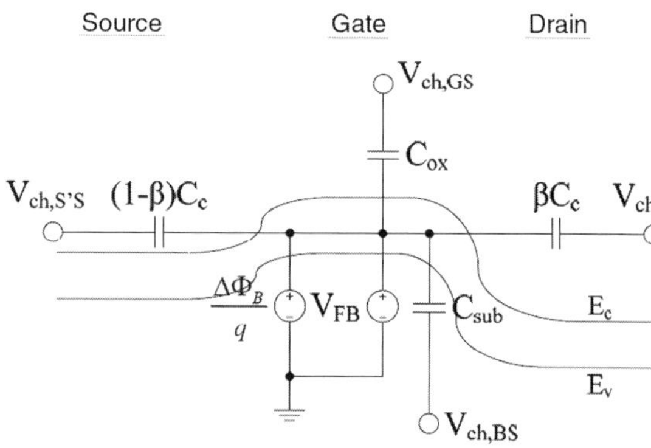

Fig. 6.10 Electrostatic capacitance model superimposed on the energy band diagram. Charge conservation is used to find the change in the channel surface potential due to a change in bias ($\Delta\Phi_{\text{B}}$). Reproduced with permission from [27]. Copyright 2007 IEEE

$$Q_{\text{cap}} = Q_{\text{CNT}}, \tag{6.19}$$

where

$$\begin{aligned} Q_{\text{cap}} = {} & C_{\text{ox}}\left(V_{\text{ch,GS}} - V_{\text{FB}}\right) + C_{\text{sub}}V_{\text{ch,BS}} \\ & + \beta C_{\text{c}}V_{\text{ch,D'S}} + (1-\beta)C_{\text{c}}V_{\text{ch,S'S}} - (C_{\text{ox}} + C_{\text{sub}} + C_{\text{c}})\frac{\Delta\Phi_{\text{B}}}{q} \end{aligned} \tag{6.20}$$

is the charge induced by the electrodes, and

$$Q_{\text{CNT}} = \frac{4q}{L_{\text{g}}}\sum_{\substack{k_m \\ m=m_0}}^{M}\sum_{\substack{k_l \\ l=0}}^{L}\left[\frac{1}{1+\mathrm{e}^{\left(\frac{E_{m,l}-\Delta\Phi_{\text{B}}}{kT}\right)}} + \frac{1}{1+\mathrm{e}^{\left(\frac{E_{m,l}-\Delta\Phi_{\text{B}}+qV_{\text{DS}}}{kT}\right)}}\right] \tag{6.21}$$

is the charge induced on the SWNT surface. $E_{m,l}$ is the carrier energy of the (m,l) quantum state relative to the intrinsic Fermi level (so $E_{m,0}$ is the half band gap of the mth subband). L_{g} is the channel length. m_0 is 1 for semiconducting SWNTs and 0 for metallic SWNTs (i.e., the metallic subband is included for metallic SWNTs and is not included for semiconducting SWNTs). The SWNT surface potential change $\Delta\Phi_{\text{B}}$ can then be computed from the equations (6.19–6.21).

Practical non-idealities such as scattering are included in the model. Three scattering mechanisms are included in the complete model: (1) acoustic phonon scattering (near-elastic scattering [30]), (2) optical phonon scattering (non-elastic scattering [31]), and (3) elastic scattering. SWNT-FET_L1 assumes near ballistic transport in the intrinsic SWNT region and only accounts for acoustic and optical phonon scattering (elastic scattering is taken into account in the next level, SWNT-FET_L2, by including a series resistance). So the transmission probabilities in (6.18), T_{LR} and T_{RL}, can be calculated as in [27] to include acoustic and optical phonon scattering effects to yield more realistic currents which are smaller than that computed under ideal, ballistic conditions.

Having found $\Delta\Phi_{\text{B}}$ and the transmission probabilities, I_{semi} can be found from (6.18). Note that a typical short-channel device (diameter $<$3 nm, L_{g} $<$100 nm) with a sub 1 V power supply, the first 2–3 subbands and the first 10–15 quantum states in a subband dominate the current and explicit summation of these quantum states is required [27]. But, for a long-channel device ($L_{\text{g}} > 100$ nm), the wave number along the axial direction approaches continuous values and (6.18) can be approximated by replacing the inner summation with an integral over k_l (also assume $T_{\text{LR}} = T_{\text{RL}} = T_{\text{M}}$):

$$I_{\text{semi}}\left(V_{\text{ch,DS}}, V_{\text{ch,GS}}\right) \approx \frac{4q^2}{h}\sum_{\substack{k_m \\ m=1}}^{M} T_m\left[V_{\text{ch,DS}} + \frac{kT}{q}\ln\left(\frac{1+\mathrm{e}^{\frac{E_{m,0}-\Delta\Phi_{\text{B}}}{kt}}}{1+\mathrm{e}^{\frac{E_{m,0}-\Delta\Phi_{\text{B}}+qV_{\text{ch,DS}}}{kt}}}\right)\right]. \tag{6.22}$$

I_{semi} is implemented as a dependent current source in the model as shown in Fig. 6.9. The two other current contributions, I_{BTBT} and I_{metal} can be found with much less effort.

Band-to-band tunneling current must be taken into account when modeling the total drain current in the SWNT in order to accurately predict the subthreshold slope and the static leakage power as I_{BTBT} can become relatively significant with negative gate bias. BTBT only occurs at the drain junction (with positive drain–source bias) and only if the following two conditions are met: (1) the conduction band at the drain side is below the valence band at the source side and (2) there are empty states at the drain side to accept the tunneling carriers. The first condition is equivalent to $V_{\text{ch,DS}} > 2E_{m,0}$, where $2E_{m,0}$ is the band gap ($E_{m,0}$ is the half band gap) of the mth subband. Thus, when calculating I_{BTBT}, only those subbands which satisfy $V_{\text{ch,DS}} > 2E_{m,0}$ should be included. The second condition is handled by using the Fermi–Dirac distribution to calculate the unoccupied states (1 – the Fermi function). The resulting I_{BTBT} model is

$$I_{\text{BTBT}} = \frac{4qkT}{h} \sum_{\substack{k_m \\ m=1}}^{M} \left[T_{\text{BTBT}} \ln \left(\frac{1 + e^{\frac{qV_{\text{ch,DS}} - E_{m,0} - E_{\text{f}}}{kt}}}{1 + e^{\frac{E_{m,0} - E_{\text{f}}}{kt}}} \right) i_m \right] \tag{6.23}$$

where E_{f} is the Fermi level of the doped source/drain region. i_m is an "accounting dummy parameter" whose value is 1 if the m th subband satisfies the first condition above for BTBT ($V_{\text{ch,DS}} > 2E_{m,0}$) and is 0 otherwise; thus, only those subbands with $V_{ch,DS} > 2E_{m,0}$ are actually summed in (6.23). T_{BTBT} is the band-to-band tunneling probability and can be found using the WKB method in [32, 33] while noting that the carriers are confined to only the axial direction (1D problem). Full derivations of T_{BTBT} are in [27].

To find the current contributed by the metallic subband in metallic SWNTs, an equation similar to (6.18) is used but both electron and hole current must be included. Only 1 subband is considered (m=0, the metallic subband), so the equation reduces to only 1 summation over k_l. Again, approximating the summation with an integral yields

$$I_{\text{metal}} = \frac{4q^2}{h} T_{\text{metal}} V_{\text{ch,DS}}. \tag{6.24}$$

The corresponding transmission probability must also be calculated as in [27]. As expected, the metallic current is independent of the channel surface potential change (and the gate potential) because the density of states of a metallic SWNT is independent of the carrier energy. This current source is implemented as a dependent conductance from drain to source (Fig. 6.9) with resistance $(4q^2/h)T_{\text{metal}}$.

In the case of a semiconducting SWNT, there is no metallic subband, so there is no metallic current. Thus, I_{metal}=0 in the case of a semiconducting SWNT and the total current for a semiconducting SWNT is I_{semi} + I_{BTBT}. On the other hand, for a metallic SWNT, I_{metal} will dominate the current. So the total current through a metallic SWNT can be approximated as just I_{metal}.

In summary, three current contributions have been modeled (Fig. 6.9) to accurately describe the SWNT-FET drive current, dynamic performance, subthreshold slope, and leakage. To further improve the accuracy in dynamic performance and transient response, a full transcapacitance network is also modeled, but not discussed here. Figure 6.9 illustrates this network and their equations and derivations can be found in [27].

It is perhaps important to note that the self-consistent solution required by the equations cannot be solved analytically without approximations. Thus, this equation form has never before been implemented in a SPICE-like circuit simulator due to runtime inefficiencies of the numerical solver approach [34]. However, for the first time, this model implements the self-consistent equations without further approximations by utilizing the well-optimized intrinsic numerical solving algorithm of the circuit simulator. Specifically, a mathematically equivalent subcircuit is used to solve for the self-consistent solution indirectly (the detailed implementation can be found in [27, 28]). The benefit of such an implementation is the much improved accuracy (while preserving reasonable runtimes), and thus this model can be used as a benchmark for assessing the accuracy of the other analytical solutions obtained through approximations of the self-consistent equations.

To model a complete SWNT-FET device for circuit-level simulations, additional levels (SWNT-FET_L2 and SWNT-FET_L3) must be included to model various other non-idealities and effects, such as those due to the source/drain extension region and contacts as well as the inter-SWNT charge screening.

6.3.3 The Full SWNT-FET Model

The Full SWNT-FET Model builds upon the Intrinsic SWNT Channel Region model (SWNT-FET_L1, first level model). The first level model accurately models the intrinsic channel portion of a SWNT-FET device. To accurately model the entire SWNT-FET, the remaining portions of the SWNT-FET (the source/drain extension regions and contacts) must be taken into account. This is accomplished in level 2, SWNT-FET_L2 (Fig. 6.8). In addition, a SWNT-FET may have multiple SWNTs per device, which results in SWNT-to-SWNT charge screening. This effect is included in the third level, SWNT-FET_L3 [27, 28, 29]. With all three levels, the SWNT-FET model is complete and can be used to accurately simulate circuits with hundreds of multiple-nanotube SWNT-FETs.

The SWNT-FET_L1 model assumed near-ballistic transport. While it accounted for phonon scattering, it did not account for elastic scattering due to the imperfect, practical SWNT fabrication. SWNT-FET_L2 accounts for this elastic scattering by modeling it as a channel resistance ($R_{ch,series}$) in series with the intrinsic SWNT channel region. The total potential drop across the channel region is the sum of the potential drop ($V_{ch,series}$) across the series resistance $R_{ch,series}$ and the potential drop

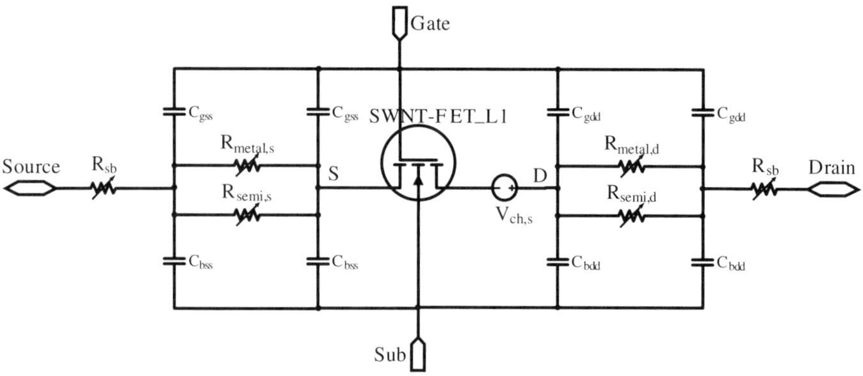

Fig. 6.11 Circuit schematic of the SWNT-FET_L2. The model includes SWNT-FET_L1 and augments it by accounting for: (1) elastic scattering in the channel (modeled as $V_{\mathrm{ch,series}}$); (2) the source/drain extension region parasitic capacitances and resistances (modeled as C_{gss}, C_{bss}, C_{gdd}, C_{bdd}, $R_{\mathrm{metal,s}}$, $R_{\mathrm{semi,s}}$, $R_{\mathrm{metal,d}}$, $R_{\mathrm{semi,d}}$); and (3) the Schottky Barrier resistance due to source/drain metal contacts (modeled as $\mathrm{R_{sb}}$). Reproduced with permission from [28]. Copyright 2007 IEEE

due to the channel quantum resistance. This $V_{\mathrm{ch,series}}$ potential drop is given by

$$V_{\mathrm{ch,series}} = \frac{L_{\mathrm{g}}}{L_{\mathrm{g}} + \frac{D_{\mathrm{CNT}}}{1.5\,\mathrm{nm}}\lambda_{\mathrm{eff}}} V_{\mathrm{DS}}, \tag{6.25}$$

where D_{CNT} is the diameter of the SWNT. It is assumed that the mean free path (MFP) is linearly proportional to the SWNT diameter, and that a SWNT with diameter 1.5 nm has a mean free path of λ_{eff}. Thus, the MFP $l_{\mathrm{eff}} = (D_{\mathrm{CNT}}/1.5\ \mathrm{nm})\ \lambda_{\mathrm{eff}}$. Figure 6.11 shows the modeled circuit for SWNT-FET_L2. The elastic scattering model is implemented by including a voltage source $V_{\mathrm{ch,series}}$ that models the voltage drop across the $R_{\mathrm{ch,series}}$ [28].

In addition, SWNT-FET_L2 also accounts for the parasitic capacitances and resistances of the heavily doped source/drain extension region (shown in Fig. 6.11). $R_{\mathrm{semi,s/d}}$ and $R_{\mathrm{metal,s/d}}$ are the resistances due to the semiconducting and metallic subbands. There are two cases to consider: (a) the source/drain region acts as local interconnect and connects the source of one SWNT-FET to the drain of another SWNT-FET (in this case, there is no metal contact); and (b) the source/drain is connected to a metal contact.

In case (a), the potential profile along the doped SWNT (extending from the source of one SWNT-FET to the drain of the next SWNT-FET) is continuous. However, in case (b), there is an additional potential drop across the doped drain to metal contact interface (as well as the metal contact to doped source interface of the next SWNT-FET) due to mode mismatch of the metal contact and the heavily doped SWNT. Figure 6.12 illustrates the Fermi level profiles for the two cases. Both of these cases are handled by the model.

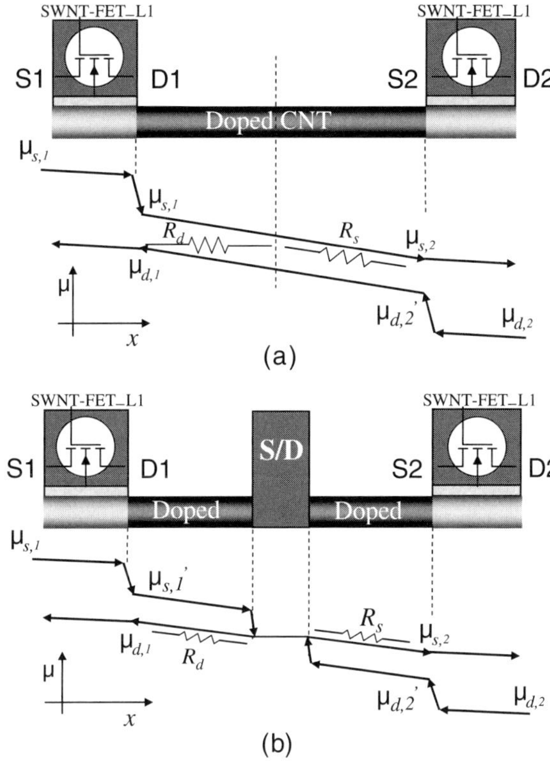

Fig. 6.12 The Fermi level profiles for case (**a**) where the heavily doped source/drain extension region is used as local interconnect to connect two SWNT-FETs and case (**b**) where the source/drain region is connected to a metal contact. These two cases results in different parasitic extension region resistances. Reproduced with permission from [28].

From [28], the source/drain extension region resistances can be found as

$$\begin{cases} R_{\mathrm{x,s}} = L_{\mathrm{s}}/(\lambda_{\mathrm{sd}} G_{\mathrm{x,c}}) & (6.26a) \\ \text{or} & \\ R_{\mathrm{x,s}} = (L_{\mathrm{s}} - \lambda_{\mathrm{sd}})/(\lambda_{\mathrm{sd}} G_{\mathrm{x,c}}) & (6.26b) \end{cases}$$

and

$$\begin{cases} R_{\mathrm{x,d}} = L_{\mathrm{d}}/(\lambda_{\mathrm{sd}} G_{\mathrm{x,c}}) & (6.27a) \\ \text{or} & \\ R_{\mathrm{x,d}} = (L_{\mathrm{d}} - \lambda_{\mathrm{sd}})/(\lambda_{\mathrm{sd}} G_{\mathrm{x,c}}) & (6.27b) \end{cases}$$

where the subscript x can denote either "semi" for the semiconducting subbands or "metallic" for the metallic subbands. Equations (6.26a) and (6.26b) describe the resistance on the source side; equations (6.27a) and (6.27b) describe the resistance on the drain side. Equations (6.26a) and (6.27a) correspond to case (a), where the SWNT is connected directly to the source/drain of another SWNT. Equations (6.26b) and (6.27b) correspond to case (b), where the SWNT is connected to a metal contact. $L_{\mathrm{s/d}}$ is the length of the source/drain extension region; λ_{sd} is the scattering mean free path in the extension region; and $G_{\mathrm{x,c}}$ is the quantum conductance of the doped SWNT. $G_{x,c}$ depends on the SWNT diameter, doping level (E_f), and the potential difference ($V_c = |\mu_s - \mu_d|/q$). Following the approach in [27, 28],

$$\begin{cases} G_{\text{semi},c}(V_c) = \frac{4q^2}{h} \sum_{\substack{k_m \\ m=1}}^{2} \left[1 + \frac{kT}{qV_c} \ln \left(\frac{1+e^{\frac{E_{m,0}-E_f-\Delta\Phi_S}{kt}}}{1+e^{\frac{E_{m,0}-E_f-\Delta\Phi_S+qV_c}{kt}}} \right) \right] & (6.28a) \\ \text{and} & \\ G_{\text{metallic},c} = \frac{4q^2}{h} & (6.28b) \end{cases}$$

where $\Delta\Phi_S$ is the change in surface potential referenced to the source. For semiconducting SWNTs, there is no metallic subband, so $G_{\text{metallic},c}$ is 0. As expected, the metallic conductance is independent of bias.

The parasitic capacitances of the source/drain extension region are also modeled:

$$\begin{cases} C_{\text{gxx}} = \frac{L_x C_{\text{of}} C_Q}{2(C_{\text{of}}+C_Q+C_{\text{sub}})} & (6.29a) \\ \text{and} & \\ C_{\text{bxx}} = \frac{L_x C_{\text{sub}} C_Q}{2(C_{\text{of}}+C_Q+C_{\text{sub}})} & (6.29b) \end{cases}$$

The subscript x can denote either "s" or "d" (e.g., C_{gdd} is the coupling capacitance from the gate to the drain extension region). C_{of} is the outer fringe capacitance from the gate to the doped source/drain region; C_Q is the quantum capacitance of the doped source/drain region; and C_{sub} is the coupling capacitance from the SWNT to the substrate. The detailed derivations are shown in [28].

From equations (6.26), (6.27), (6.28), and (6.29), all the parasitics can be computed. These parasitic resistances and capacitances are included in SWNT-FET_L2 in the form of a π-model for the source/drain extension region (see Fig. 6.11).

Lastly, SWNT-FET_L2 also models the Schottky Barrier (SB) resistance for the source/drain metal contacts, if any. The following simplifying assumptions are made when calculating the SB resistance: (1) the doped SWNT region is long enough such that there is no surface potential modulation due to the quantum confinement effects; (2) the dipole effects are neglected; (3) there are no pinning effects; and (4) the depletion region is approximated as a step function. As done in [28], the SB resistance R_{sb} can be expressed as

$$R_{\text{sb}} = \frac{1}{G_{\text{semi},c}} \left(\frac{1}{T_{\text{sb}}} - 1 \right), \qquad (6.30)$$

where T_{sb} is the transmission probability through the Schottky Barrier. In the case where the source/drain is not connected to a metal contact, R_{sb} is set to 0 in the model (Fig. 6.11).

Next, SWNT-FET_L3 completes the model by addressing two important effects: (1) SWNT-to-SWNT charge screening when there are multiple nanotubes under a single gate and (2) the gate-to-neighboring-contacts parasitic capacitances. The circuit schematic is shown in Fig. 6.13.

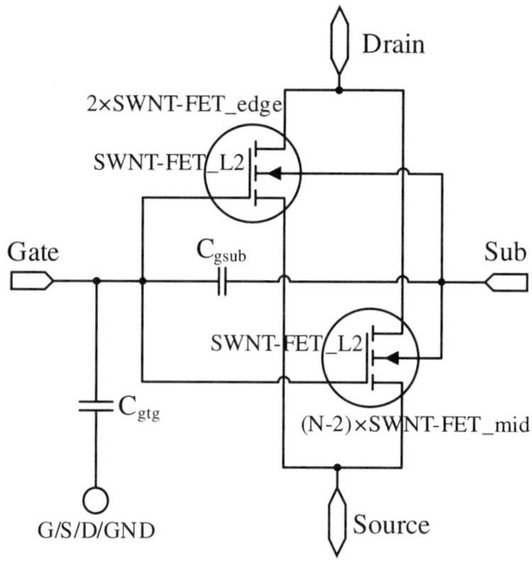

Fig. 6.13 Circuit schematic of the SWNT-FET_L3. The model includes SWNT-FET_L2 and augments it by accounting for: (1) inter-SWNT charge screening by considering two cases (edge and middle) of screening effects, and (2) parasitic gate capacitances C_{sub} and C_{gtg} to the substrate and adjacent gate/source/drain contacts respectively. Reproduced with permission from [28]. Copyright 2007 IEEE

When there are multiple SWNTs per device, the SWNTs can be grouped into two categories: (i) the two SWNTs near the edge of the device and (ii) the remaining (*N*–2) SWNTs in the middle of the device, where *N* is the total number of SWNTs. The SWNTs at the edge only experience charge screening from one side (there is only 1 adjacent SWNT); while the SWNTs in the middle experience charge screening from two sides (there are two adjacent SWNTs, one to each side). The SWNTs within each category are treated identically. The screening effect is taken into account by calculating the effective gate to channel capacitance C_{gc} for each category. As in [29]:

$$C_{edge} = \frac{C_{scr}C_{inf}}{C_{scr} + \eta C_{inf}} \tag{6.31}$$

and

$$C_{middle} = \frac{2}{\eta}C_{edge} + \left(1 - \frac{2}{\eta}\right)C_{inf}, \tag{6.32}$$

where C_{edge} and C_{middle} are the effective gate-to-channel capacitances for CNTs in the edge and middle region, respectively. C_{inf} is the gate-to-channel capacitance if no charge screening effects were present; C_{scr} is the net equivalent capacitance due to charge screening effects from nearby SWNTs; and lastly, η (described fully in [29]) is parameter dependent on the geometry and the number of SWNTs under the same gate. However, the electric field is well confined within the gate dielectric for typical gate geometries (wide gate relative to the thin gate dielectric); thus, the SWNTs far away will have negligible charge screening effects on each other and negligible effects on each other's gate-to-channel capacitance. In other words, the effective gate-to-channel capacitance of a given SWNT will not depend on the number of SWNTs under the gate. (It can be approximated that charge screening is dominated by the immediately adjacent SWNTs to each side, thus the effective gate-to-channel capacitance will depend on whether the SWNT is an "edge SWNT" or a "middle SWNT".) In this case, η reduces to 1, and (6.31) and (6.32) simplifies to

$$C_{\mathrm{edge}} = \frac{C_{\mathrm{scr}} C_{\mathrm{inf}}}{C_{\mathrm{scr}} + C_{\mathrm{inf}}} \tag{6.33}$$

and

$$C_{\mathrm{middle}} = 2C_{\mathrm{edge}} - C_{\mathrm{inf}} = \frac{C_{\mathrm{scr}} C_{\mathrm{inf}} - C_{\mathrm{inf}}^2}{C_{\mathrm{scr}} + C_{\mathrm{inf}}}. \tag{6.34}$$

In the case where there is only one nanotube present, the gate-to-channel capacitance is C_{inf}. This is approximated by the model as simply $C_{\mathrm{inf}} \approx C_{\mathrm{edge}}$ evaluated at a large inter-SWNT pitch so that C_{scr} becomes very large. For two SWNTs per gate, both SWNTs each have an effective gate-to-channel capacitance of C_{edge}. For three or more SWNTs, two SWNTs (the ones at the edge) each have an effective gate-to-channel capacitance of C_{edge}, while the middle SWNT(s) has an effective gate-to-channel capacitance of C_{middle}. Detailed derivations of SWNT charge screening are in [29]. The model handles multiple nanotube SWNT-FETs automatically.

There is also a parasitic coupling capacitance directly from the gate metal to the substrate (C_{sub}) and a lumped parasitic coupling capacitance from the gate metal to the adjacent gate/source/drain contact C_{gtg} (shown in Fig. 6.8 SWNT-FET_L3). Due to the typical aspect ratios of these contacts, C_{gtg} can be rather significant and must be modeled for accurate AC simulations. These additional parasitic capacitances can be derived as in [28].

The SWNT-FET model is now complete, with all three levels of the hierarchy. The model includes many practical non-idealities from scattering in transport to contact parasitics. The full SWNT-FET model can be used in a variety of ways, including the study of a single SWNT-FET device to exploring the design space for optimal SWNT density/pitch. A few example applications are illustrated in the next section. The SWNT-FET model can also be used to simulate large, complementary circuits with hundreds of p- and n-type SWNT-FETs.

The most up-to-date version of the model can be found on the Stanford Nano website (http://nano.stanford.edu) [35].

6.3.4 Validation of the SWNT Compact Model

The Intrinsic SWNT Channel Region model (SWNT-FET_L1) is simulated alone to check that the results and implications are consistent with fundamental physics since this is the core component in the full model. Figure 6.14 shows simulations with various incremental non-idealities. Under the near-ballistic transport assumption, the

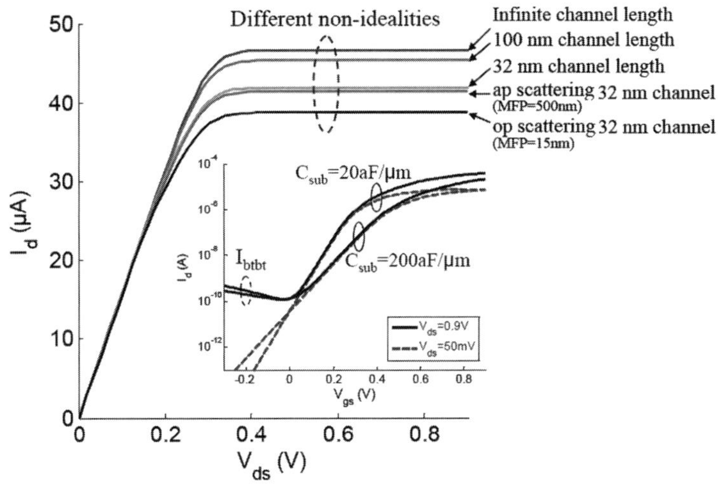

Fig. 6.14 I_D-V_D curve for a SWNT-FET with incremental non-idealities under the near-ballistic assumption. V_{GS}=0.9 V, V_{FB}=0 V, chirality=(19,0), 3 nm HfO_2 top gate dielectric, 10 um SiO_2 insulating layer (between the SWNT and the Si substrate). The inset shows the drain current as a function of V_{GS} and illustrates the band-to-band tunneling effect. Reproduced with permission from [27]. Copyright 2007 IEEE

saturation current in this figure shows the theoretical (with the noted non-idealities) upper limit for the SWNT-FET current. As expected with the near-ballistic assumption (i.e., no scattering in the nanotube channel), there is little difference (<3%) between 100 nm channel length and an infinitely long channel length. However, for short channel devices (<32 nm), drain current decreases with channel length (at L_g=32 nm, the on current is ~90% of the ideal, long channel case). This is because as the channel length decreases, there is increasingly larger energy quantization (k_l) in the axial direction, reducing the number of conducting modes. Further including the phonon scattering effects (optical and acoustic), the on current reduces by another 7% (Pessimistic assumptions are used to obtain the upper bound on the impact of phonon scattering on the DC characteristics. For real devices with >100 nm channel length, non-ballistic transport, and operated under lower V_{dd}, OP scattering is not expected to be the main detractor of DC performance.) The effect of band-to-band tunneling current is also noticeable for high V_{DS} and negative V_{GS} (Fig. 6.14 inset) [27].

Figure 6.15 illustrates more clearly the dependence of drain current with channel length. The figure shows the acoustic and optical phonon scattering mechanisms and the regions where they dominate, respectively. Given a fixed bias (V_{GS}=V_{DS}=0.9 V), for shorter channel lengths (L_g~<2 μm), current is limited by optical phonon scattering because of the shorter optical phonon scattering mean free path (~15 nm as oppose to ~500 nm for acoustic phonons). For longer channel lengths (L_g~>2 μm) and V_{DS}=0.9 V, the current becomes limited by the acoustic phonon scattering mainly because the carriers undergo multiple acoustic phonon scattering events in the channel, therefore, preventing them from gaining enough

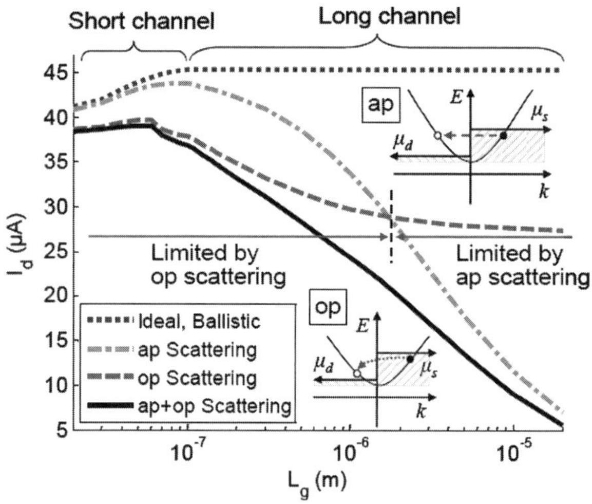

Fig. 6.15 CNT drain current as a function of channel length with and without scattering effects. I_D is taken at $V_{GS}=V_{DS}=0.9$ V. With phonon scattering, the current becomes a strong function of the channel length. Reproduced with permission from [27]. Copyright 2007 IEEE

energy to emit optical phonons. It should be noted, however, that at sufficiently high electric fields (high V_{DS}), the optical phonon emission is activated even for long channel devices, at which point the current becomes limited by the optical phonon scattering. In the ideal, ballistic case, current is mostly independent of the channel length; but, with the phonon scattering effects taken into account, the current then becomes a strong function of channel length [27].

The SWNT-FET_L1 model also implements a transcapacitance network to model the electrostatic coupling between the nodes. Figure 6.16 shows a SWNT-FET connected similarly to a MOS capacitor. The capacitances are shown in Fig. 6.16 as a function of the channel surface potential. The capacitances are non-linear and these

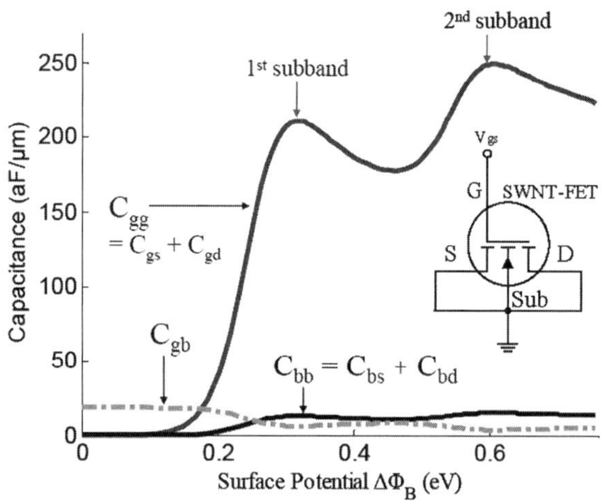

Fig. 6.16 C–V curves for various capacitance components in a SWNT-FET connected as a MOSCAP. V_{FB} is set to 0 and T=300 K. The two peaks in the gate capacitance (C_{gg}) correspond to the first and second subbands. Reproduced with permission from [27]. Copyright 2007 IEEE

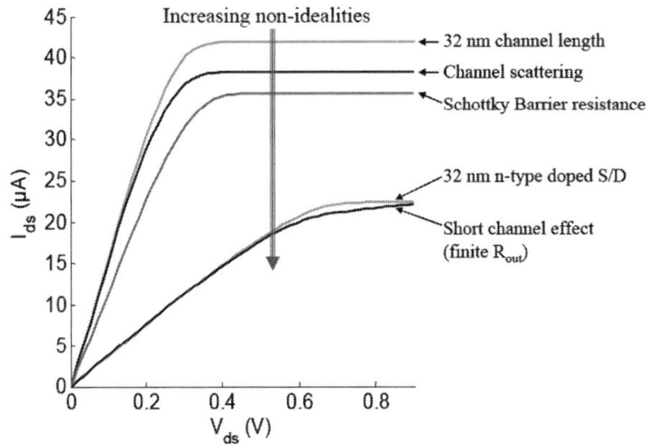

Fig. 6.17 I_{DS}–V_{DS} curve for a SWNT-FET with incremental non-idealities in the full SWNT-FET model. The device parameters are: $V_{GS} = 0.9$ V, $V_{FB} = 0$ V, chirality = (19,0), 3 nm HfO_2 top gate dielectric, 10 μm SiO_2 insulating layer (between the SWNT and the Si substrate). The metal work function and SWNT work function are 4.6 and 4.5 eV, respectively. Reproduced with permission from [28]. Copyright 2007 IEEE

CVs are extremely important for analog small signal applications. Also, the two peaks in the total gate capacitance correspond to the first and second subbands [27].

Now we take a look at the entire SWNT Compact Model to validate the full model. Similar to Fig. 6.14, Fig. 6.17 shows I_D vs V_{DS} as non-idealities are incrementally added to the model. For a 32 nm channel length SWNT-FET, phonon scattering (from SWNT-FET_L1) and elastic scattering (from SWNT-FET_L2) result in a total current degradation of about 10%. The Schottky Barrier resistance from the source/drain metal contacts further reduces current by about 5%. Next, by including the parasitic resistances of the doped source/drain extension regions, the current is significantly reduced by about 40%. This illustrates the importance of placing care into the design and fabrication of the source/drain extension regions.

The effects of inter-SWNT charge screening is also very important, as evident in Fig. 6.18. As the SWNT density is increased (the SWNT pitch is decreased), there is more SWNT charge screening and the current decrease. This suggests that increasing the SWNT density results in a tradeoff between higher drive current (more SWNTs in the channel) and higher current per SWNT (due to charge screening).

6.3.5 Applications of the SWNT-FET Compact Model

Figure 6.19 further explores the tradeoff of increasing SWNT density by simulating a SWNT-FET complementary inverter with varying number of SWNTs within the inverter (the inverter gate width is fixed at 32 nm). The performance, in terms of Fanout-of 4 (FO4) delay, of the SWNT-FET inverter is compared to that of a Si CMOS at the 32 nm technology node. The results suggest that the optimal tradeoff

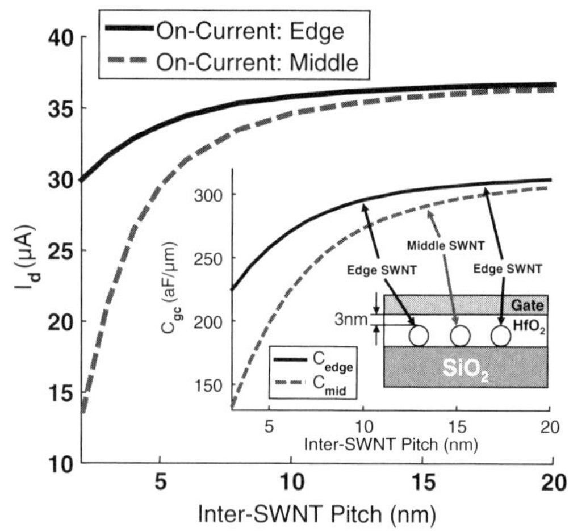

Fig. 6.18 I_D as a function of the SWNT pitch. SWNTs experience more charge screening as the pitch is reduced. The SWNTs within a device are grouped into edge SWNTs (*solid line*) and middle SWNTs (*dashed line*). The inset shows the effective gate-to-channel capacitance for each SWNT group. The middle SWNTs experience SWNT charge screening from both sides, and thus exhibit significantly reduced I_D. Reproduced with permission from [27]. Copyright 2007 IEEE

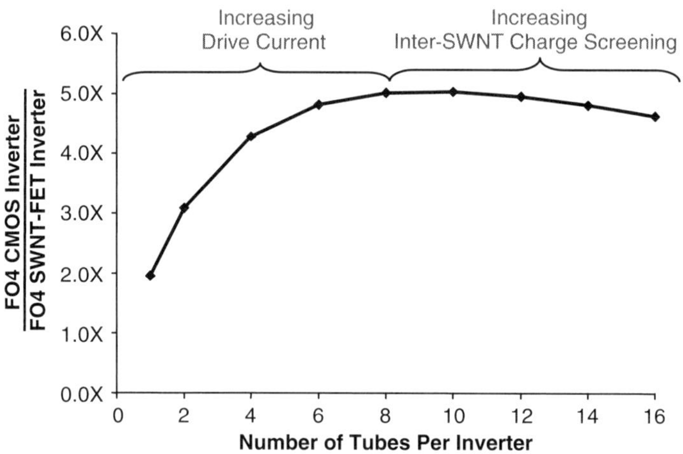

Fig. 6.19 Tradeoff between increasing drive current and increasing inter-SWNT charge screening for various SWNT densities. The inverter under test is 32 nm wide. The optimal performance occurs at 8 SWNTs/32 nm = 4 nm SWNT pitch when increasing drive current from more SWNTs balances out the diminishing individual SWNT current from increased charge screening. Reproduced with permission from [36]. Copyright 2007 IEEE

occurs at a SWNT pitch of 4 nm (i.e., SWNT density of ~250 SWNTs/μm) [36], which yields about 5x better performance for SWNT-FETs over Si CMOS. Fewer than ~250 SWNTs/μm density results in better current per SWNT (less SWNT-to-SWNT charge screening), but much less total current (fewer SWNTs in total). On the other hand, greater than ~250 SWNTs/μm density results in much less current per SWNT (more SWNT-to-SWNT charge screening), and net less total current though there are more SWNTs in total.

Table 6.1 Comparison of gate capacitance, on current, and off current between SWNT-FETs and Si MOSFETs. Note the currents are in units of current per capacitance. The off currents for SWNT-FET and MOSFET have been set to the same value. The last column shows the (SWNT-FET I_{on})/(MOSFET I_{on}) ratio. Reproduced with permission from [28]. Copyright 2007 IEEE

$L_{Channel}$ = 18 nm	Gate C_{eff}	I_{off} (nA/fF)	I_{on} (mA/fF)	I_{on}/I_{off}	SWNT-FET/MOS
nMOS	1.1 fF/μm	383	1.198	3128	N/A
nSWNT-FET	3.6 aF/FET	383	7.236	18863	6.03
pMOS	1.1 fF/μm	253	0.5229	2066	N/A
pSWNT-FET	3.6 aF/FET	253	7.172	28389	13.74

Additionally, n-type and p-type SWNT-FET devices were compared against Si MOSFET devices at the 32 nm technology node. The results show that n-type and p-type SWNT-FETs have ~ 6x and 13x, respectively, better drive current (in current per unit capacitance) performance than Si nFET and pFET (Table 6.1) [28].

J. Deng et al. [36] has also reported the use of this Circuit Compatible SWNT-FET Model to investigate the effects on energy and FO_4 of various variations and imperfections on SWNT-FET inverters. Diameter, doping level, and the percentage of metallic SWNTs were varied to study the impact of these

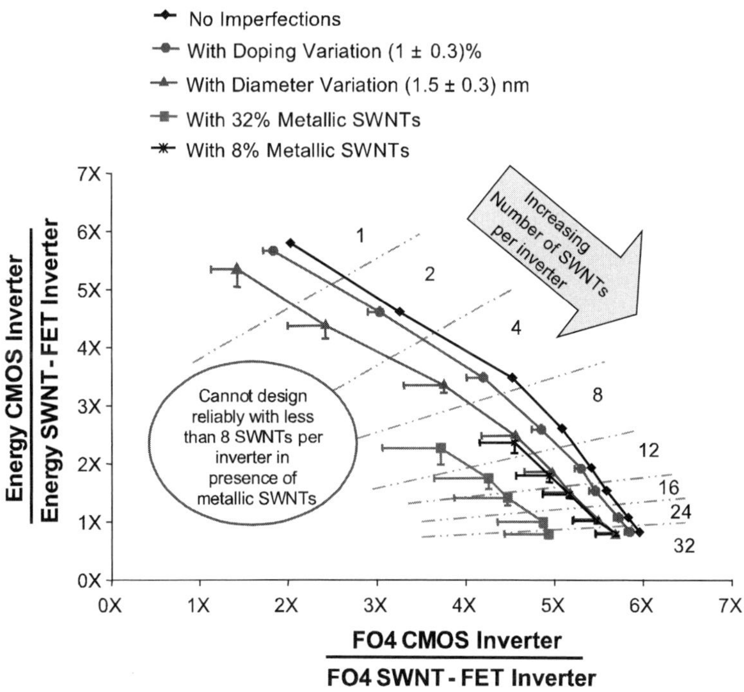

Fig. 6.20 Energy per cycle and FO_4 delay improvement for SWNT-FET inverter at 3σ points (σ is the standard deviation) compared to 32 nm CMOS inverter. Error bars indicate 6σ variation. Reproduced with permission from [36]. Copyright 2007 IEEE

variations on circuit reliability. Reference [36] notes that of these three sources of variation, metallic SWNTs appear to be the most problematic. Metallic SWNTs in a SWNT-FET (roughly 1/3 of SWNTs are metallic) can cause the source and drain to short, as the gate voltage has no control over metallic SWNT conduction. But even assuming that all metallic SWNTs can be removed (to leave only semiconducting SWNTs for the SWNT-FET channels), this results in a random number of semiconducting SWNTs in the SWNT-FET (given a fixed number of SWNTs per SWNT-FET, there is a random number of metallic SWNTs and thus also a random number of semiconducting SWNTs). The simulations show that this results in significantly degraded performance. Furthermore, the high probability of metallic SWNTs can result in "defective" inverters since some inverters may have no semiconducting SWNTs at all after metallic CNT removal. Hence, the high probability of metallic SWNTs is a primary problem for SWNT-FET large scale circuits. Figure 6.20 summarizes the study [36].

6.4 Summary

SWNT-FETs are a promising extension to Si CMOS beyond the International Technology Roadmap for Semiconductors (ITRS) [37] into the nanoscale regime. As SWNT-FET technology continues to mature, a spectrum of modeling tools will be required to aid in design and verification. In this chapter, we introduced two SWNT-FET device models. The first model, the Schottky Barrier SWNT-FET model, is a device-level model that can be used to rapidly explored device designs. The second model, the CNFET Compact model, is a circuit-compatible model used for quickly simulating large-scale, CMOS-like SWNT-FET circuits to evaluate both DC and AC performance and functionality. In addition, several applications of these models were also presented, illustrating the power of these tools in revealing important design considerations for the future.

Acknowledgments We thank Prof. M. Lundstrom (Purdue) and Prof. J. Guo (U. Florida) for discussions on the modeling of the Schottky barriers. The support and encouragement of Dr. Jim Hutchby (SRC) and Dr. Wilfried Haensch (IBM) for the development of the compact model for the carbon nanotube transistor is greatly appreciated. We would like to thank our collaborators, Prof. Subhasish Mitra and Nishant Patil, who have significantly contributed to the work described here. This work was supported in part by the Charles Powell Foundation, the National Science Foundation (ECS-0501096), the Semiconductor Research Corporation (SRC), and the MARCO FENA and C2S2 Focus Research Center Programs.

References

1. J. Guo, S. Datta, M. Lundstrom, and M.P. Anantram, "Toward multi-scale modeling of carbon nanotube transistors," *The International Journal on Multiscale Computer Engineering*, vol. 2, pp. 257–277, 2004.
2. S. Beebe. F. Rotella, Z. Sahul, D. Yergeau, G. McKenna, I. So, Z. Yu, K.C. Wu, E. Kan, J. McVittie, and R.W. Dutton, "Next generation Stanford TCAD – PISCES 2ET and SUPREM 007," *International Electron Devices Meeting*, pp. 213–216, 1994.

3. L.W. Nagel, "SPICE2: A computer program to simulate semiconductor circuits," *Technical Report ERL-M520*, University of California, Berkeley, 1975.
4. S. Scheffer, L. Lavagno, and G. Martin, "EDA for IC system design, verification, and testing," *CRC Taylor and Francis*, Boca Raton, Florida, 2006.
5. A. Hazeghi, T. Krishnamohan, H.-S.P. Wong, "Schottky-barrier carbon nanotube field effect transistor modeling," *IEEE Transactions on Electron Devices*, vol. 54, pp. 439–445, 2007.
6. M. Lundstrom and J. Guo, *Nanoscale Transistors: Device Physics, Modeling, and Simulation,* Springer Publishing, 2006.
7. J. Deng and H.-S.P. Wong, "A circuit-compatible SPICE model for enhancement mode carbon nanotube field effect transistors," *International Conference on Simulation of Semiconductor Devices and Processes* (*SISPAD*), pp. 166–169, 2006.
8. K. Natori, Y. Kimura, and T. Shimizu, "Characteristics of a carbon nanotube field-effect transistor analyzed as a ballistic nanowire field-effect transistor," *Journal of Applied Physics*, vol. 97, p. 34306, 2005.
9. J. Guo, M. Lundstrom, and S. Datta, "Performance projections for ballistic carbon nanotube field-effect transistors," *Applied. Physics Letters*, vol. 80, p. 3192, 2002.
10. J. Guo, S. Datta, and M. Lundstrom, "A numerical study of scaling issues for Schottky-barrier carbon nanotube transistors," *IEEE Transactions on Electron Devices*, vol. 51, pp. 172–177, 2004.
11. A. Javey, J. Guo, Q. Wang, M. Lundstrom, and H. Dai, "Ballistic carbon nanotube field-effect transistors," *Nature*, vol. 424, p. 654, 2003.
12. S.-H. Oh, D. Monroe, and J. M. Hergenrother, "Analytic description of short-channel effects in fully-depleted double-gate and cylindrical, surrounding-gate MOSFETs," *IEEE Electron Device Letters*, vol. 21, p. 445, 2000.
13. A. Hazeghi, T. Krishnamohan, and H.-S.P. Wong, "Schottky-Barrier carbon nanotube field effect transistor modeling," *6th IEEE Conference on Nanotechnology*, pp. 238–241, Cincinnati, OH, June 17–20, 2006.
14. Ph. Avouris, J. Appenzeller, R. Martel, and S. Wind, "Carbon nanotube electronics," *Proceedings of the IEEE*, vol. 91, pp. 1772–1784, 2003.
15. A. Javey, R. Tu, D.B. Farmer, J. Guo. R.G. Gordon, and H. Dai, "High performance n-type carbon nanotube field-effect transistors with chemically doped contacts," *Nano Letters*, vol. 5, pp. 345–348, 2005.-
16. Y.M. Lin, J. Appenzeller, and Ph. Avouris, "High-performance carbon nanotube field-effect transistor with tunable polarities," *IEEE Transactions on Nanotechnology*, vol. 4, p. 481, 2005.
17. M. Radosavljevic, J. Appenzeller, and Ph. Avouris, "High performance of potassium n-doped carbon nanotube field-effect transistors," *Applied Physics Letters*, vol. 84, p. 3693, 2004.
18. J. Appenzeller, M. Radosavljevic, J. Knoch, and Ph. Avouris, "Tunneling versus thermionic emission in one-dimensional semiconductors," *Physics Review Letters*, vol. 92, p. 48301, 2004.
19. S. Datta, *Electronic Transport in Mesoscopic Systems*, Cambridge University Press, 1995.
20. J. W. Mintmire and C. T. White, "Universal density of states for carbon nanotubes," *Physics Review Letters*, vol. 81, p. 2506, 1998.
21. F. Leonard and J. Tersoff, "Role of Fermi-level pinning in nanotube schottky diodes," *Physics Review Letters*, vol. 84, p. 4693, 2000.
22. W. Kim, A. Javey, R. Tu, J. Cao, Q. Wang, and H. Dai, "Electrical contacts to carbon nanotubes down to 1 nm in diameter," *Applied Physics Letters*, vol. 87, p. 173101, 2005.
23. J. Deng and H.-S.P. Wong, "Metrics for performance benchmarking of nanoscale Si and carbon nanotube FETs including device non-idealities," *IEEE Transactions on Electron Devices*, pp. 1317–1322, June, 2006.
24. BSIM3 and BSIM4 are developed by the Device Research Group of the Department of Electrical Engineering and Computer Science, University of California, Berkeley and copyrighted by the University of California. www-device.eecs.berkeley.edu/~bsim3/bsim˙ent.html

25. A. Raychowdhury, S. Mukhopadhyay, and K. Roy, "A circuit-compatible model of ballistic carbon nanotube field-effect transistors," *IEEE Transactions on Computer-Aided Design of Integrated Circuits and Systems*, vol. 23, pp. 1411–1420, 2004.
26. C. Dwyer, M. Cheung, and D. J. Sorin, "Semi-empirical SPICE models for carbon nanotube FET logic," *4th IEEE Conference on Nanotechnology*, pp. 386–388, 2004.
27. J. Deng and H.-S.P. Wong, "A compact SPICE model for carbon nanotube field effect transistors including nonidealities and its application – Part I: model of the intrinsic channel region," *IEEE Transactions on Electron Devices*, vol. 54, pp. 3186–3194, 2007.
28. J. Deng and H.-S.P. Wong, "A compact SPICE model for carbon nanotube field effect transistors including nonidealities and its application – Part II: full device model and circuit performance benchmarking," *IEEE Transactions on Electron Devices*, vol. 54, pp. 3195–3205, 2007.
29. J. Deng and H.-S.P. Wong, "Modeling and analysis of planar gate capacitance for 1-d fet with multiple cylindrical conducting channels," *IEEE Transactions on Electron Devices*, vol. 54, pp. 2377–2385, 2007.
30. D. Mann, A. Javey, J. Kong, Q. Wang, and H. Dai, "Ballistic transport in metallic nanotubes with reliable Pd ohmic contacts," *Nano Letters*, vol. 3, pp. 1541–1544, 2003.
31. Z. Yao, C.L. Kane, and C. Dekker, "High-field electrical transport in single-wall carbon nanotubes," *Physics Review Letters*, vol. 84, pp. 2941–2944, 2000.
32. E. O. Kane, "Zener tunneling in semiconductors," *Journal of Physics and Chemistry of Solids*, vol. 12, pp. 181–188, 1959.
33. E.O. Kane, "Theory of tunneling," *Journal of Applied Physics*, vol. 32, pp. 83–91, 1961.
34. A. Raychowdhury, S. Mukhopadhyay, and K. Roy, "A circuit-compatible model of ballistic carbon nanotube field-effect transistors," *IEEE Transactions on Computer-Aided Design of Integrated Circuits and Systems,* vol. 23, pp. 1411–1420, 2004.
35. Stanford Nano Website. http://nano.stanford.edu
36. J. Deng, N. Patil, K. Ryu, A. Badmaev, C. Zhou, S. Mitra, and H.-S.P. Wong, "Carbon nanotube transistor circuits: circuit-level performance benchmarking and design options for living with imperfections," *International Solid State Circuits Conference (ISSCC)*, pp. 70–71, 2007.
37. International Technology Roadmap for Semiconductors. http://www.itrs.net/

Chapter 7
Performance Modeling for Carbon Nanotube Interconnects

Azad Naeemi and James D. Meindl

7.1 Introduction

Since their discovery in 1991, carbon nanotubes (CNT) have received tremendous research interest as they have many unique mechanical, electrical, thermal and chemical properties [1]. A single-walled carbon nanotube (SWNT) is a graphene roll with a diameter of 0.5 to a few nanometers that depending on its chirality can be either metallic or semiconductor. Multi-walled carbon nanotubes (MWNT), on the other hand, are concentric graphene tubes that may have diameters from a few to a hundred nanometers. The in-plane sp^2 bonding in graphene is even stronger than the sp^3 bonding in diamond [2], and carbon nanotubes, therefore, have very high mechanical strengths.

A SWNT is close to an ideal one-dimensional system of electrons that gives rise to many unique electrical and thermal properties, some of which were discussed in Chapter 1. Since electrons can move in one dimension only, the phase space for scattering in nanotubes is very limited; electrons can be scattered only backward. The mean free path in high-quality nanotubes, therefore, is in the micron range (when the bias voltage is low and there is no high energy phonon scattering) [3]. This is in contrast to a three-dimensional metallic wire in which electrons can be backscattered by various small-angle scatterings, and the mean free paths are in the range of a few tens of nanometers. In addition, carbon nanotubes have the potential of being used as both transistors and interconnects since they can be either metallic or semiconducting depending on their chirality.

Interconnects are considered as one of the grandest challenges that gigascale integration faces because of the delay they add to the critical paths, the power they dissipate, the noise and jitter they induce on one another, and their vulnerability to electromigration. As will be demonstrated in this chapter, carbon nanotubes can potentially address these challenges if they are optimally utilized. This chapter aims at quantifying the physical limits of carbon nanotubes and comparing them with

A. Naeemi (✉)
Microelectronics Research Center, Georgia Institute of Technology, Atlanta, GA 30332, USA

A. Javey, J. Kong (eds.), *Carbon Nanotube Electronics*, Series on Integrated Circuits and Systems, DOI 10.1007/978-0-387-69285-2_7,

those of copper interconnects to demonstrate their viability and also developing some guidelines for the nature of CNT technology needed to fully exploit their potentials.

This chapter starts with deriving a physical circuit model for SWNTs that is accurate for all lengths and voltages. This circuit model is useful for compact modeling, and can be easily used in circuit simulators like SPICE. SWNT-bundles are then modeled in Section 7.3, and their conductivity is compared with that of copper wires. Qualitative physical explanations and analogies are also provided to offer insight and clarify some misconceptions about SWNTs. In Section 7.4, compact physical models are developed for conductivity of MWNTs, and their performance is compared with SWNT-bundles and copper wires. In Section 7.5, a comprehensive study of the application of carbon nanotubes in a multi-level interconnect network at the local, semi-global and global levels is presented. The results are summarized in Section 7.6.

7.2 Circuit Models for SWNTs

7.2.1 Kinetic Inductance

Inductance has conventionally been defined as the resistance to current change due to Faraday's law, and it represents the energy stored in the magnetic field generated by current, $1/2L_M I^2$, where L_{M} is the inductance and I is the current. The mechanical counterpart for inductance is mass as it opposes velocity change, and gives rise to kinetic energy $(1/2mv^2)$. Electric current itself is the flow of carriers that have nonzero mass and therefore nonzero kinetic energy. The total energy associated with electric current is [4]

$$E = \int\limits_{all\,space} \frac{1}{2}\mu H^2 dV + \int\limits_{conductor} \frac{1}{2} n m v^2 dV \tag{7.1}$$

where μ is permeability, H is the magnetizing force, and m, v and n are mass, speed, and density of charged particles, respectively. In normal wires, the energy stored in the magnetic field is significantly larger than the kinetic energy of electrons, and the second integral is negligible. However, there are cases in which the kinetic energy of electrons is comparable or even larger than the energy in the magnetic field. In those cases, a kinetic inductance needs to be defined to represent the kinetic energy of electrons. The classic model for current is

$$I = qnvA, \tag{7.2}$$

where A is the cross-sectional area and q is the charge of the carrier particles. Assuming constant current density in a conductor, kinetic inductance per unit length becomes equal to [5]

$$l_k = \frac{1}{A}\left(\frac{m}{nq^2}\right) \tag{7.3}$$

which is inversely proportional to the cross-sectional area. Qualitatively, as the cross-sectional area decreases, fewer carriers will be available, and to maintain a constant current, electrons need to move faster. But, because of the quadratic dependency of kinetic energy on speed, the kinetic energy and hence kinetic inductance increases as the cross-sectional area becomes smaller. Magnetic inductance of a wire, on the other hand, depends mainly on its distance to a return path and weakly on the cross-sectional area. Thus, as the cross-sectional dimensions of a wire become smaller, the ratio of kinetic inductance to magnetic inductance becomes larger. For a normal wire, this kinetic inductance is very difficult to observe no matter how small the cross-sectional area is, because its reactive impedance is always going to be negligible compared to its resistance. Resistance per unit length is

$$r = \frac{1}{A}\left(\frac{m}{nq^2}\right)\frac{1}{\tau}, \tag{7.4}$$

where τ is the average collision time for carriers. Even in the best metals, $1/\tau$ is in the order of 10–100 THz [6]; therefore, for virtually all practical frequencies $\omega l_k << r$. For superconductors, however, there is no resistance, and kinetic inductance has been modeled and measured for many years. In thin film superconductors, it is shown that kinetic inductance can be up to four orders of magnitude larger than its magnetic counterpart [5]. Far infrared detectors [7] and delay lines with phase velocities 100 times smaller than the speed of light [5] are two examples of practical devices that operate based on kinetic inductance in thin film superconductors.

Kinetic inductance for a carbon nanotube becomes important as the mean free path of electrons and hence the collision time for electrons can be very large. The models described above offer a good insight, even though this would be a classical description for kinetic inductance. From the quantum mechanical point of view, one can look at the available energy states in an ideal quantum wire that is shown in Fig. 7.1 [8]. These states can be simply calculated by solving the Schrödinger equation in one dimension and applying the boundary conditions. Due to degenerative approximation, one can assume that all energy states below the Fermi level are occupied and the ones above it are empty. For zero current, the number of electrons moving from left to right are the same as those moving in opposite direction, thus canceling the impact of one another. To generate a current from left to right, some of the left movers must be converted to right movers. But due to the finite density of states in 1D conductors together with the requirement of the Pauli Exclusion Principle, the converted electrons have to fill in the higher energy levels. Hence, as current increases the total kinetic energy of electrons increases by $N^2/D(\varepsilon_F)$ where

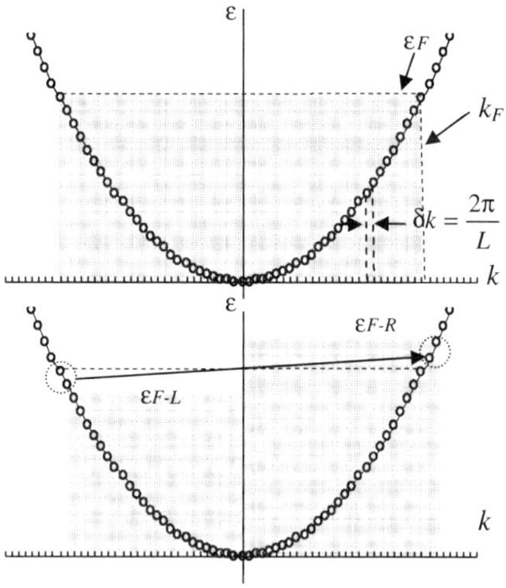

Fig. 7.1 Electron energy versus wave-vector in a quantum wire. Allowable states are shown by open circles. Top: current equal to zero, the numbers of right-mover electrons (positive k) and left-mover electrons (negative k) are the same. Bottom: a net current from left to right. Some of the left-movers are converted to right-movers because of which the kinetic energy of the system has increased. The axes are not to scale since $k_F >>> \delta k$. Reproduced with permission from [8]. Copyright 2007 IEEE

N is the number of converted electrons and $D(\varepsilon_F)$ is the density of states near the Fermi level. The density of states in a quantum wire is $D(\varepsilon_F) = L/(hv_F)$, and current is equal to $I = 2Nev_F/L$, where v_F is the Fermi velocity, the speed of electrons at the Fermi level ($mv_F{}^2/2 = \varepsilon_F$), and L is length. The kinetic inductance per unit length can therefore be written as [9, 10, 11]

$$l_k = \frac{h}{2v_F e^2} \tag{7.5}$$

For graphene and hence carbon nanotubes, $v_F = 8\times10^5$ m/s [10]. There are four parallel channels due to spin and sublattice degeneracy (see Chapter 1), and the overall kinetic inductance per unit length of carbon nanotubes is therefore around 4 nH/μm, more than four orders of magnitude larger than its magnetic counterpart.

Like most fundamental limits, it is difficult to observe kinetic inductance as there are other limits that are more visible and dominant. For instance, the reactive impedance of a SWNT at 10 GHz due to its kinetic inductance is 251 Ω/μm whereas its resistance is in the order of 6.5 kΩ/μm. That is the reason why the attempts to observe kinetic inductance in SWNTs at this frequency range [12] have not been successful. Note that the commercial microwave test equipments used in [12] are designed for loads with close to 50 Ω impedances whereas tested carbon nanotubes had resistances above 200 kΩ. Such a large impedance mismatch severely limits the resolution of the measurements [13].

7.2.2 Capacitance

To add electric charge to a quantum wire, one must add electrons to available states above the Fermi level (Pauli Exclusion Principle). The required energy to add electric charge Q to a quantum wire is [9, 10, 11]

$$E = \frac{1}{2}\frac{Q^2}{C_E} + \frac{(Q/e)^2}{D(\varepsilon_F)}, \tag{7.6}$$

where the first term is the energy stored in the electric field (C_E is the electrostatic capacitance) and e is electron charge. One can therefore define a quantum capacitance in series to the electrostatic capacitance as

$$c_Q = 2e^2/(h\nu_F) \tag{7.7}$$

and has a value in the order of 100 aF/μm, in the same order of the electrostatic capacitance of a typical wire above a ground plane. More detailed discussion on quantum capacitance can be found in Chapter 3.

7.2.3 Resistance

There is a minimum resistance of h/e^2 associated to an ideal quantum wire neglecting spin degeneracy [14]. A simple qualitative explanation comes from Fig. 7.1 in which electrons moving from left to right having a higher Fermi level compared to the ones moving in the opposite direction. The electrons in the left contact should have a higher energy to be able to enter the quantum wire. Hence, the contact on the left should have a lower potential compared to the right contact, and this leads to the quantum resistance. Considering a fourfold spin and sublattice degeneracy of a SWNT, the quantum resistance becomes 6.45 kΩ.

The quantum resistance is the minimum resistance of a quantum wire if there are no scatterings at the contacts or along the quantum wire. Scatterings at the contacts give rise to the contact resistance that may be up to hundreds of kilo-ohms if the contacts are poor. There have been many reports, however, indicating contact resistances in the order of a few kilo-ohms or even hundred ohms [15–18] by using appropriate metal materials (refer to Chapter 3) that show that there is no fundamental limitation in lowering the contact resistance to the level that can be ignored compared to the quantum resistance.

Electrons moving along a CNT can get scattered by defects and also phonons, and their mean free paths are finite (see Chapter 1 for more detail). Resistance, therefore, is a function of length. Both linear and exponential length dependencies have been reported. The nanotubes with resistances linearly proportional to their length normally have large mean free paths (~1 μm, ignoring the high energy phonons) [19, 20], and the ones that show exponential dependency normally have smaller mean free paths in the order of a few hundred nanometers [21]. One likely explanation is that electrons in nanotubes with high defect density get scattered multiple times before they loose their phase coherence. The incident and scattered electron waves can therefore interfere and become localized. This phenomenon is proven to give rise to exponential dependency of resistance on length [14]. For the rest of this chapter, it is assumed that the defect density is low and the resistance is a linear function of length.

At small bias voltages, electrons in SWNTs get backscattered by defects and acoustic phonons only, and electron mean free path (mfp) can be as large as 1.6 μm in high-quality nanotubes [19]. At higher voltages, electrons get backscattered by optical and zone-boundary phonons that have energies around $\hbar\Omega \approx 0.16$ eV [15, 19, 22] as discussed in Chapter 1. An electron with energy E can emit a phonon of energy $\hbar\Omega$ only if there is an available state with energy $E - \hbar\Omega$. The length through which an electron must be accelerated by electric field E to attain this energy is $\ell_{\Omega} = \hbar\Omega/eE$, where e is electron charge. Once it attains this energy, it travels on average $\ell_0 = 15$ nm before it scatters, and emits an optical or zone boundary phonon [19, 22]. The effective mfp can be obtained via Matthiessen's rule given by

$$\frac{1}{\ell_{eff}} = \frac{1}{\ell_e} + \frac{1}{\ell_{\Omega} + \ell_0} \tag{7.8}$$

where ℓ_e is the low-bias mfp. Resistance of a SWNT is given by [15]

$$R = R_c + R_Q(1 + L/\ell_{eff}) \tag{7.9}$$

where R_C is the contact resistance, and L is the nanotube length. The electric field along a nanotube may vary (e.g., it is larger near the contacts), and (7.9) should be written in an integral form:

$$R = R_c + R_Q + R_Q \int_{x=0}^{x=L} \frac{dx}{\ell_{eff}(x)}, \tag{7.10}$$

$$= R_c + R_Q + \left(R_Q \frac{L}{\ell_e} + \int_{x=0}^{x=L} \frac{dx}{\frac{I_0}{E(x)} + \frac{\ell_0}{R_Q}} \right),$$

where $I_0 \equiv h/(4e(\hbar\Omega)) = 25$ μA. The term inside the bracket shows the distributed resistance whereas the rest are lumped resistances split between the two ends.

7.2.4 Equivalent Circuit

Based on resistance, capacitance, and inductance values described above, an equivalent circuit for the distributed resistance of a metallic SWNT is proposed in Fig. 7.2 that is valid for all bias voltages and lengths assuming that only the two subbands that cross the Fermi level conduct the current. In the equivalent circuit in Fig. 7.2, R_V is a voltage-dependent resistor corresponding to ℓ_{Ω}, R_{shunt} is a shunt resistor corresponding to ℓ_0, and R_e represents elastic scatterings by acoustic phonons. The bandgap of the first non-crossing subband is equal to ~2.6(eV)/diameter(nm) [17]. For SWNTs with typical diameters of 1–2 nm, non-crossing subbands may conduct

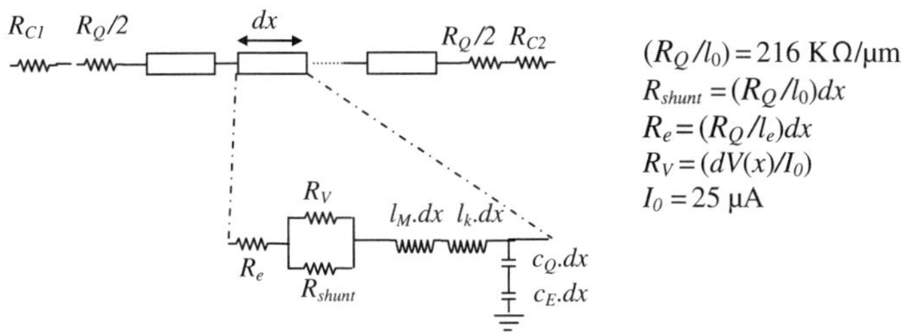

Fig. 7.2 The equivalent circuit for resistance of metallic SWNTs based on physical models for electron–phonon scattering. This circuit can be used in HSPICE simulations using a voltage-dependent resistance element (VCR) for R_V. R_{c1} and R_{c2} are the contact resistances at each end. Magnetic inductance (l_M) and electrostatic capacitance (c_E) values depend on geometry. Reproduced with permission from [8]. Copyright 2007 IEEE

only at voltages above 1.3–2.6 V [17]. The power supply voltage in current digital chips is close to 1.2 V, and is projected to scale down to ~0.5 V at the end of the ITRS [23]. For virtually all interconnect applications, therefore, the equivalent circuit in Fig. 7.2 is adequate. This circuit can be used in HSPICE simulations using a voltage-dependent resistance element (VCR) for R_V [24].

It should be noted that for most interconnect applications, $R_{\text{shunt}} > 10R_V$; hence, R_{shunt} can be neglected unless a nanotube is so short and/or the bias voltage is so large that the electric field becomes larger than 0.54 V/μm. Ignoring R_{shunt} simplifies (7.10) to

$$R = R_{\text{low-bias}} + \frac{V}{I_0} \tag{7.11}$$

where $R_{\text{low-bias}} \equiv R_Q(1+L/\ell_e)$. Equation (7.11) is the same as the model presented in [15] that has been verified experimentally. Note that the total resistance given by equation (7.11) is useful for steady state analyses, and the distributed circuit shown in Fig. 7.2 should be used for transient analysis even when R_{shunt} can be ignored. The voltage-dependent component in equation (7.11) is independent of the nanotube length, quality, and contacts whereas the low-bias resistance is sensitive to these parameters. Bias voltage, therefore, affects short high-quality nanotubes the most. For instance, if $\ell_e = 1$ μm and the bias voltage is 1 V, the resistance is 307%, 56%, and 6.25% larger than the low-bias resistance for the lengths of 1 μm, 10 μm, and 100 μm, respectively. Another important point about (7.11) is that the resistance increases monotonically with voltage. Hence, a piecewise model for resistance based on a critical voltage as suggested by [25] does not appear to be valid.

The equivalent circuit shown in Fig. 7.2 is a distributed R(V)LC circuit or a transmission line in which the wave propagation speed is

$$v = 1/\sqrt{lc} = \sqrt{\frac{4}{l_k}\left(\frac{1}{c_E} + \frac{1}{4c_Q}\right)}, \tag{7.12}$$

where l and c are inductance and capacitance per unit length, respectively. Assuming that the electrostatic and quantum capacitances are equal, the wave speed is $\sqrt{5}v_F = 1.78\times10^6$ m/s, which is 162 times smaller than the speed of light in free space. To see the impact of this limitation, one can compare the step responses of an ideal SWNT (ballistic) and a copper interconnect implemented at the end of the ITRS as shown in Fig. 7.3. It can be seen that even an ideal SWNT will have a delay larger than the minimum size copper wire for two reasons, the large quantum resistance and the large signal travel time. For copper wires, the cross-sectional dimensions are assumed to be the minimum wire width at the 22 nm node projected for the year 2016 [23], and the specularity parameter, p (an empirical parameter representing the fraction of electrons reflected specularly at the surfaces of wires) and the reflection coefficient at grain boundaries, R are both pessimistically assumed to be 0.5. Reflectivity coefficients as small as 0.2 are reported [26].

In a conventional transmission line, the wave propagation speed is always very close to the speed of light, independent of geometry. This is mainly due to two properties associated with the magnetic inductance. First, magnetic inductance of a wire is a function of its distance to its return path. One may lower the capacitance between a wire and a ground plane by increasing the distance between them. However, this increases magnetic inductance, and the wave propagation speed remains approximately unchanged. The second property is mutual inductance. The equivalent inductance of two similar inductors in parallel is $l_s/2+l_m/2$ where l_s and l_m are the self and mutual inductances. For two adjacent lines, mutual inductance is close to self inductance, and connecting lines in parallel does not lower the inductance considerably which is why inductance changes very slowly as cross-sectional area increases.

Kinetic inductance, however, is independent of the distance to return path, and there is no mutual kinetic inductance either. Hence unlike conventional transmission

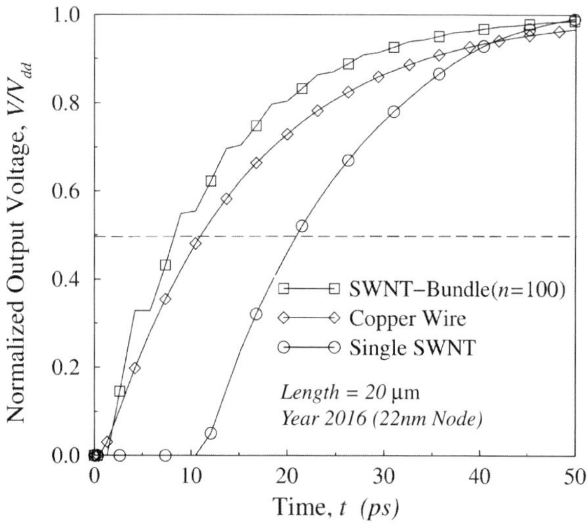

Fig. 7.3 Step responses of copper, ideal single SWNT and ideal SWNT-bundle interconnects versus time. Interconnects are 20 μm long. Copper and SWNT-bundles are 27 nm wide, the minimum feature size projected for the 22 nm technology node (year 2016). Reproduced with permission from [8]. Copyright 2007 IEEE

lines in which wave propagation is independent of geometry, the wave propagation speed in a bundle of carbon nanotubes depends on the number of metallic nanotubes and their distance to their nearby ground plane. Having adequate numbers of metallic nanotubes in a bundle makes the wave propagation speed large enough such that the RC charge up time becomes dominant compared to the travel time. This is shown in Fig. 7.3.

7.3 Circuit Models for SWNT–Bundles

Knowing that nanotube bundles are needed to outperform copper wires, it is important to understand the physical properties of such bundles. Some major issues regarding nanotube bundles are reviewed here that lead to important guidelines regarding the way these bundles need to be made.

One important question is whether or not each nanotube retains its physical properties once it is placed in a bundle. A rigorous analysis for a regular array of SWNTs with chirality of (10,10) shows that a bandgap appears in the band structure of nanotubes [27]. This is due to the interaction between nanotubes in a bundle that causes a broken symmetry, and induces a pseudo-gap of about 0.1 eV. This can change the physical properties of a bundle of carbon nanotubes dramatically. In current synthesis methods, however, nanotubes with random chiralities are produced, and adjacent nanotubes in a bundle have different chiralities. The eigen-states for electrons in neighboring nanotubes are hence different. This severely limits the inter-nanotube coupling, and nanotubes largely retain their properties. This has been proven both theoretically and experimentally [28, 29].

The weak inter-nanotube coupling, however, has major implications. To utilize all nanotubes within a bundle, they all should have the same length as the bundle, and there should be good ohmic connections to all of them at both ends of the bundle. There are reports of electrical connections to all nanotubes within vertical bundles. Vertical bundles potentially can be used as vias in multi-level interconnect networks [30]. Making electrical connections to all nanotubes within horizontal bundles is a remaining challenge. For horizontal bundles, only connections to outer nanotubes have been demonstrated [29].

7.3.1 Conductivity

The resistance of a nanotube bundle is determined by the number of metallic nanotubes that are well connected, n_{m}, and the quality of nanotubes and contacts. Conductivity of a SWNT-bundle is therefore

$$\sigma_{SWNT} = (n_m/A_{bundle})/\left(\frac{R_c}{L} + R_Q\left(\frac{1}{L} + \frac{1}{\ell_{eff}}\right)\right) \tag{7.13}$$

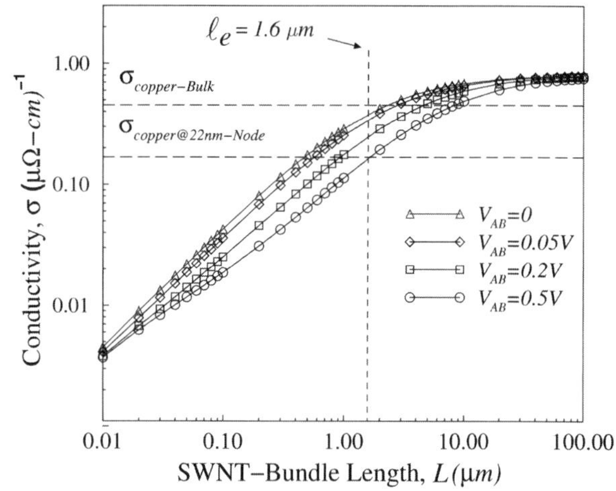

Fig. 7.4 Conductivity of Densely packed SWNT-bundles versus length for various bias voltages. One third of nanotubes are assumed to be metallic, mean free path is 1.6 μm and nanotubes are 1 nm in diameter. Conductivities of bulk copper and copper lines with 27 nm width are also shown for reference. Reproduced with permission from [8]. Copyright 2007 IEEE

where A_{bundle} is the cross-sectional area of the bundle and R_c is the contact resistance. Statistically, 1/3 of SWNTs are metallic and the rest are semiconductor [1]. A typical value for the diameter of SWNTs is 1 nm [1]. The ultimate conductivity of SWNT-bundle therefore corresponds to 1 metallic nanotube per 3 nm^2 and $R_c << R_Q/L$, which is plotted in Fig. 7.4 versus length for various bias voltages assuming low-bias mean free path of 1.6 μm [19]. Bulk copper conductivity and copper conductivity at the 22 nm node ($W = 27$ nm) are also shown for reference, assuming $R = 0.3$ and $p = 0.25$, which are taken as average values reported or used in literature [23, 31]. It can be seen that there is a critical length below which conductivity of nanotube-bundles becomes smaller than that of copper wires.

In reality, the nanotubes in a bundle may not be densely packed. Porous bundles simply have smaller conductivities, proportional to the fraction of the cross-sectional area occupied by nanotubes for which Fig. 7.4 can be modified accordingly. Although bias voltage can increase the resistance of interconnects, a worst case analysis for signal interconnects shows that using the low-bias resistance has less than 20% error if adequate number (>20) of metallic nanotubes are used in a bundle [32]. This is because for long interconnects the relative increase in resistance due to the bias voltage is small as equation (7.11) shows. For short interconnects, on the other hand, most of the voltage drop is across the driver since driver resistance is typically much larger than interconnect resistance if adequate metallic SWNTs are used. Furthermore, total delay is not very sensitive to the interconnect resistance for short interconnects.

7.3.2 Capacitance

To calculate the electrostatic capacitance of a SWNT-bundle, one should note that nanotubes in a bundle have the same potential because they are connected in parallel and they are also capacitively coupled [33]. The electrostatic capacitances among nanotubes can therefore be neglected, and only the electrostatic capacitances to ground and neighboring bundles should be considered.

The quantum capacitance depends on the total density of states; hence, the quantum capacitance of a SWNT-bundle is the sum of the quantum capacitances of all metallic nanotubes. For more than 10 metallic SWNTs in a bundle, the electrostatic capacitance becomes more than 40 times smaller than the quantum capacitance. Since the two capacitances are in series, the quantum capacitance can be neglected.

For a densely packed bundle of all metallic SWNTs, capacitance is equal to the capacitance of a metallic wire with the same cross-sectional dimensions (Fig. 7.5). The only difference between an ideally packed SWNT-bundle and a perfectly smooth metallic wire is due to the 1 nm surface roughness for bundles. However, since this roughness is much smaller than the spacing to ground and also to neighboring bundles, its impact on capacitance is less than 3% based on the simulations performed by the field solver RAPHAEL [34]. This is consistent with the physical models presented in [35], which proves that a rough surface can be approximated with a smooth surface at the peak height if the lateral period of fluctuations is small compared to the spacing between the two electrodes of a capacitor. This case of

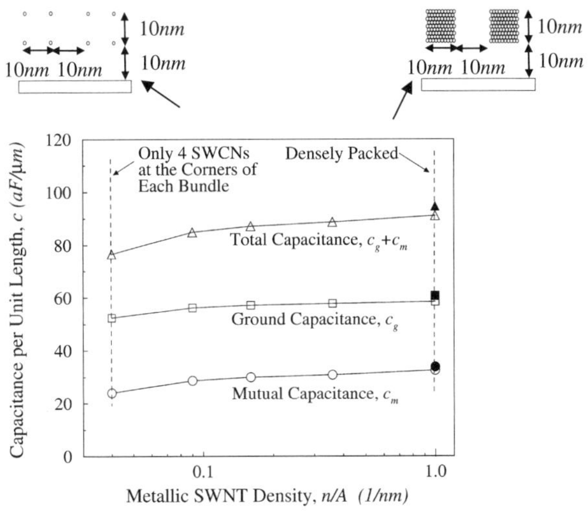

Fig. 7.5 Capacitance per unit length for SWNT-bundles versus density of metallic SWNTs. Solid points show the capacitance per unit length values for ideally smooth copper wires with the same cross-sectional dimensions. The 10 nm by 10 nm cross-sectional dimensions are chosen arbitrarily for the ease of simulations. Two major conclusions: (1) densely packed bundles have capacitance values the same as those of copper wires and (2) capacitance values decrease very slowly as bundles become more porous. Quantum capacitance is ignored in this plot because even for the case of only four SWNTs per bundle it is more than 20 times larger than the electrostatic capacitance. Reproduced with permission from [8]. Copyright 2007 IEEE

"fast fluctuations" gives the upper limit for capacitance of rough surfaces [35]. It is worthwhile to note that even copper wires are not perfectly smooth and a 1 nm surface roughness can be quite common.

As the porosity within a bundle increases, capacitance decreases very slowly as Fig. 7.5 shows in which capacitance is calculated using RAPHAEL field solver. Even for a very porous bundle with only four nanotubes at the corners, capacitance is only 20% smaller than that of a densely packed bundle. Such a small change in capacitance for such a large change in nanotube density is due to fringing effect. Therefore, it is safe to assume that the capacitance of a SWNT-bundle is the same as a copper wire with the same cross-sectional dimensions even if it is not densely packed.

7.3.3 Inductance

While kinetic inductance of a SWNT-bundle is inversely proportional to the number of metallic nanotubes, magnetic inductance depends mainly on the distance to the return path. Total inductance is the summation of kinetic and magnetic inductances. For $n < 10^4$, the kinetic inductance is dominant, and increasing the number of metallic nanotubes in a bundle decreases the overall inductance linearly. For $n > 10^4$, the kinetic inductance becomes comparable with or even smaller than the magnetic inductance as more and more nanotubes are packed in a bundle, and the overall inductance asymptotically reaches the magnetic inductance.

The impact of inductance on the delay of an interconnect can be ignored if the resistances associated with the driver and/or the interconnect are large compared to the characteristic impedance of the interconnect [36]:

$$0.4rL + 0.7R_{tr} > Z_0, \tag{7.14}$$

where R_{tr} is driver resistance, r is resistance per unit length, and Z_0 is the characteristic impedance. In this case, the interconnect is considered to be RC limited. Knowing that $Z_0 = \sqrt{l/c}$, where l and c are inductance and capacitance per unit length of the interconnect, respectively, this condition for a SWNT-bundle can be written as

$$0.7R_{tr} > \sqrt{\frac{l_k/(4n) + l_m}{c}} - 0.4R_Q(1 + L/l_{eff})/n \tag{7.15}$$

The minimum driver resistance needed for a SWNT-bundle to be RC-limited has been plotted versus the number of metallic SWNTs in a bundle for various lengths in Fig. 7.6. It has been assumed that all cross-sectional dimensions of bundles (width, thickness, and spacing to ground and adjacent bundles) scale proportionally as n varies because of which the capacitance and magnetic inductance values remain constant. It can be acquired from Fig. 7.6 that for most practical cases, SWNT-bundles will be operating in the RC regime as for short interconnects normally small

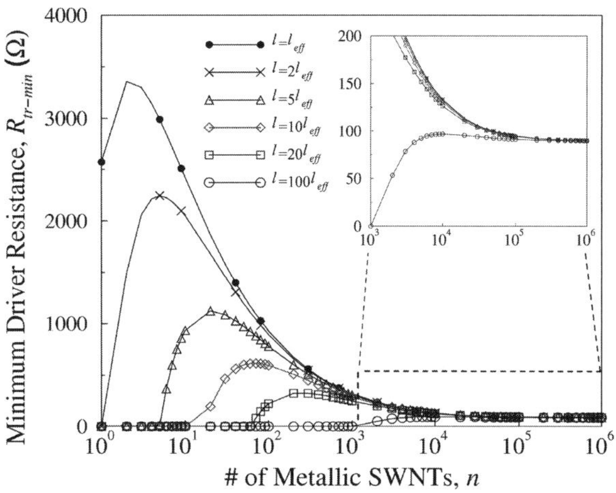

Fig. 7.6 Minimum driver resistance for which SWNT-bundles are RC-limited

drivers with large resistances are used. On the other hand, for long interconnects, a large number of SWNTs is needed to have a small overall resistance and hence a reasonable delay.

7.4 Circuit Models for MWNTs

Diameters of MWNTs may vary in a wide range of a few to hundreds of nanometers, and they may have from a few to many shells. Initially, most experiments indicated that only the outer shell in a MWNT conducts. Recently, however, it has been confirmed that all shells can conduct if they are properly connected to the electrical contacts [30, 37–39]. Early experiments made contacts to the outer shells only, and due to the weak inter-shell coupling, the inner shells had a small impact on the overall conduction. In [37], researchers have grown a 25 μm long MWNT with an outer diameter of 100 nm. They have reported an overall resistance of 35 Ω that has been achieved through welding the inner and outer shells to a tungsten probe using an electrical discharge. In [30], vertical MWNTs with diameters of about 10 nm are grown. The average resistance of these nanotubes is 1700 Ω. To achieve such low resistances reported in [30, 37] not only most shells need to conduct, but also the contact resistance for each shell should be relatively small.

The experiments reported in [30, 37] are yet to be adopted for large scale integration. They, however, prompt the question of whether or not MWNTs can potentially outperform copper wires or even SWNTs in terms of conductivity. In this section, compact physical models are derived to determine the ultimate potential performances of MWNTs with various diameters and lengths and compare them with copper wires and SWNT-bundles. The results offer important guidelines regarding which kind of nanotubes needs to be developed for various interconnect applications.

7.4.1 Number of Conduction Channels per Shell

It is easier to write the band structure for a zigzag nanotube shell with an axis along the x direction even though the results are going to be general [14]. A zigzag nanotube has a chirality of (n, 0), and the peaks and the valleys of its valence and conduction bands with respect to the Fermi level are [14, 40]

$$E_{\nu} \equiv E_{(k_x=0)} = \pm\frac{3ta_0}{d}\left|\nu - \frac{2n}{3}\right|, \tag{7.16}$$

where a_0 is the length of carbon–carbon bonds, t is the Hamiltonian matrix element between neighboring carbon atoms, d is the shell diameter, and ν is an integer less than n. The plus and minus signs correspond to the valleys of the conduction bands and the peaks of the valence bands, respectively. If n is a multiple of 3 then the shell becomes metallic as there would be a ν for which $E_{\nu} = 0$. Although (7.16) is derived for isolated shells, it is valid for all the shells in MWNTs as in practice adjacent shells have different chiralities and thus they do not have considerable coupling with each other, and the band structures of the shells remain the same [40]. This is also true for SWNTs in bundles when they have random chiralities [28].

At zero temperature ($T = 0$ K), all energy states below the Fermi level are occupied and the ones above it are free. Completely full or empty subbands can not contribute to the conduction; hence, only those that cross the Fermi level determine the number of channels [14]. At temperatures above 0 K electrons distribute according to the Fermi–Dirac distribution, thus various subbands contribute to conduction accordingly. The total number of channels for each shell can be written as [37]

$$N_{chan/shell} = \sum_{All\ SubBands} \frac{1}{\exp(|E_{\upsilon}|/k_B T) + 1} \tag{7.17}$$

By looking at (7.16) and (7.17), one can see that the number of channels per shell depends on whether it is a metal or semiconductor and is a function of ta_0/dk_BT. Increasing the temperature has exactly the same effect as increasing the diameter because the contribution of each subband is determined by its distance to the Fermi level, E_{ν}, normalized to the thermal energy, k_BT, where E_{ν} is inversely proportional to the diameter. The number of channels per shell at $T = 300$ K is plotted versus the diameter in Fig. 7.7 for metallic and semiconductor shells.

Assuming that the shells have random chiralities, statistically one third of the shells are going to be metallic and the rest semiconductor [14, 40]. Based on this concept, the average number of channels per shell is also plotted which is constant for small diameters and increases linearly for large diameters [41]. The average number of channels per shell can therefore be approximated by

$$\begin{aligned} \tilde{N}_{chan/shell}(d) &\approx ad + b \quad d > 3\text{nm} \\ &\approx \frac{2}{3} \qquad\quad d < 6\text{nm} \end{aligned}, \tag{7.18}$$

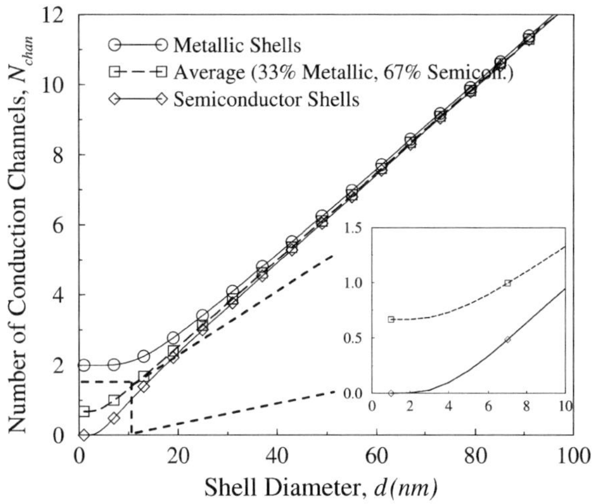

Fig. 7.7 Number of conduction channels per graphene shell versus shell diameter for metallic and semiconductor shells. The average number of conduction channels is also plotted assuming that statistically one third of the shells are metallic. The region inside the dashed square is magnified in the inset image

where a is $0.14\ \text{nm}^{-1}$ and b is 0.2.[1] The error of (7.18) is less than 15% for all values of d. Note that the two regions in (7.18) have an overlap, and for $3\ \text{nm} < d < 6\ \text{nm}$ both constant and linear functions can be used without any considerable error. This makes the later mathematical derivations easier.

It is important to note that (7.17) is valid for small bias voltages, and as the voltage increases, more subbands contribute to the conduction [42]. For interconnect applications, we are more interested in the low-bias conductance as the electric field along interconnects is typically small [32]. Equation (7.18) is, therefore, accurate enough for nanotube interconnects.

7.4.2 Total Conductance

The total number of channels in a MWNT is simply

$$N_{chan} = \sum_{all\ shells} \tilde{N}_{chan/shell}(d) \tag{7.19}$$

The spacing between the shells in a MWNT corresponds to the van der Waals distance between the graphene layers in graphite, which is $\delta = 0.34$ nm [37, 40]. The ratio of $d_{\min}/d_{\max}$ varies in different MWNTs, and values between 0.35 to 0.8 have been observed [30, 37, 40]. Assuming an average value of 0.5 [37], the number of channels per MWNT is

[1] In [41], a factor of 2 corresponding to the contributions of both conduction and valence subbands was missed in the code written to calculate the number of conduction channels in a graphene shell. The value of a was consequently underestimated by a factor of 2.

$$N_{chan} = \left(1 + \left[\frac{(d_{\max} - d_{\min})}{2\delta}\right]\right)\left(\frac{1}{2}a(d_{\max} + d_{\min}) + b\right) \quad for \quad d_{\max} > 6\,nm \tag{7.20}$$

and

$$N_{chan} = \frac{2}{3}\left(1 + \left[\frac{(d_{\max} - d_{\min})}{2\delta}\right]\right) \quad for \quad d_{\max} < 6\,nm \tag{7.21}$$

where the bracket sign represents the integer part. Number of channels per unit area is the critical parameter which has been plotted in Fig. 7.8 assuming that $d_{\min} \approx d_{\max}/2$.

Electrons in a carbon nanotube have large but finite mfps. It has been shown that mfp is linearly proportional to diameter [43, 44] as

$$\ell_{eff} = \frac{\sqrt{3}\pi t^2 d}{2\text{var}^2(\varepsilon) + 9\text{var}^2(t)} \tag{7.22}$$

where var(ε) and var(t) are the variances of on-site energy ε and the matrix element between nearest neighbors, t, respectively. Conductance per channel is [3,14]

$$G = G_0/(1 + L/\ell_{eff}) \tag{7.23}$$

where G_0 is the quantum conductance 1/12.9 kΩ, L is the nanotube length, and ℓ_{eff} can be written as $l_{\text{eff}} = Kd$. For a 1 nm shell, a typical value for ℓ_{eff} is 1 μm [3], and K is, therefore, in the order of 1000. Total conductance of a MWNT is

$$G_{total} = \sum_{All\,shells} \frac{G_0 \tilde{N}_{chan/shell}(d)}{(1 + L/Kd)}. \tag{7.24}$$

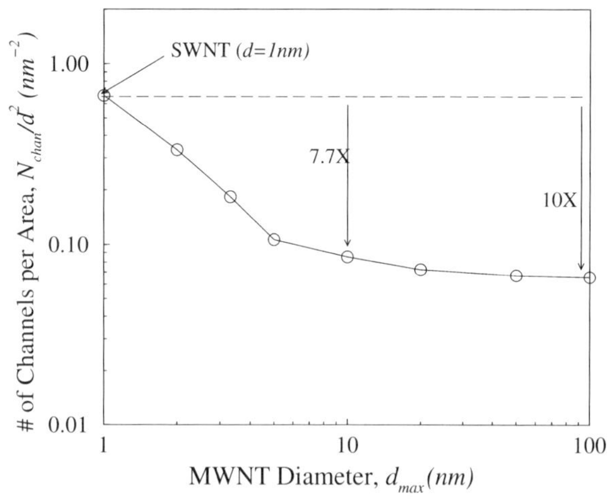

Fig. 7.8 Number of conduction channels per cross-sectional area in a MWNT

Since the spacing between shells is considerably smaller than the nanotube diameter, (7.24) can be approximated with an integral. For $d_{\max} > 6$ nm, conductivity of a MWNT becomes equal to

$$\begin{aligned} \sigma &\equiv G_{total} L/A \\ &= \left[\left(1-\frac{d_{\min}^2}{d_{\max}^2}\right)\frac{a}{2} + \left(b - \frac{aL}{K}\right)\left(\left(\frac{1}{d_{\max}} - \frac{d_{\min}}{d_{\max}^2}\right) - \frac{L}{d_{\max}^2 K}\ln\frac{d_{\max}+\frac{L}{K}}{d_{\min}+\frac{L}{K}}\right)\right]\frac{LG_0}{2\delta}. \end{aligned} \tag{7.26}$$

At the length at which $b - aL/K = 0$ ($L = Kb/a$), conductivity becomes independent of the diameter. For $L < Kb/a$, conductivity decreases as diameter increases whereas for $L > Kb/a$ it increases with diameter. This is an important point as it highlights the need for nanotubes with smallest possible diameters for short interconnects and nanotubes with largest possible diameters for long interconnects.

For $D_{\max} < 6$ nm,

$$\sigma = \left(\left(\frac{1}{d_{\max}} - \frac{d_{\min}}{d_{\max}^2}\right) - \frac{L}{d_{\max}^2 K}\ln\frac{d_{\max}+\frac{L}{K}}{d_{\min}+\frac{L}{K}}\right)\frac{LG_0}{3\delta} \tag{7.27}$$

that always increases as diameter decreases regardless of the length. Conductivity of MWNTs is plotted in Fig. 7.9 versus length for various outer diameters.

For long lengths that $L >> Kd = \ell_{eff}$ and also assuming that $d_{\min} = d_{\max}/2$, (7.25) simplifies to

$$\sigma = \frac{G_0 K}{2\delta}\left(\frac{7}{24}a d_{\max} + \frac{3}{8}b\right). \tag{7.28}$$

Figure 7.9 plots the conductivity of MWNTs (various diameters) and SWNTs versus length. For SWNT-bundles, conductivity is independent of bundle size. It is worthwhile to mention that in Fig. 7.9, SWNT-bundles are assumed to be densely

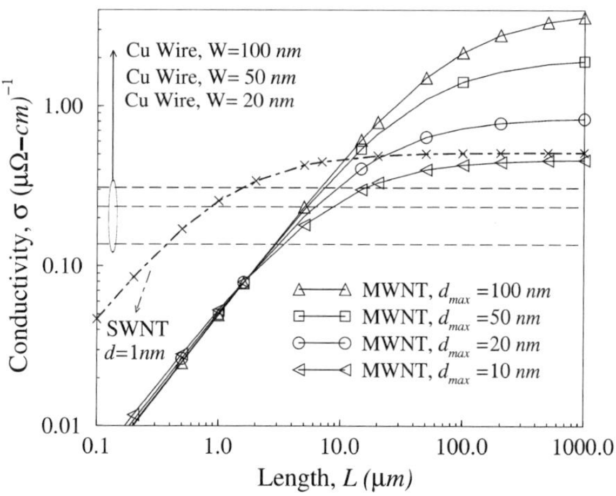

Fig. 7.9 Conductivity of MWNTs with various diameters and bundles of densely packed SWNTs versus length. SWNTs are assumed to be 1 nm in diameter and have random chiralities and a 1 μm mean free path. As a reference, the conductivity of copper wires with various widths are also shown by the *dashed lines*. The specularity parameter and reflection coefficient for copper wires are assumed to be 0.25 and 0.3

packed so that their ultimate potential conductivity can be calculated. Porous bundles simply have smaller conductivities proportional to the fraction of the cross-sectional area occupied by nanotubes that make large MWNTs even more favorable.

7.4.3 Inductance and Capacitance

Like a SWNT-bundle, a MWNT has both kinetic and magnetic components of inductance. The kinetic inductance is 8 nH/μm (considering spin degeneracy) for each conduction channel and its overall value is inversely proportional to the number of conduction channels. Magnetic inductance, however, has both the self and mutual components, and its value depends on the distance of the MWNT to its return path.

Likewise, there is a quantum capacitance of 200 aF/μm per conduction channel (considering spin degeneracy) in series to the electrostatic capacitance. To calculate the electrostatic capacitance, only the outer shell needs to be taken into account as the inner shells are shielded.

7.5 Carbon Nanotube Interconnects

Having the equivalent circuit models for carbon nanotubes, it is important to identify the best ways of fully exploiting their unique properties to enhance the performance of interconnects. Multi-level interconnect networks in high-performance chips are complicated with up to 10 metal levels with various pitches and thicknesses. They accommodate interconnects with a very wide range of lengths and cross-sectional dimensions that are generally categorized as local, semi-global or intermediate, and global. These enormous sets of interconnects have different and sometimes opposing requirements, and it is critical to study the application of nanotubes for each type separately.

7.5.1 Local Interconnects

Short local interconnects are typically routed in the lowest metal levels with the minimum wiring pitch that the current technology allows. Capacitance, not resistance, is the main concern for these interconnects (Fig. 7.10). This is because a minimum-size interconnect must be roughly several hundred gate pitches long so that its resistance becomes comparable to that of its driver, whereas it can have a capacitance larger than a typical logic gate if it is only ten gate pitches long. This

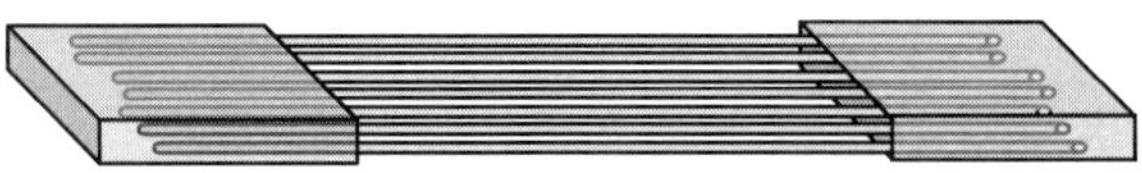

Fig. 7.10 A monolayer SWNT interconnect. Reproduced with permission from [8]. Copyright 2007 IEEE

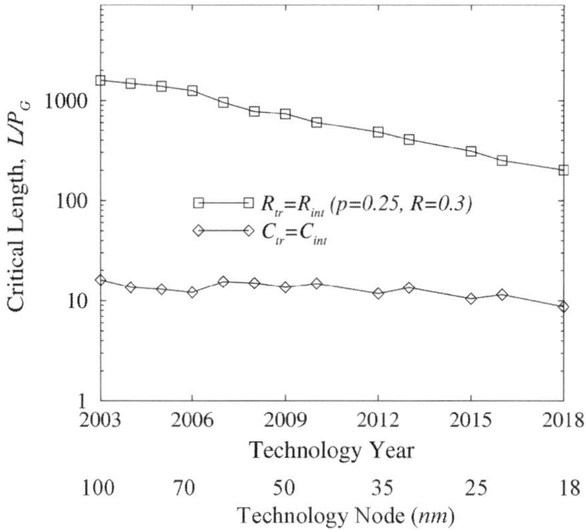

Fig. 7.11 Critical lengths beyond which interconnect resistance (*upper curve*) or capacitance (*lower curve*) become larger than the output resistance (*upper curve*) or input capacitance (*lower curve*) of a typical driver or receiver. Critical lengths are in gate pitches and the typical driver is assumed to have channel length to width ratio of 10 for the n-FET transistors. p-FET transistors are assumed to be twice larger. Minimum size copper wires are considered and surface and grain boundary scatterings are also taken into account. Reproduced with permission from [8]. Copyright 2007 IEEE

is shown in Fig. 7.11 in which the lengths at which resistance and capacitance of interconnects become equal to those of their drivers and receivers are plotted versus the technology generation. The critical length is shown in gate pitch (P_G) because the lengths of local interconnects scale linearly with the gate pitch as the technology advances.

The large gap between the two curves in Fig 7.11 indicates that for a large number of interconnects, any decrease in interconnect capacitance with a moderate increase in resistance can improve the performance. This can be achieved by using interconnects with smaller aspect ratios (thickness to width ratio). In most processors, however, designers have to use interconnects with aspect ratios of height to width larger than 1.5, even for the lowest metal levels to avoid electromigration [45]. Even if electromigration is by some means mitigated (e.g., through sub-ambient cooling), the interconnect aspect ratio cannot be made very small because of the thickness variation that is caused during chemical-mechanical-polishing (CMP).

Carbon nanotubes can conduct orders of magnitude larger current densities compared to copper [15] without any observable electromigration. Mono- or a few-layer SWNT interconnects with nanometer thicknesses as shown in Fig. 7.10 can offer a significant reduction in lateral capacitance while having a reasonable resistance. The cross-sections of copper interconnects, monolayer nanotube interconnects above a thick dielectric layer, and monolayer nanotube interconnects above a thin dielectric layer are shown in Fig. 7.12. The width and spacing of the interconnects in all three

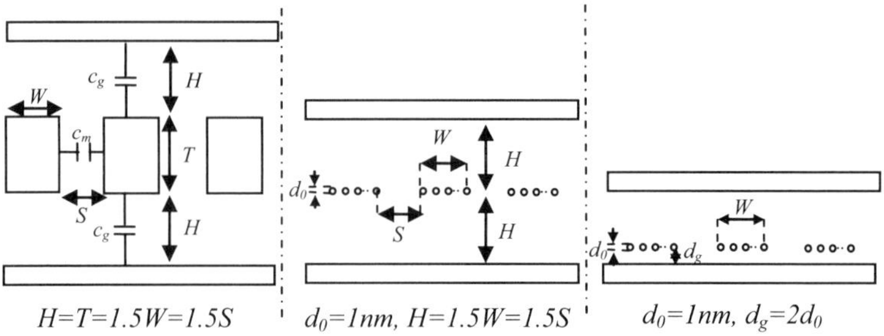

Fig. 7.12 The three interconnect configurations considered in this subsection. The *left picture* shows copper interconnects with aspect ratio of 1.5. The *middle one* shows monolayer nanotube interconnects above a thick dielectric and the *right picture* shows monolayer nanotube interconnects above a thin dielectric layer. Reproduced with permission from [8]. Copyright 2007 IEEE

cases are assumed to be equal to the minimum feature size at the 22 nm technology node. Typically, the inter-level dielectric layer is roughly as thick as the copper wires. For Fig. 7.12(a) and (b), therefore, the dielectric thicknesses are assumed to be 1.5 times the wire width. In Fig. 7.12(c), the dielectric thickness has been arbitrarily chosen to be 2 nm to highlight the impact of dielectric thickness on the performance of nanotube interconnects.

Per unit length values of capacitance to ground, c_g, and capacitance between adjacent interconnects, c_m, for each case are also shown in Table 7.1, which are calculated by RAPHAEL [34]. When an interconnect switches, its adjacent lines may stay quiet ($c = 2c_m + 2c_g$), may switch in the same direction ($c = 2c_g$) or in opposite direction ($c = 4c_m + 2c_g$). The average capacitance per unit length for each interconnect can therefore be considered to be $2c_m + 2c_g$, which is also given for each case in Table 7.1. It can be seen that the average capacitance for a monolayer nanotube above a thick dielectric layer is 50% smaller than that of the copper interconnects. This would have a major impact on gigascale integration (GSI) chips that are mainly

Table 7.1 Capacitance values for the structures shown in Fig. 7.12. Reproduced with permission from [8]. Copyright 2007 IEEE

	Copper wires	Monolayer SWNT above a thick dielectric	Monolayer SWNT above a thin dielectric
Capacitance to ground, c_g (aF/μm)	35.6	27.2	238
Line-to-line capacitance, c_m (aF/μm)	38.6	9.9	0.9
Average capacitance, $2c_m+2c_g$ (aF/μm)	148.5	74.5	476
Capacitance variation, $(4c_m+2c_g)/2c_g$	3.16	1.73	1.007

power limited. In a high-performance chip, 70% of total capacitance is due to interconnects, the majority of which are short local interconnects [46]. A nanotube interconnect above a thin dielectric layer, however, has a capacitance more than three times larger than that of a copper wire.

Dynamic delay variation due to different switching patterns of the neighbors is also an important issue. The worst-case capacitance corresponds to when a line and its neighbors switch anti-phase ($c = 4c_m + 2c_g$), and the minimum capacitance corresponds to when they switch in-phase ($c = 2c_g$). Because of smaller lateral capacitances, the maximum variation in capacitance of monolayer nanotube interconnects is much smaller than that of copper interconnects (73% versus 216% as indicated in Table 7.1).

The aforementioned advantages of monolayer nanotube interconnects with thick dielectrics in terms of capacitance values remain constant at various generations of technology as long as the cross-sectional dimensions scale proportionally with technology. Interconnect resistance, however, changes with technology. For the 22 nm node projected for the year 2016 with minimum interconnect width of 27 nm, resistance is 42 Ω/μm assuming aspect ratio of 1.5 and $p = 0.25$, $R = 0.3$. For a mono- or a few-layer SWNT interconnects, resistance will depend on the number of metallic nanotubes, their mean free paths, and the contact resistance. Since these interconnects are short, the contact resistance is very important. A mono- or a few-layer nanotube interconnect, however, can be partially buried in metallic contacts from both sides which facilitates achieving small contact resistances (Fig. 7.10). For this configuration, there are many reports of small contact resistances ($<<R_Q$) [3, 15–17]. Having 20 metallic nanotubes can offer a resistance of 325 Ω +190 Ω/μm, and are hence more resistive compared to copper wires. Despite this increase in the resistance, mono- or a few-layer SWNT interconnects can be faster than copper wires of the same width if they are not very long (<10 to 20 μm). This is shown in Fig. 7.13 in which delay is plotted versus length. This improvement in latency is

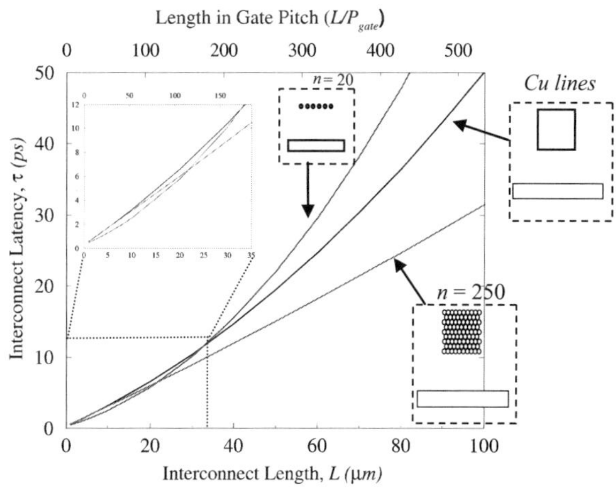

Fig. 7.13 Delay versus interconnect length for a few-layer SWNT (20 metallic SWNTs) interconnects, minimum size copper wires (W = 27 nm), and SWNT-bundle with 250 metallic SWNTs. For short lengths, thin few-layer SWNT interconnects are faster because of smaller capacitance. For long lengths, SWNT-bundles are faster because of their lower resistance

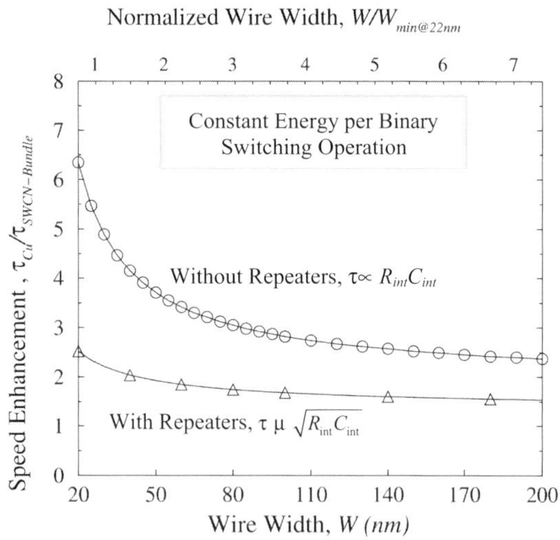

Fig. 7.14 The improvement in interconnect RC delay for semi-global interconnects versus wire width (lower horizontal axis) or wire width normalized to minimum wire width at the 22 nm node (upper horizontal axis). Density of metallic SWNTs and mean free path of electrons are assumed to be 1 per 3 nm^2 and 1.6 μm, respectively. For lower densities, the improvements scale accordingly. Energy per binary switching remains constant. Reproduced with permission from [8]. Copyright 2007 IEEE

in addition to the major reduction in power dissipation, which is the most important advantage of these thin nanotube interconnects.

In terms of fabrication, the mono or a few-layer nanotube interconnects have many advantages. They can be grown easier and faster since they are relatively short. Making connections to all nanotubes is also going to be easier compared to SWNT-bundles. This is very important for local interconnects as they are very sensitive to the contact resistance due to their short lengths.

Semi-global interconnects are longer than the local interconnects, and their delay is mainly determined by their RC product because of which they are normally routed in metal levels above the local levels with cross-sectional dimensions larger than the minimum feature sizes. In some cases, repeaters are also used to improve their speed as their delays become proportional to the square root of their RC products.

SWNT-bundles can replace the copper wires at the semi-global level, and improve their latency because of their smaller resistivity. While for local interconnects contact resistance was very important, semi-global interconnects are less sensitive to it as they are typically many mean free paths long (>>1 μm). Mean free path and density of metallic nanotubes in bundles are the key parameters that determine the performance improvement. The improvement in delay that SWNT-bundles may offer are plotted in Fig. 7.14 versus wire width, assuming interconnects are long enough that the contact resistance can be ignored ($R_c < R_Q L/\ell_e$). Both cases that repeaters are used or are not have been considered. It has been assumed that there is one metallic nanotube per 3 nm^2. If there are fewer nanotubes, the plots simply scale accordingly.

Replacing copper wires with nanotube-bundles of the same cross-sectional dimensions has no major impact on energy per binary switching as it depends on interconnect capacitance. Even capacitance associated with repeaters does not

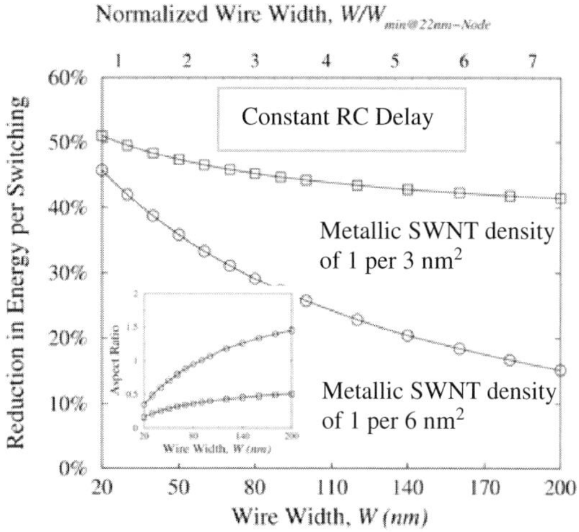

Fig. 7.15 The improvement in the binary switching operation (or interconnect capacitance) versus wire width when the RC delays of SWNT interconnects are equal to those of copper wires with a thickness to width ratio of 2. Inset Plot: The corresponding aspect ratio versus wire width. The mean free path of electrons is assumed to be 1.6 μm. Reproduced with permission from [8]. Copyright 2007 IEEE

change considerably because lowering resistance requires fewer larger repeaters, and the total capacitance remains the same [48]. If power dissipation is the main concern, one can use SWNT-bundles with smaller thicknesses while keeping the RC delays the same as those of copper wires. This lowers the lateral capacitance and hence energy per binary switching.

The improvement in average capacitance, which is the same as the improvement in energy per binary switching, is plotted in Fig. 7.15 for metallic nanotube densities of 1 per 3 nm^2 and 1 per 6 nm^2. The inset plot shows the corresponding aspect ratios. It can be seen in Figs. 7.14 and 7.15 that even a porous SWNT bundle can offer significant improvements in delay or power dissipation for semi-global interconnects. Low-cost ASIC chips normally use close to minimum feature size wire widths for all metal levels to minimize the number of levels [47]. Those designs would benefit most from replacing copper wires with SWNT-bundles.

7.5.2 Global Interconnects

Replacing copper wires with SWNT or MWNT-bundles increases the conductivity; hence, lowers the delay. However, since global interconnects normally have large cross-sectional dimensions, they normally operate in the shallow RLC region, which is when interconnect inductance has a moderate impact on delay. A large reduction in their resistance can push them deep into the RLC region where the delay becomes time-of-flight limited and decreases very slowly with lowering the resistance. Furthermore, crosstalk noise can become very large as mutual inductance is a far reaching effect because of which many aggressors can affect a victim line [33].

Global interconnects are typically data buses between macrocells consisting of several hundred thousand to a few million gates. This allows greater flexibility in the design of the inter-macrocell global interconnects. For instance, two low bandwidth interconnects can be used instead of one high bandwidth interconnect, or vice versa. Therefore, unlike intra-macrocell interconnects, the global interconnect routing need not be restricted to a fixed netlist. This flexibility permits optimization of the global interconnections for achieving a large bandwidth density and a small latency simultaneously.

The optimal wire width is defined as the width at which bandwidth per unit width-reciprocal latency product is maximized [33, 48]. Assuming optimal repeater insertion, the optimal wire width becomes equal to [33]

$$W_{opt} = 2.53 c_0 \sqrt{\frac{c_g}{\varepsilon_r} \frac{\rho}{A_r} R_0 C_0}, \tag{7.28}$$

where c_0 is the speed of light in free space, ρ is conductor resistivity, A_{r} is the interconnect thickness to width ratio, c_{g} is capacitance per unit length, and $R_0 C_0$ is the intrinsic delay of repeaters. The optimal wire width is in the shallow RLC region, where the difference between RC and RLC model latencies is only 10% [48]. Using the optimal wire width also offers the best trade-off between energy dissipation and bandwidth density, requires a small repeater area for global interconnects (less than 1% of the chip area), reduces via blockage considerably [48], and guarantees small and constant crosstalk noise in all technology generations [33].

By re-optimizing the interconnect width for SWNT or MWNT-bundles one can improve bandwidth density without entering the deep RLC region. The optimal wire width given by (7.28) scales proportional to the square root of resistivity. Improving resistivity increases bandwidth density without any latency penalty or any increase in crosstalk. The optimal global wire width is plotted versus technology generation in Fig. 7.16 for copper, and SWNT- and MWNT-bundles. It can be seen that bundles

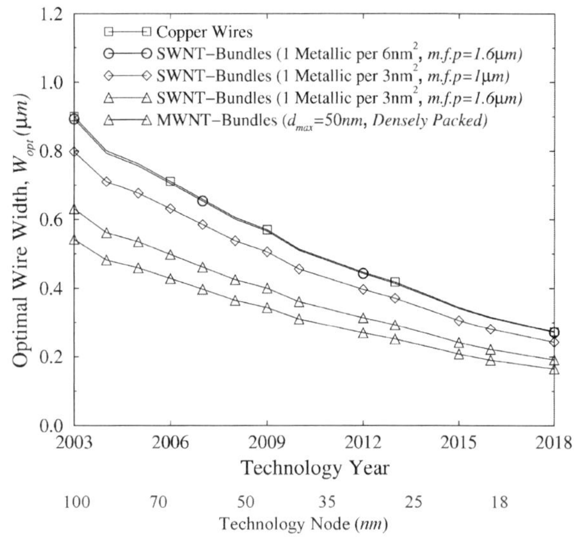

Fig. 7.16 Optimal global wire width versus technology generation for copper wires and SWNT-bundles and MWNT-bundles. The mfp to diameter ratio (K) is assumed to be 1000 for MWNTs. Bandwidth density is inversely proportional to the optimal wire width

of MWNTs ($D = 50$ nm) can potentially lower the optimal wire width by up to 40%; hence, improve the bandwidth density by up to 60%. SWNT-bundles, however, can lower the optimal wire width considerably only if they are densely packed (1 metallic SWNT per 3 nm^2) and have mean free paths as large as 1.6 μm. This is because global wires have resistivities close to bulk copper and SWNT-bundles can outperform them only if they are dense and have large mean free paths.

7.6 Conclusions

Equivalent circuit models are developed for metallic SWNTs, SWNT-bundles and MWNTs. These models are then used to quantify their potential performance as interconnects in future gigascale systems. Based on physical models, an R(V)LC model is proposed for SWNTs which is accurate for all voltages and lengths, and can be used for both compact modeling and SPICE simulations. Because of the large quantum resistance and kinetic inductance that rise from the one dimensionality of SWNTs, a single SWNT above a ground plane is going to be much slower than minimum size copper wires, even at the end of the ITRS. By putting SWNTs in a bundle, however, resistance and kinetic inductance both decrease linearly with the number of metallic nanotubes. This is in contrast with magnetic inductance that changes very slowly with cross-sectional dimensions. Having adequate number of metallic SWNTs, a bundle of SWNTs can outperform copper wires in terms of resistance, and can have adequately small kinetic inductance. Capacitance of a densely packed SWNT-bundle is determined by its cross-sectional dimensions, and is very close to that of a copper wire with the same dimensions ($<$3% difference). Capacitance changes very slowly as porosity increases in a bundle. For instance, a bundle with only four SWNTs in its corners has a capacitance of only 20% smaller compared to a densely packed bundle.

Conductivity of MWNTs is also modeled assuming good ohmic connection to all shells and a constant level of disorder in graphene shells with various diameters. For MWNTs with diameters larger than 6 nm, there is a critical length at which conductivity becomes independent of diameter ($\sim$2 μm). For nanotubes shorter than the critical length, increasing the diameter lowers the conductivity whereas for nanotubes longer than the critical length, increasing the diameter increases the conductivity. This is because for long nanotubes electron mean free path, which increases linearly with diameter, is important, whereas for short nanotubes quantum resistance and hence the total number of conduction channels per unit area are the key parameters. For long lengths (a few hundreds of microns), MWNTs can have conductivities several times larger than those of copper or SWNT-bundles. For short lengths ($<$10 μm), however, dense SWNT-bundles offer more than two times higher conductivities as compared to MWNTs. For vertical applications (vias), therefore, SWNT-bundles or bundles of MWNTs with small diameters are needed as these interconnects are normally shorter than a few microns.

By optimally customizing SWNT interconnects at the local, semi-global, and global levels, significant improvements in power dissipation, delay, crosstalk, and bandwidth density can be achieved. For local interconnects, mono- or few-layer SWNTs interconnects offer 50% reduction in capacitance, 48% reduction in capacitance coupling between adjacent lines, and up to 20% reduction in delay. Contact resistance is important for these interconnects that are short. For this configuration, however, there are many reports of small contact resistances. For semi-global interconnects, bundles of SWNTs can be used to achieve higher conductivities. Having higher conductivity enables either lowering latency by using the same cross-sectional dimensions as those of copper wires, or lowering power dissipation by using smaller aspect ratios while delay remains constant. Since copper resistivity for semi-global interconnects is larger than its bulk resistivity, even a moderate number of metallic nanotubes per cross-sectional area (1 per 6 nm^2) can offer a considerable reduction in power dissipation (more than 35% for twice minimum size at 22 nm node) or delay (more than 70% if repeaters are used and more than 40% otherwise assuming wiring width equal to twice minimum size at the 22 nm node). Low-cost ASIC chips can hence benefit significantly from SWNT interconnects as they normally use minimum size wires for all interconnects to minimize the number of metal levels. For global interconnects, bundles of MWNTs with large diameters (e.g., 50 nm) can offer higher conductivities compared to copper wires (2.5$\times$). Replacing the copper interconnects with MWNT-bundles therefore allows using smaller interconnect dimensions (40% smaller for 50 nm MWNTs) while delay and crosstalk noise remain constant. This would increase bandwidth density of global interconnects (by 60% for 50 nm MWNTs). SWNT-bundles may also be used for the same purpose only if they are very dense (1 metallic SWNT per 3 nm^2) and have mean free paths as large as 1.6 μm.

References

1. Topics in Applied Physics, "Carbon Nanotubes: Synthesis, Structure, Properties and Applications," M.S. Dresselhaus, G. Dresselhaus, and Ph. Avouris (Eds.), Berlin, New York, Springer, 2000.
2. A. P. Graham, et al., "How do carbon nanotubes fit into the semiconductor roadmap?" *Appl. Phys. A*, vol. 80, pp. 1141–1151, 2005.
3. P. L. McEuen, M. S. Fuhrer, and H. Park, "Single-walled carbon nanotube electronics," *IEEE Trans. Nanotech.*, vol. 1, pp. 78–85, March 2002.
4. R. Meservey and P. M. Tedrow, "Measurement of the kinetic inductance of superconducting linear structures," *J. Appl. Phys.*, vol. 40, pp. 2028–2034, April 1969.
5. J. M. Pond, J. H. Claassen, and W. L. Carter, "Measurement and modeling of kinetic inductance microstrip delay lines," *IEEE Trans. Microwave Theory and Techniques*, vol. MTT-35, pp. 1256–1262, Dec. 1987.
6. H. Ehrenreich and H. R. Philipp, "Optical properties of Ag and Cu," *Phys. Rev.*, vol. 128, pp. 1622–1629, Nov. 1962.
7. A. Porch, P. Mauskopf, S. Doyle, and C. Dunscombe, "Calculation of the characteristics of coplanar resonators for kinetic inductance detectors," *IEEE Trans. Appl. Superconductivity*, vol. 15, pp. 552–555, June 2005.

8. A. Naeemi and J. D. Meindl, "Design and performance modeling for single-wall carbon nanotubes as local, semi-global and global interconnects in gigascale integrated systems," *IEEE Trans. Electron Devices*, vol. 54, pp. 26–37, Jan. 2007.
9. M. W. Bockrath, "Carbon Nanotubes: Electrons in One Dimension," Ph.D. dissertation, University of California, Berekeley, CA 1999.
10. P. J. Burke, "Lüttinger liquid theory as a model of the gigahertz electrical properties of carbon nanotubes," *IEEE Trans. Nanotech.* vol. 1, pp. 129–144, Sept. 2002.
11. S. Salahuddin, M. Lundstrom, and S. Datta, "Transport effects on signal propagation in quantum wires," *IEEE Trans. Electron Devices*, vol. 52, pp. 1734–1741, Aug. 2005.
12. Z. Yu and P. J. Burke, "Microwave transport in metallic single-walled carbon nanotubes," *Nano Letters*, vol. 5, pp. 1403–1406, June 2005.
13. J. R. Juroshek, C. A. Hoer, and R. F. Kaiser, "Calibrating network analyzers with imperfect test ports," *IEEE Transactions on Instrumentation and Measurement,* vol. 38, pp. 898–901, 1989.
14. S. Datta, "Electronic Transport in Mesoscopic Systems," Cambridge University Press, 1995.
15. Z. Yao, C. L. Kane, and C. Dekker, "High-field electrical transport in single-wall carbon nanotubes," *Phys. Rev. Lett.*, vol. 84, pp. 2941–2944, 2000.
16. W. Liang, M. Bockrath, D. Bozovic, J. H. Hafner, M. Tinkham, and H. Park, "Fabry-perot interference in nanotube electron waveguide," *Nature*, vol. 411, pp. 665–669, June 2001.
17. A. Javey, J. Guo, M. Paulsson, Q. Wang, D. Mann, M. Lundstrom, and H. Dai, "High-field quasiballistic transport in short carbon nanotubes," *Phys. Rev. Lett.*, vol. 92, p. 106804, March 2004.
18. O. Hjortstam, P. Isberg, S. Söderholm, and H. Dai, "Can we achieve ultra-low resistivity in carbon nanotube-based metal composites?" *Appl. Phys. A*, vol. 78, pp. 1175–1179, Jan. 2004.
19. J. Y. Park, S. Rosenbelt, Y. Yaish, V. Sazonova, H. Üstunel, S. Braig, T. A. Arias, and P. L. McEuen, "Electron–phonon scattering in metallic single-walled carbon nanotubes," *Nano Lett.*, vol. 4, pp. 517–520, 2004.
20. S. Li, Z. Yu, C. Rutherglen, and P. J. Burke, "Electrical properties of 0.4 cm long single-walled carbon nanotubes," *Nano Lett.*, vol. 4, pp. 2003–2007, 2004.
21. P. J. de Pablo, et al., "Nonlinear resistance versus length in single-walled carbon nanotubes," *Phys. Rev. Lett.* Vol. 88, p. 036804, Jan. 2002.
22. A. Javey, P. Qi, Q. Wang, and H. Dai, "10- to 50-nm-long quasi-ballistic carbon nanotube devices obtained without complex lithography," *Proc. Natl. Acad. Sci. USA*, vol. 101, pp. 13408–13410, 2004.
23. International Technology Roadmap for Semiconductors (ITRS), ed., Semiconductor Industry Association, CA, 2003.
24. HSPICE Simulation and Analysis User Guide, Release U-2003.03-PA, Synopsys, Inc., 2003.
25. A. Raychowdhury and K. Roy, "A circuit model for carbon nanotube interconnects: comparative study with Cu interconnects for scaled technologies," *Int. Conf. Comp. Aided Design*, pp. 237–240, 2004.
26. W. Steinhogl, G. Schindler, G. Steinlesberger, M. Traving, and M. Engelhardt, "Comprehensive study of the resistivity of copper wires with lateral dimensions of 100 nm and smaller," *J. Appl. Phys.*, vol. 97, pp. 023706-7, 2005.
27. P. Delaney, H. J. Choi, J. Ihm, S. G. Louie, and M. L. Cohen, "Broken symmetry and pseudogaps in ropes of carbon nanotubes," *Phys. Rev. B*, vol. 60, pp. 7899–7904, Sept. 1999.
28. A. A. Maarouf, C. L. Kane, and E. J. Mele, "Electronic structure of carbon nanotube ropes," *Phys. Rev. B*, vol. 61, pp. 11156–11165, April 2000.
29. H. Stahl, J. Appenzeller, R. Martel, and P. Avouris, "Intertube coupling in ropes of single-wall carbon nanotubes," *Phys. Rev. B*, vol. 61, pp. 11156–11165, April 2000.
30. M. Nihei, D. Kondo, A. Kawabata, Sh. Sato, H. Shioya, M. Sakaue, T. Iwai, M. Ohfuti, and Y. Awano, "Low-resistance multi-walled carbon nanotube vias with parallel channel conduction of inner shells," *IEEE Int. Interconnect Technol. Conf.*, pp. 234–236, June 2005.

31. S. M. Rossnagel and T. S. Kuan, “Alteration of Cu conductivity in the size effect regime,” *J. Vac. Sci. Technol. B*, vol. 22, pp. 240–247, Jan./Feb. 2004.
32. A. Naeemi and J. D. Meindl, “Impact of electron–phonon scattering on the performance of carbon nanotube interconnects for gigascale integration (GSI),” *IEEE Electron Device Lett.*, pp. 476–478, July 2005.
33. A. Naeemi, J. A. Davis, and J. D. Meindl, “Compact physical models for multilevel interconnect crosstalk in gigascale integration (GSI),” *IEEE Trans. Electron. Devices*, vol. 51, pp. 1902–1912, Nov. 2004.
34. RAPHAEL, “Interconnect Analysis Program”, TMA Inc, 1996.
35. N. C. Bruce, A. García-Valenzuela, and D. Kouznetsov, “Rough-surface capacitor: approximations of the capacitance with elementary functions,” *J. Phys. D: Appl. Phys.*, vol. 32, pp. 2692–2702, 1999.
36. R. Venkatesan, J. A. Davis, and J. D. Meindl, “Compact distributed RLC interconnect models – part IV: unified models for time delay, crosstalk, and repeater insertion,” *IEEE Trans. Electron Devices*, vol. 50, pp. 1094–1102, 2003.
37. H. J. Li, W. G. Lu, J. J. Li, X. D. Bai, and C. Z. Gu, ”Multichannel ballistic transport in multiwall carbon nanotubes,” *Phys. Rev. Lett.*, vol. 95, pp. 086601-4, 2005.
38. Q. Yanm J. Wu, G. Zhou, W. Duan, and B. Gu, “Ab initio study of transport properties of multiwalled carbon nanotubes,” *Phys. Rev. B*, vol. 72, p. 155425, Oct. 2005.
39. J. Y. Huang, S. Chen, S. H. Jo, Z. Wang, D. X. Han, G. Chen, M. S. Dresselhaus, and Z. F. Ren, “Atomic-scale imaging of wall-by-wall breakdown and concurrent transport measurements in multiwall carbon nanotubes,” *Phys. Rev. Lett.*, vol. 94, pp. 236802-4, 2005.
40. L. Forró and C. Schönenberger, “Physical properties of multi-wall nanotubes,” Topics in Applied Physics, Carbon Nanotubes: Synthesis, Structure, Properties and Applications, M.S. Dresselhaus, G. Dresselhaus, and Ph. Avouris (Eds.), Springer, 2000.
41. A. Naeemi and J. D. Meindl, “Compact physical models for multiwall carbon-nanotube interconnects,” *IEEE Electron Device Lett.*, vol. 27, pp. 338–340, May 2006.
42. A. Svizhenko, M. P. Anantram, and T. R. Govindan, “Ballistic transport and electrostatics in metallic carbon nanotubes,” *IEEE Trans. Nanotech.*, vol. 4, pp. 557–562, Sept. 2005.
43. C. T. White and T. N. Todorov, “Carbon nanotubes as long ballistic conductors,” *Nature*, vol. 393, pp. 240–242, May 1998.
44. J. Jiang, J. Dong, H. T. Yang, and D. Y. Xing, “Universal expression for localization length in metallic carbon nanotubes,” *Phys. Rev. B*, vol. 64, p. 045409, July 2001.
45. M. Inohara et al., “High performance copper and low-k interconnect technology fully compatible to 90 nm-node SOC application (CMOS4),” *IEDM*, pp. 77–80, 2002.
46. T. Sakurai, “Perspectives on power-aware electronics,” *IEEE ISSCC Dig, Tech.* Papers, pp. 26–29, Feb. 2003.
47. P. Zarkesh-Ha, P. Wright, S. Lakshminarayanan, C.-C. Cheng, W. Loh, and W. Lynch, “Back-end process optimization for 90 nm high-density ASIC chips,” *IEEE Int. Interconnect Technol. Conf.*, pp. 123–125, June 2003.
48. A. Naeemi, R. Venkatesan, and J. D. Meindl, “Optimal global interconnects for GSI,” *IEEE Trans. Electron. Devices*, vol. 50, pp. 980–987, April 2003.

Chapter 8
Chemical Sensing with SWNT FETs

Kyeong-Jae Lee and Jing Kong

8.1 Introduction

SWNTs possess unique properties that make them excellent candidates for sensing technology. Because the properties of a SWNT depend sensitively on its structure and because a SWNT is composed entirely of surface atoms, a slight variation of its environment tends to have a noticeable effect on its properties. Many types of sensors have been demonstrated using nanotubes, such as chemical [1–3], biological [4], flow [5], strain [6], pressure [7, 8], thermal [9], and mass [10] sensors. In this chapter, we will focus on the chemical sensing using SWNT FET devices.

Chemical sensors based on individual SWNTs were first demonstrated in year 2000 [11, 12]. The devices were constructed in the field effect transistor scheme. In Ref. [11], a constant bias was applied between the source and the drain electrodes and the current of the SWNT was monitored while gas molecules were introduced into the chamber. It was found that the electrical conductance of a semiconducting SWNT dramatically increases and decreases upon exposure to gaseous molecules of NO_2 and NH_3, respectively, as shown in Fig. 8.1. These SWNT sensors have many advantages, including fast responses, small sizes (therefore high packing density), high sensitivities (sub-ppm levels) and room-temperature operation, etc. As a result, there has been tremendous interest in using SWNT FETs as chemical and biological sensors.

Two different sensing mechanisms have been proposed in literature. The first one suggests the charge transfer between the SWNTs and the analyte molecules that adsorb on the SWNT surface [13], which then gives rise to the carrier density, and thus channel conductance change. The second mechanism resorts to the modification of the Schottky barriers (SB) at the metal-SWNT contacts due to the adsorption of molecules on both the metal and the SWNT. Depending on the specific analyte and the contact material, reports vary on which effect dominates the

K.-J. Lee (✉)
Department of Electrical Engineering and Computer Science Massachusetts Institute of Technology Cambridge MA 02139 USA

A. Javey, J. Kong (eds.), *Carbon Nanotube Electronics*, Series on Integrated Circuits and Systems, DOI 10.1007/978-0-387-69285-2_8,

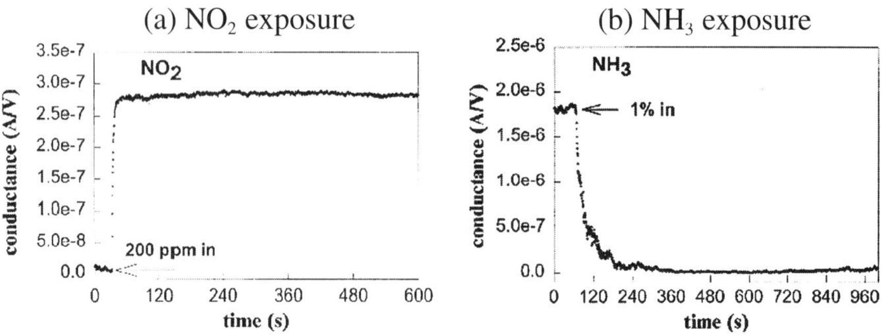

Fig. 8.1 Electrical responses of SWNT chemical sensors upon exposure of analyte molecules. (**a**) Conductance (under $Vg = +4$ V) versus time in a 200-ppm NO_2 flow. (**b**) Conductance ($Vg = 0$) versus time recorded with the same S-SWNT sample as in (a) in a flow of Ar containing 1% NH_3. Reprinted with permission from [11]. Copyright 2001 AAAS

sensing response. Bradley et al. propose that NH_3 has a larger impact on the nanotube channel than the SBs formed by Ti/Au contacts [14], while O_2 and H_2 seem to change the metal work function of Ti and Pd [15, 16]. The effects of NO_2 have been widely studied but experimental results still suggest conflicting conclusions [11, 17–20]. In most literature of SWNT FET sensor research, qualitative pictures of the sensing response were provided. It is the aim of this chapter to develop a quantitative understanding of the sensing mechanisms, particularly, to differentiate the two contributing mechanisms. NO_2 is used as an example system throughout the chapter; however, the formulism can be directly applied to different analyte molecules.

This chapter is organized as follows. Section 8.2 gives an analytical expression of the conductance changes due to the two mechanisms. Section 8.3 presents the model which links the gas adsorption process with the electrical response of the SWNT devices, and suggests different approximations under different conditions. Section 8.4 applies the derived model to experimental data and demonstrates that useful information, such as binding energy and adsorption rate of the gas molecule on SWNT surface, can be obtained from fitting experimental data with the model. Section 8.5 briefly discusses various performance aspects of SWNT sensors and both conductance and capacitance-based techniques using the same type of SWNT devices. At last, Section 8.6 summarizes this chapter.

8.2 Source of Conductance Change

In the gas sensing process, the electrical signal (most often current) changes due to the gas adsorption on the SWNT and metal surface. The key parameter connecting the electrical response to the physics of gas adsorption is the surface coverage θ. Surface coverage is defined as the occupation probability of a particular gas molecule on a surface [21, 22]. The majority of the SWNT sensor papers associate the conductance change (ΔG) directly to the surface coverage (i.e., $\Delta G \propto \theta$). This assumption fits certain sensing profiles fairly well but fails to account for all effects. In the following, the surface coverage on the metal contact and the nanotube are treated separately.

8.2.1 Schottky Barrier Modulation Due to Gas Adsorption

Experimental work shows that gas adsorption on the metal contacts can alter the work function of the metal, thus affecting the energy alignments at the junction leading to the change in SB height [16, 23]. For example, when Pd-contacted SWNT FETs are exposed to H_2, the Pd work function decreases and SB height increases for hole transport in SWNTs as discussed in Chapter 3 [16]. Doping can also be used to modify the SBs. Javey et al. have demonstrated SWNT FETs with potassium (K) doped source and drain regions. Since Fermi level pinning is small or nearly absent in SWNTs, they conclude that n-type contacts are formed as a result of SB height reduction as well as SB width thinning [24]. A general approach of relating the SB height to the conductance is taken here.

For a p-type SWNT FET, the difference between the metal work function and SWNT valence band in reference to vacuum level determines the SB height for holes. The "nominal" SB height Φ_0 is [25, 26]:

$$\Phi_0 = \left(\phi_{\mathrm{SWNT}} + \frac{E_{\mathrm{g}}}{2} \right) - \phi_{\mathrm{M}}, \tag{8.1}$$

where $\phi_{\mathrm{SWNT}} = 4.7$ eV [23] is the SWNT work function, and E_{g} and ϕ_{M} are the band gap and metal work function, respectively. Since little Fermi level pinning exists at the metal–nanotube interface [16, 27], this classical formalism is able to qualitatively describe how the SB height varies with SWNT diameter. However, for nanoscale quasi-1D systems, the depletion width depends exponentially on doping [28], and band realignment is weak due to the limited available depletion width. In SWNTs, this leads to a relatively small and slowly varying SB height with SWNT diameters, which has important consequences when modeling the SB height. Thus, a more accurate approximation presented by Leonard and Talin is used here [29]. The SB height is given as:

$$\Phi_{\mathrm{SB}} \approx \frac{kT}{\beta} \ln \left(\frac{\alpha \sqrt{E_{\mathrm{g}}/2kT}}{\ln \alpha \sqrt{E_{\mathrm{g}}/2kT} - \Phi_0/kT} \right), \tag{8.2}$$

where Φ_0 is given in Eq. (8.1), and after simplifying the expressions in [29], $\alpha = \frac{e^2(2/\pi)^{3/2}}{3\sqrt{\beta} a \gamma d C}$, ($\beta = 0.7$, $a = 0.142$ nm C–C bond length, $d =$ SWNT diameter, $\gamma =$ 2.5 eV is tight-binding overlap integral, $e =$ the electron charge, $k =$ Boltzmann factor, $T = 300$ K, and C = capacitance per unit area between metal-SWNT). Assume an initial SB height ($\Phi_{\mathrm{SB,i}}$) and a final SB height ($\Phi_{\mathrm{SB,f}}$) before and after analyte adsorption and that the band gap of the CNT does not change after adsorption. Then the following expression becomes:

$$
\begin{aligned}
\exp\left(\frac{\Delta\Phi_{\mathrm{SB}}}{kT}\right) &= \exp\left[\frac{\Phi_{\mathrm{SB,f}} - \Phi_{\mathrm{SB,i}}}{kT}\right] \\
&= \exp\left[\frac{1}{\beta}\ln\left(\frac{\alpha\sqrt{E_{\mathrm{g}}/2kT}}{\ln\alpha\sqrt{E_{\mathrm{g}}/2kT} - \Phi_{0,\mathrm{f}}/kT}\right)\right. \\
&\quad \left. - \frac{1}{\beta}\ln\left(\frac{\alpha\sqrt{E_{\mathrm{g}}/2kT}}{\ln\alpha\sqrt{E_{\mathrm{g}}/2kT} - \Phi_{0,\mathrm{i}}/kT}\right)\right] \\
&= \exp\left[\frac{1}{\beta}\ln\left(\frac{\ln\alpha\sqrt{E_{\mathrm{g}}/2kT} - \Phi_{0,\mathrm{i}}/kT}{\ln\alpha\sqrt{E_{\mathrm{g}}/2kT} - \Phi_{0,\mathrm{f}}/kT}\right)\right] \\
&= \left[\frac{\ln\alpha\sqrt{E_g/2kT} - \Phi_{0,i}/kT}{\left(\ln\alpha\sqrt{E_g/2kT} - \Phi_{0,i}/kT\right) - \left(\Phi_{0,f} - \Phi_{0,i}\right)/kT}\right]^{1/\beta} \\
&= \left[1 - \frac{\Delta\Phi_0/kT}{\left(\ln\alpha\sqrt{E_{\mathrm{g}}/2kT} - \Phi_{0,\mathrm{i}}/kT\right)}\right]^{-1/\beta}.
\end{aligned}
\tag{8.3}
$$

As can be seen, the SB height change can be attributed to the change in Φ_0. The work function of metals is known to change when gas molecules are adsorbed on the surface. A surface dipole layer forming on the metal surface causes polarization of the molecules. Evidence suggests that there is a linear relation between ϕ_{M} and surface coverage for certain molecules, while extreme non-linearity exists for others [21, 22, 30, 31]. Understanding details of surface orientation of the adsorbate and metal contact geometry is important to accurately model the gas–surface interaction. However, little data is found in literature on the nature of NO_2 adsorption, and detailed studies of NO_2 adsorption on metal surfaces is beyond the scope of this chapter's discussion. To first order, a reasonable approximation is to assume a simple linear relation between $\Delta\Phi_0$ and the coverage at the contact area (θ_{M}). This approach is taken here, which then simplifies Eq. (8.3) to:

$$
\exp\left(\frac{\Delta\Phi_{\mathrm{SB}}}{kT}\right) = (1 - b\theta_{\mathrm{M}})^{-1/\beta} \tag{8.4}
$$

where b is a constant, which can be either positive or negative depending on the polarizability and dipole moment of the adsorbate and is proportional to

$\frac{1}{kT\left(\ln \alpha\sqrt{E_g/2kT}-\Phi_{0,i}/kT\right)}$. The conductance will either increase or decrease depending on the sign of b.

8.2.2 Charge Transfer to Nanotube

The charge transfer between the SWNT and adsorbed molecules vary to a large extent depending on the nature of their interactions. Certain gas molecules have been calculated to interact weakly with minimal charge transfer to CNTs [32, 33], such as acetone, H_2, and CH_4. Other molecules are predicted by calculation to have weak interactions, such as NH_3, but have experimentally demonstrated noticeable charge transfer. It has been proposed that in the case of NH_3, the charge transfer is mediated by a surface H_2O layer [34] or defect sites [35]. The CNT sensor response to NO_2 has been heavily researched, and the acceptor nature of NO_2 has been accepted by most researchers.

Charge transfer doping is much stronger for SWNTs than planar devices due to quasi-1D electrostatics [28]. With strong charge transferring analytes, reports show that ΔG can be dominated by charge transfer from the adsorbate molecules [36]. In such cases, a simple phenomenological model can be applied which relates the conductance change to the adsorption of molecules on the SWNT [37]. This is described by a site-binding picture, where each atom along the nanotube surface is modeled as a binding site for molecular adsorption. Assuming that the charge carrier density (n) is proportional to the available binding sites, the surface coverage on the SWNT (θ_{NT}) dictates how much of those binding sites are occupied giving rise to the conductance change. Thus, for a final and initial carrier density of n_f and n_0, respectively:

$$\theta_{NT} = \frac{\text{Occupied Binding Sites}}{\text{Total Binding Sites}} \propto \frac{n_f - n_0}{n_0}$$

Therefore $\delta\theta_{NT} = \frac{n_f - n_0}{n_0}$, where δ is a proportionality constant, and

$$\frac{n_0}{n_f} = \frac{1}{\delta\theta_{NT} + 1} \tag{8.5}$$

8.3 Modeling Gas Adsorption

Most papers found in literature attribute the change in conductance (ΔG) simply to the surface coverage on the SWNT [17, 36, 37]. While this approach may be viable in certain cases, the effect of the SB modulation is overlooked. Suehiro et al. used Al-contacted CNT FETs that form large SBs [19]. High contact-resistance devices elucidate the nature of SB modulation, which suggest that the total resistance needs to be modeled as the sum of the contact resistance (R_M) and the SWNT channel resistance (R_{NT}), instead of only considering the SWNT channel conductance.

The effects discussed in the previous section are directly used here. Assume an initial and final resistance as R_0 ($= R_M + R_{NT}$) and R_f ($= R_{M,f} + R_{NT,f}$), respectively, then the channel resistance will be inversely proportional to the carrier density ($R_{NT} \propto 1/n_0$). Under a fixed bias voltage, the current density through a SB will be exponentially dependent on the SB height and is proportional to the carrier density. Thus, $R_M \propto \frac{1}{n_0} \exp\left(\frac{\Phi_{SB,0}}{kT}\right)$. Then, using Eq. (8.4) and (8.5), the total resistance change becomes:

$$\begin{aligned} \Delta R = R_f - R_0 &= R_M\left(\frac{R_{M,f}}{R_M} - 1\right) + R_{NT}\left(\frac{R_{NT,f}}{R_{NT}} - 1\right) \\ &= R_M\left(\frac{n_0}{n_f}\exp\left(\frac{\Delta\Phi_{SB}}{kT}\right) - 1\right) + R_{NT}\left(\frac{n_0}{n_f} - 1\right) \\ &= R_M\left[(\delta\theta_{NT} + 1)^{-1}(1 - b\theta_M)^{-1/\beta} - 1\right] + R_{NT}\frac{-\delta\theta_{NT}}{\delta\theta_{NT} + 1}. \end{aligned} \tag{8.6}$$

Therefore the ratio between ΔG and initial conductance is:

$$\begin{aligned} \frac{\Delta G}{G_0} &= \frac{G_f - G_0}{G_0} = \frac{R_0 - R_f}{R_f} = \frac{-\Delta R}{R_0 + \Delta R} \\ &= \frac{R_M\left[1 - (\delta\theta_{NT} + 1)^{-1}(1 - b\theta_M)^{-1/\beta}\right] + R_{NT}\dfrac{\delta\theta_{NT}}{\delta\theta_{NT} + 1}}{R_M(\delta\theta_{NT} + 1)^{-1}(1 - b\theta_M)^{-1/\beta} + R_{NT}\dfrac{1}{\delta\theta_{NT} + 1}}. \end{aligned} \tag{8.7}$$

Now, the surface coverage, θ_M and θ_{NT} are obtained through the Langmuir model. A Langmuir isotherm assumes a monolayer coverage of gas molecules on the surface. The site-binding model presented in [37] exactly resembles a Langmuir isotherm and shares the same mathematical form. This interpretation is consistent with experimental data showing that conductance change eventually saturates with increasing concentration of the gas molecules [35]. The steady-state response of the surface coverage in a Langmuir isotherm is readily given as [21, 22]:

$$\theta_M = \frac{K_M P}{K_M P + 1} \tag{8.8}$$

$$\theta_{NT} = \frac{K_{NT} P}{K_{NT} P + 1}, \tag{8.9}$$

where P is the partial pressure of the adsorbate (Pascal), and K_M, K_{NT} are the Langmuir constant (Pascal^{-1}), and θ_M and θ_{NT} are the final surface coverage. For the case of NO_2, using parameters found in [17], the adsorption and desorption equilibrium of NO_2 results in:

$$K_M = \frac{S\sigma}{\sqrt{2\pi mkT\nu}} \exp\left(\frac{E_{b,M}}{kT}\right) \tag{8.10}$$

$$K_{NT} = \frac{S\sigma}{\sqrt{2\pi mkT\upsilon}} \exp\left(\frac{E_{b,NT}}{kT}\right), \tag{8.11}$$

where $E_{b,M}$ and $E_{b,NT}$ are the NO_2 binding energies to the metal and CNT surfaces, respectively, and m is the molecular mass, S is the sticking coefficient, $\sigma \sim 10^{-19}\text{m}^2$ is the molecular cross-section, and $\upsilon \sim 10^{12}$/s is the molecular vibration frequency.

Now, by combining Eq. (8.7) with Eqs. (8.8, 8.9, 8.10 and 8.11), experimental data can be understood using this model. If the transient form of $\theta_M(t)$ and $\theta_{NT}(t)$ are used, the equation can be used to fit the transient sensing response (as long as the chamber can be filled by the analyte gas in a much shorter time scale). The transient response is described in Section 8.4.3. Using the steady state form of θ_M and θ_{NT}, the equation can be used to analyze the concentration or partial pressure dependent response.

8.4 Application to Data

8.4.1 Contributions from the Metal Contact Versus the Channel

The relative strengths of the surface interaction between the gas molecule and the metal/SWNT will determine which effect dominates the sensor response. If the sensing response is dominated by the metal contact, one can assume that charge transfer effect is negligible (i.e., $\delta \to 0$). Conversely, if the surface interactions at the CNT surface dominate the sensing response, the dipole moment or the polarizability at the metal contact can be ignored (i.e., $b \to 0$). From Eq. (8.7), the sensing response at either limit simplifies to:

$$\text{Metal contact dominated: } \frac{\Delta G}{G_0} \cong \frac{R_M\left[1-(1-b\theta_M)^{-1/\beta}\right]}{R_M\,(1-b\theta_M)^{-1/\beta}+R_{NT}}. \tag{8.12}$$

$$\text{Charge transfer to SWNT dominated: } \frac{\Delta G}{G_0} = \delta\theta_{NT}. \tag{8.13}$$

Note that in the case of charge transfer dominated, it does not matter whether the total resistance is dominated by R_{NT} or R_M, $\Delta G/G_0$ will be the same form. In addition, this result indicates that this normalized conductance change is roughly constant at a given concentration. In other words, the conductance change is directly proportional to the initial conductance, which is consistent with the site-binding model presented in [36, 37]. The initial conductance is assumed to be proportional to the number of SWNTs within each device. A larger number of SWNTs relates to a larger number of binding sites and hence a higher chance of surface binding and charge transfer to occur.

If neither effect can be neglected, i.e., b and δ are non-zero, but the initial resistance is dominated by metal contact, i.e., $R_M >> R_{NT}$, then we can obtain the third approximation result:

$$\frac{\Delta G}{G_0} \cong (\delta\theta_{NT} + 1))(1 - b\theta_M)^{1/\beta} - 1. \tag{8.14}$$

The exact composition of R_M and R_{NT} is difficult to quantify and will vary severely from device to device. Nonetheless, it has been suggested that the metal contact resistance can vary over a much wider dynamic range [26, 38] than the channel resistance. The SWNT channel resistance is on the order of 4~6 kΩ/μm at low bias [39], and devices with multiple nanotubes have lower effective SWNT channel resistance. Figure 8.2 assumes a fixed value of $R_{NT} = 30\ k\Omega$ (i.e., $R_0 = R_M + 30\ k\Omega$) and plots the change in resistance/conductance as a function of the initial resistance/conductance. From Fig. 8.2, it can be seen that when the charge transfer to SWNT dominates (i.e., $b \rightarrow 0$), both ΔR_R_0 and ΔG_G_0 has a linear response (*red dashed* lines in Fig 8.2(a) and (b)). On the other hand, if the sensing response is dominated by change in the metal contact, the log–log plot of ΔG_G_0 will be highly non-linear (*blue dashed line* in Fig. 8.2(b)). By comparing the experimental sensing results with these curves, the dominating mechanism of sensing response for different analyte molecules can be identified.

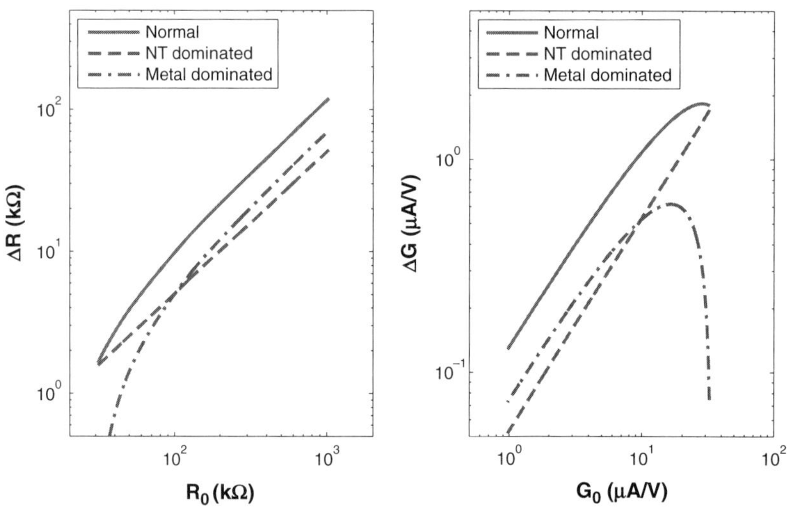

Fig. 8.2 Resistance (conductance) change as a function of initial resistance (conductance). Assumes a fixed $R_{NT} = 30k\Omega$ and exposure to 100 ppm NO_2. Normal conditions use values of $E_{b,M} = 0.4$ eV, $E_{b,NT} = 0.4$ eV, $b = -0.5$, and $\delta = 0.5$. Either b or δ is set to zero in the limiting case

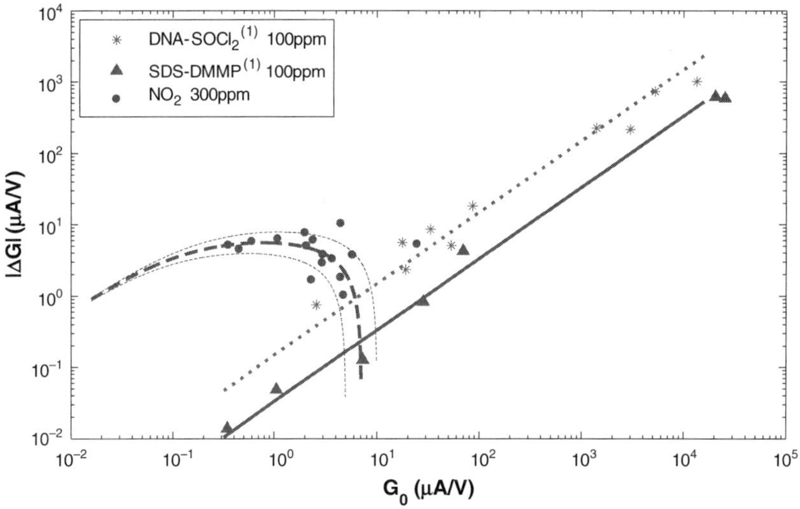

Fig. 8.3 Experimental data on ΔG vs G_0. Lines represent data fitting with the previously derived equations. NO_2 fitting assumes values of $E_{b,M} = 0.48$ eV, $E_{b,NT} = 0.73$ eV, $b = -20.32$, and $\delta = 0.0027$. [(1)]Data from [36]

Figure 8.3 shows the log–log plot of conductance response of several NO_2 sensor devices together with $SOCl_2$ and DMMP responses for comparison. The NO_2 sensing responses are measured by SWNT FETs with Cr/Au metal contact. The $SOCl_2$ and DMMP sensing responses are from [36]. Lee et al. used devices made by dielectrophoresis with hundreds of parallel SWNTs in between Ti/Au electrodes to detect thionyl chloride ($SOCl_2$) and dimethyl methylphosphonate (DMMP). Two types of CNT samples are prepared in a suspension using sodium dodecyl sulfate (SDS) or DNA ($d(GT)_{15}$). These sensors fit an exact linear profile and are indicative of a charge transfer to SWNT-dominated sensing mechanism, which is consistent with their report. The NO_2 sensing curve exhibits non-linearities indicating a metal contact-dominated sensing mechanism. The data points were fit to a curve with a higher metal-NO_2 interaction: $b = -20.32$ and $\delta = 0.0027$. This result demonstrates that, in NO_2 sensing, even though charge transfer doping has been clearly indicated [11], the response is dominated by the SB change at the metal contact.

8.4.2 Partially Exposed Devices

Partially exposed SWNT FET sensors devices have been fabricated by researchers to differentiate the roles of metal contacts and charge transfer doping. These devices are mostly covered by a passivation layer (polymer or oxide) with either only the SWNT channel or the metal/SWNT contact region being exposed [14, 20, 40].

However, due to the differences in the devices (such as the size of the exposed area versus the FET channel length, or nature of metal-NT contact), seemingly contradictory results have been reported. In Ref. [40], responses to NO_2 from both channel-exposed and contact-exposed devices were observed, indicating SB variation and charge transfer doping are playing equally important roles in the NO_2 sensing. Moreover, modulations of the sub-threshold swing of the FET devices in opposite directions for channel-exposed and contact-exposed devices were found. On the other hand, in Ref. [20], Zhang et al. conducted similar experiments, however, only immediate responses with contact-exposed devices were observed. The slow responses in channel-exposed devices were attributed to NO_2 diffusing through the passivation resist layer to the metal contact [20]. Several possible reasons could be contributing to the discrepancies. The most likely one is the differences regarding the distance from the edge of exposed region to the edge of the metal contact in the channel-exposed devices among these studies. Since the depletion length in nanotube–nanotube hetero-junctions decay logarithmically with distance [28], a decay length on the order of few hundred nanometers is anticipated by normal SWNT FET devices [14]. Therefore, if the edge of the exposure region is less than 1 μm away from the metal contact, even though the contact region is protected by the resist, the doping level of the nanotube at the contact region will be close to the exposed region. Therefore, the device can not be considered as three differently doped regions (edge-doped center-edge), but need to be considered as one channel

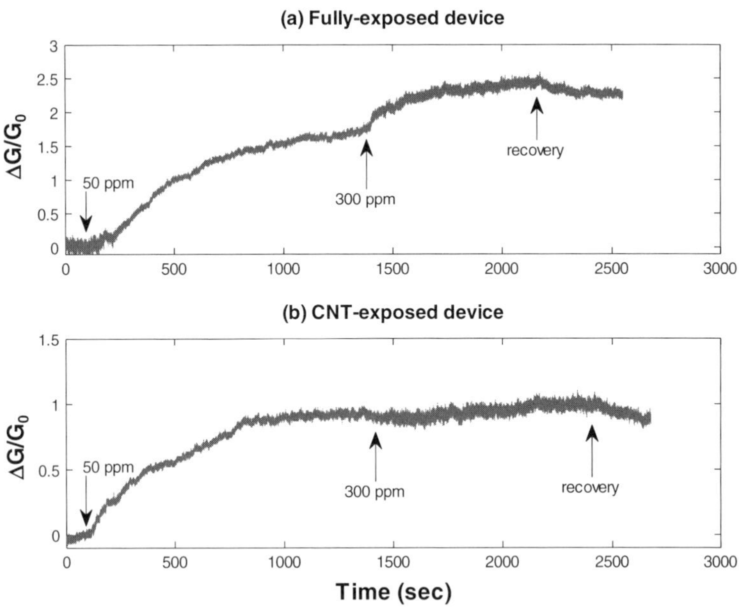

Fig. 8.4 Transient response of a (**a**) fully exposed and (**b**) channel-exposed device upon exposure to 50 ppm and 300 ppm NO_2

with a weakly varying Fermi level. Figure 8.4 shows the transient response of a fully exposed device and a channel-exposed device (4 μm channel and 3.5 μm exposed center region) to NO_2. The channel-exposed device shows a similar response to the fully exposed device in this case, which is consistent with the hypothesis. However, in order to obtain a definite confirmation of such an effect, a systematic study on devices with various distances from the exposure region edge to the metal contact is needed.

8.4.3 Transient Response

Understanding the dynamics of surface coverage is important in modeling the transient response of SWNT FET sensors. If the adsorption rate constants associated with the metal contact and CNT are k_M and k_{NT}, then the desorption rate is k_M/K_M and k_{NT}/K_{NT}, respectively, where K_M and K_{NT}, are the Langmuir constants in the steady state response in Eqs. (8.8) and (8.9). Balancing the adsorption and desorption rates yields the following rate equations [21, 22, 37]:

$$\frac{d\theta_M}{dt} = k_M P\,(1 - \theta_M) - \frac{k_M}{K_M}\theta_M \tag{8.15}$$

$$\frac{d\theta_{NT}}{dt} = k_{NT} P\,(1 - \theta_{NT}) - \frac{k_{NT}}{K_{NT}}\theta_{NT}. \tag{8.16}$$

which are valid when the gas molecules are present in the sensing chamber. Given that the SWNT FET sensor has no analyte molecules attached in the beginning (i.e., $\theta_M\,(0) = \theta_{NT}\,(0) = 0$) and that the surface coverage is at most equal to one assuming no multi-layer stacking of analytes, the transient form for the surface coverage results in:

$$\theta_M(t) = \frac{K_M P}{K_M P + 1}\left(1 - \exp\left(-k_M \frac{1 + K_M P}{K_M} t\right)\right) \tag{8.17}$$

$$\theta_{NT}(t) = \frac{K_{NT} P}{K_{NT} P + 1}\left(1 - \exp\left(-k_{NT} \frac{1 + K_{NT} P}{K_{NT}} t\right)\right). \tag{8.18}$$

At steady-state ($t \to \infty$), the above equations approach Eqs. (8.8) and (8.9). When the gas molecule is removed from the chamber, then the adsorption terms are erased from the rate balance equations (8.15) and (8.16). Thus, only the desorption terms (second term on right side of Eqs. (8.15) and (8.16)) determine the overall rate of change. During this recovery phase, the surface coverage assumes the transient form of a simple exponential decay.

Figure 8.5 shows the transient response of two similar devices, denoted as Dev #1 and Dev #2. Dev #1 and Dev #2 presented here refer to the Cr/Au-contacted, fully exposed devices. Both devices consist of single semiconducting SWNTs, have

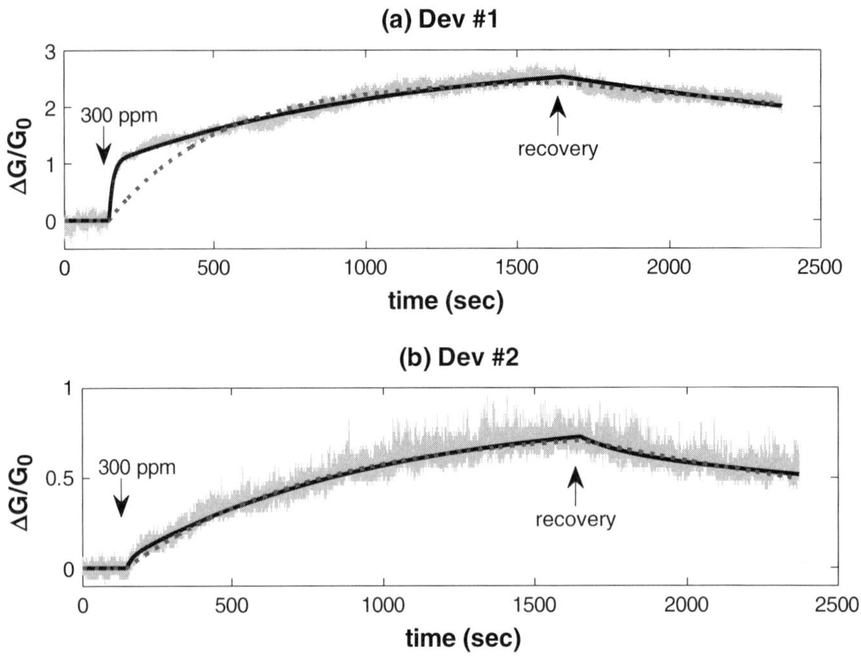

Fig. 8.5 Transient response to 300 ppm NO_2 for semiconducting devices (**a**) Dev #1 and (**b**) Dev #2. *Dark solid lines* are fitted curves using the model presented in Eq. (8.7). *Dashed lines* are fitted curves using a simple exponential rise/decay equation

similar DC resistance values (416.2 kΩ and 433.3 kΩ), and were measured simultaneously while being exposed to 300 ppm NO_2. In this figure, the dark-colored solid lines are fitted curves by combining Eq. (8.7) with (8.15, 8.16, 8.17 and 8.18) assuming a channel resistance of $R_{CNT} = 28$ kΩ. The fitting parameters are listed in Table 8.1.

The dashed lines in Fig. 8.5 are fitted curves to a generic exponential rise/decay equation (i.e., $\Delta G/G_0 \sim \theta$), which does not model the initial sharp increase in conductance for Dev #1. The same is true for Dev #2, but the difference is hardly noticeable. While the commonly used form of $\Delta G/G_0 \sim \theta$ cannot explain certain features, the model presented here combines sensing effects along the channel and the metal contact separately and fits both devices very well along all parts of the curve. Recall

Table 8.1 Fitting parameters for (dark solid) curves in Fig. 8.5

	δ	b	k_{NT} ($\text{Pascal}^{-1}\ \text{s}^{-1}$)	k_M ($\text{Pascal}^{-1}\ \text{s}^{-1}$)	K_{NT} (Pascal^{-1})	K_M (Pascal^{-1})
Dev #1	0.936	−0.7253	3.432e^{-5}	2.50e^{-3}	804.8	5.414
Dev #2	1.001	−0.0468	3.034e^{-5}	1.59e^{-5}	0.1059	0.2377

Table 8.2 Langmuir constants and binding energies for curves in Figs. 8.5 and 8.6

	K_{NT} (Pascal^{-1})	K_M (Pascal^{-1})	$E_{b,NT}$ (eV)	$E_{b,M}$ (eV)
Dev #1	804.8	5.414	0.688	0.559
Dev #2	0.1059	0.2377	0.457	0.478
Dev #3	1.594	1.796	0.527	0.530
Dev #4	0.2139	0.1054	0.475	0.457
Dev #5	0.0174	0.0184	0.410	0.412
PEI-coated[1]	5247	30.75	0.736	0.603

[1]Data from [17].

that b and δ can vary from device to device since the effective band gap, number of nanotubes, and mixture of metallic and semiconducting nanotubes are different. However, one can expect that the CNT adsorption rate constants (k_{NT}) will be more or less similar for both devices. Table 8.1 shows that the SWNT adsorption rate is roughly 3×10^{-3} Pascal^{-1}s^{-1} for both devices.

8.4.4 Surface Binding

The surface binding energies of NO_2 to the metal and CNT can be found by fitting curves to $\Delta G/G_0$ as a function of partial pressure of NO_2. The steady-state form of the surface coverage Eqs. (8.8) and (8.9) should be used here. Dev #3, #4, and #5 are similar devices as Dev #1 and Dev #2 presented in the previous section using Cr/Au as contact metal. Qi et al. also present SWNT FETs but with Mo metal contacts [17]. They report that coating the devices with polyethyleneimine (PEI) turns them into n-type FETs and effectively increases the sticking coefficient by 2 orders of magnitude. The conductance response is shown in Fig. 8.6. Note that the range of partial pressure is 3–4 orders of magnitude less than devices presented here (Dev #1–5), indicating higher sensitivity in [17]. Table 8.2 lists the Langmuir constants and respective binding energies for the devices in Figs. 8.5 and 8.6. The binding energies are calculated from Eqs. (8.10) and (8.11) assuming a unity sticking coefficient ($S = 1$).

Although there is no consistent interpretation of the Langmuir constants K_M and K_{NT} in Table 8.2, a few observations can be made. First, PEI coating does indeed yield very high values of K_M and K_{NT}, indicating either higher binding energy or sticking coefficient. Assuming a unity sticking coefficient ($S = 1$), the metal–NO_2 binding energies can vary over a wide range, possibly indicating that gas adsorption on metal surfaces is sensitive to other factors such as surface roughness, geometry, and contact quality. Similarly, the NO_2–SWNT binding energy can vary depending on the type of the nanotube. Theoretical calculations predict NO_2–SWNT binding energies of 0.3–0.8 eV [33, 41]. Systematic variations of metal contact and CNTs are needed to further elucidate the nature of surface binding of gas molecules on SWNT FET devices.

8.5 Additional Aspects of SWNT FET Sensors

In the previous sections, the sensing mechanisms were discussed at length, using a gas adsorption model to fit the experimental results. In this section, more practical issues regarding the performance of a sensor are discussed, such as sensor response time, sensitivity and specificity, reversibility, etc. The final section of this chapter briefly mentions another type of signal transduction using these SWNT FET devices – capacitance based sensing, and the combination of capacitance/conductance sensing technology.

8.5.1 Response Time

Using the sensing response curve in Fig. 8.1 as an example, we can see that if the sensor response time is defined as the resistance/conductance change of 90% of the overall change, the response time is several minutes. In fact, comparing SWNT FET sensors fabricated by different researchers, this is a typical value for response time, which is also comparable to metal oxide or conducting polymer sensors [42, 43]. This value is a direct result of the adsorption rate.

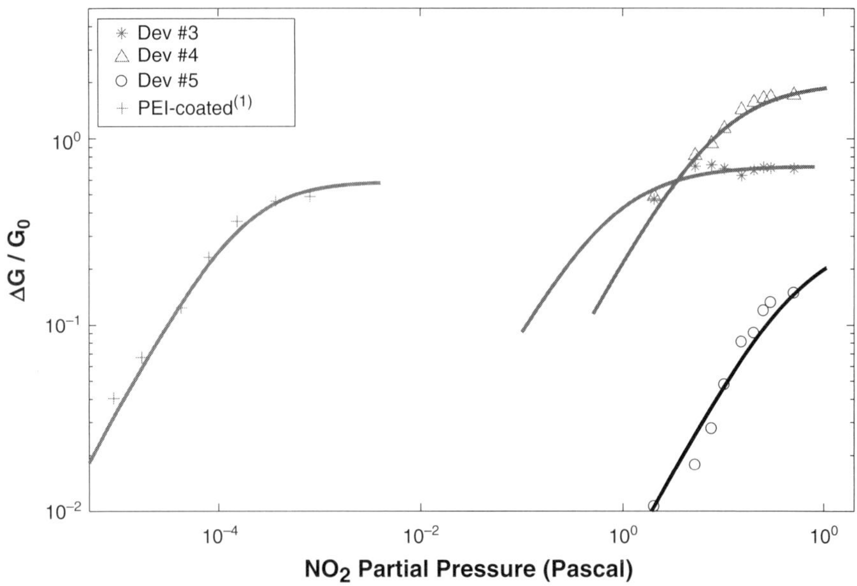

Fig. 8.6 Pressure dependence of standard (Cr/Au, fully exposed) devices and a [(1)]PEI-coated device from [17]. Point markers are measured data and *solid lines* represent fitted curves

8.5.2 Chemical Specificity

The SWNT FET sensors have only two types of responses, either increasing or decreasing in conductance, depending on the electron withdrawing or donating nature of the analyte and the metal contact response due to the analyte adsorption. As an example, NO_2 and SO_2 molecules give the same type of response and thus can not be easily differentiated using the simple devices, which limits the SWNT sensors chemical specificity. Furthermore, for gas molecules such as CO which lack specific interactions with the nanotubes and metal contact, there are no sensing responses, limiting the applicable ranges of these SWNT sensors.

Chemical modifications to the devices have been used to impart selectivity to the analytes. The modification can be on the metal contact, for example, when Pd is used as the metal contacts for SWNT FETs, the conductance of the devices change upon the exposure to H_2 gas due to SB modification [16]. More often, the modification consists of a layer of coating on both the metal contact and the surface of the SWNTs which can adsorb specific analyte molecules. As a similar example, by evaporating Pd nanoparticles on the surface of SWNT devices, molecular H_2 sensors can be enabled with excellent sensing response. This occurs because H_2 dissociates on the surface of Pd and atomic hydrogen dissolves in Pd and lowers the work function of Pd [44], so that extra electron charges are donated to the SWNT that is in contact with the Pd, and the conductance of the SWNT device decreases. By coating the SWNT devices with polyethyleneimine (PEI), SWNT sensors have been shown to respond only to NO_2 when a gas mixture of NH_3 and NO_2 was introduced to the environment. This NH_3 insensitivity is attributed to low binding affinity of NH_3 on the electron-rich (due to high-density amines on PEI) SWNTs. On the other hand, when SWNT devices are coated with another type of polymer, nafion, the detection of NO_2 was blocked and only NH_3 can be recognized from the mixture. Other polymer coatings, such as chlorosofonated polyethelene and hydroxypropyl cellulose have been utilized to implement sensing for Cl_2 and HCl species [45]. SS-DNA coated SWNT devices have been implemented to sense methanol, propionic acid, trimethylamine (TMA), dinitrotoluene (DNT), and dimethyl methylphosphonate (DMMP; a simulant for the nerve agent sarin) [46]. Although the sensing mechanism due to this modification was not fully understood, the sensors were found to be self-regenerating, i.e., samples maintain a constant response with no need for sensor refreshing for approximately 50 gas exposure cycles.

8.5.3 Sensitivity

Carbon nanotube gas sensors are among the only materials to electrically transduce molecules binding to their surface directly at analyte concentrations well below the ppb level [17, 47]. Conventional solid-state sensors operate by detecting the loading of a sensor material with the analyte. For low concentrations such as 1 ppb there are insufficient vapor molecules to load the active material to detectable levels. Because

SWNTs have small sizes and all of their atoms are on the surface, exceptionally high sensitivity is enabled. Furthermore, as mentioned in Section 8.5.2, by decorating the SWNT sensor surface with specific coating, such as in [17], higher binding energies/tendencies can be achieved, giving rise to the enhanced sensitivity. Robinson et al. have also shown that the presence of defects can also greatly enhance the sensitivity of SWNT sensors [35].

8.5.4 Recovery

Since the operation principle of the SWNT sensors are based on molecular adsorption on surface, due to the relatively large binding energy the adsorbate molecules tend to remain attached to the SWNTs long after the analyte is removed from the surrounding atmosphere. In order to achieve quick reversibility, various methods have been demonstrated to be effective, including heat treatment [11], UV irradiation [17, 48] or applying a gate bias [49].

8.5.5 Capacitance-Based Sensing

In parallel with the conductance based chemical sensing with SWNT electronic devices, during the past two years Snow et al. have demonstrated capacitance-based chemical vapor detection using the same type of SWNT devices [3, 50]. The signal transduction of the capacitance-based sensing relies on the polarization of the molecular adsorbates. Under an applied gate voltage a large radial electric field emanates from the SWNT surface. This electric field polarizes molecular adsorbates on the SWNT surface, producing an increase in capacitance. This capacitance change provides a fast, sensitive, low-noise transduction mode to detect a wide range of chemical vapors. Weakly interacting chemicals which can not produce strong enough conductance response can be detected with this approach. For the conductance-based detection, it is desired to have only semiconducting SWNTs to achieve large sensing response; however, for the capacitance-based scheme, metallic and semiconducting SWNTs contribute similarly to the response. Since the capacitance of an individual SWNT is small, on the order of 10 aF, capacitance-based detection with individual SWNTs is not very accurate. Rather, devices with SWNT mats are desirable for this type of detection. Since both the conductance- and capacitance-based schemes use the same type of SWNT devices, a simultaneous sensing scheme can be carried out to enhance the selectivity and accuracy. Similar gas adsorption/desorption picture can be applied to understand the capacitance responses during sensing; however, it has been found that the conductance response saturates at a certain pressure presumably due to the full coverage of activated sites on the surface, whereas capacitance response continuously increases, indicating the possible adsorption of analyte molecules beyond the first monolayer [51].

8.6 Summary

In this chapter, we have shown that the sensing response of SWNT FET sensors can be understood in the picture of the gas adsorption kinetics and gas molecule interactions with the nanotube and metal contacts. There are two mechanisms contributing to the sensing response, one is the modulation of the metal contacts due to the analyte adsorption, the other is the charge transfer to the SWNTs. Depending on the specific analyte molecule and the contact material, one mechanism could be dominant over the other, or both could be contributing equally. Approximations under different scenarios were investigated, and using this model, parameters such as binding energy and absorption rates can be derived from experimental results, both obtained by us and from literature.

SWNT sensors offer great promise for compact, low-power chemical detectors. Compared to other sensor technologies, SWNT sensors have the unique capability of detecting extremely low concentrations, and thus are very suitable for trace amount chemical agent detection, such as nerve agents and explosives. Although the chemical specificity is still a challenge at this stage, pattern recognition techniques using arrays of sensors can be used to provide further discrimination [45]. Snow et al. have also envisioned that a micro gas chromatograph can be incorporated in front of the SWNT sensors to detect and distinguish minute target analytes [51]. SWNT sensors have already been commercialized for specific gas detection such as H_2 and CO_2 by Nanomix, Inc, and further development in the chemical specificity and/or low cost fabrication of SWNT devices will extend the SWNT sensors to a much wider application range.

Acknowledgments Part of the work presented in this chapter was based on Dr. Jing Kong's thesis with Prof. Hongjie Dai at Stanford University. We deeply appreciate the valuable advices and contribution from Prof. Dai. The authors would also like to thank Prof. Anantha Chandrakasan at MIT for the encouragement and support for this work. This work was funded in part by the MARCO IFC Focus Research Center Program and Intel Corporation.

References

1. J. Kong, N. Franklin, C. Zhou, M. Chapline, S. Peng, K. Cho, and H. Dai, "Nanotube molecular wires as chemical sensors," *Science*, vol. 287, pp. 622–625, 2000.
2. A. Modi, N. Koratkar, E. Lass, B. Q. Wei, and P. M. Ajayan, "Miniaturized gas ionization sensors using carbon nanotubes," *Nature*, vol. 424, pp. 171–174, 2003.
3. E. S. Snow, F. K. Perkins, E. J. Houser, S. C. Badescu, and T. L. Reinecke, "Chemical detection with a single-walled carbon nanotube capacitor," *Science*, vol. 307, pp. 1942–1945, 2005.
4. R. J. Chen, S. Bangsaruntip, K. A. Drouvalakis, N. W. S. Kam, M. Shim, Y. M. Li, W. Kim, P. J. Utz, and H. J. Dai, "Noncovalent functionalization of carbon nanotubes for highly specific electronic biosensors," *Proceedings of the National Academy of Sciences of the United States of America*, vol. 100, pp. 4984–4989, 2003.
5. S. Ghosh, A. K. Sood, and N. Kumar, "Carbon nanotube flow sensors," *Science*, vol. 299, pp. 1042–1044, 2003.
6. P. Dharap, Z. L. Li, S. Nagarajaiah, and E. V. Barrera, "Nanotube film based on single-wall carbon nanotubes for strain sensing," *Nanotechnology*, vol. 15, pp. 379–382, 2004.

7. R. J. Grow, Q. Wang, J. Cao, D. W. Wang, and H. J. Dai, "Piezoresistance of carbon nanotubes on deformable thin-film membranes," *Applied Physics Letters*, vol. 86, p. 093104, 2005.
8. J. A. Wu, J. Zang, B. Larade, H. Guo, X. G. Gong, and F. Liu, "Computational design of carbon nanotube electromechanical pressure sensors," *Physical Review B*, vol. 69, p. 153406, 2004.
9. C. Y. Kuo, C. L. Chan, C. Gau, C. W. Liu, S. H. Shiau, and J. H. Ting, "Nano temperature sensor using selective lateral growth of carbon nanotube between electrodes," *IEEE Transactions on Nanotechnology*, vol. 6, pp. 63–69, 2007.
10. D. H. Wu, W. T. Chien, C. S. Chen, and H. H. Chen, "Resonant frequency analysis of fixed-free single-walled carbon nanotube-based mass sensor," *Sensors and Actuators A-Physical*, vol. 126, pp. 117–121, 2006.
11. J. Kong, N. R. Franklin, C. W. Zhou, M. G. Chapline, S. Peng, K. J. Cho, and H. J. Dai, "Nanotube molecular wires as chemical sensors," *Science*, vol. 287, pp. 622–625, 2000.
12. P. Collins, K. Bradley, M. Ishigami, and A. Zettl, "Extreme oxygen sensitivity of electronic properties of carbon nanotubes," *Science*, vol. 287, pp. 1801–1804, 2000.
13. S. Peng and K. J. Cho, "Chemical control of nanotube electronics,"n *Nanotechnology*, vol. 11, pp. 57–60, 2000.
14. K. Bradley, J. C. P. Gabriel, A. Star, and G. Gruner, "Short-channel effects in contact-passivated nanotube chemical sensors," *Applied Physics Letters*, vol. 83, pp. 3821–3823, 2003.
15. S. Heinze, J. Tersoff, R. Martel, V. Derycke, J. Appenzeller, and P. Avouris, "Carbon nanotubes as Schottky barrier transistors," *Physical Review Letters*, vol. 89, p. 106801, 2002.
16. A. Javey, J. Guo, Q. Wang, M. Lundstrom, and H. J. Dai, "Ballistic carbon nanotube field-effect transistors," *Nature*, vol. 424, pp. 654–657, 2003.
17. P. F. Qi, O. Vermesh, M. Grecu, A. Javey, O. Wang, H. J. Dai, S. Peng, and K. J. Cho, "Toward large arrays of multiplex functionalized carbon nanotube sensors for highly sensitive and selective molecular detection," *Nano Letters*, vol. 3, pp. 347–351, 2003.
18. S. Santucci, S. Picozzi, F. Di Gregorio, L. Lozzi, C. Cantalini, L. Valentini, J. M. Kenny, and B. Delley, "NO_2 and CO gas adsorption on carbon nanotubes: Experiment and theory," *Journal of Chemical Physics*, vol. 119, pp. 10904–10910, 2003.
19. J. Suehiro, H. Imakiire, S. Hidaka, W. D. Ding, G. B. Zhou, K. Imasaka, and M. Hara, "Schottky-type response of carbon nanotube NO_2 gas sensor fabricated onto aluminum electrodes by dielectrophoresis," *Sensors and Actuators B-Chemical*, vol. 114, pp. 943–949, 2006.
20. J. Zhang, A. Boyd, A. Tselev, M. Paranjape, and P. Barbara, "Mechanism of NO_2 detection in carbon nanotube field effect transistor chemical sensors," *Applied Physics Letters*, vol. 88, p. 123112, 2006.
21. G. A. Somorjai, *Introduction to surface chemistry and catalysis*. New York: John Wiley & Sons, Inc., 1994.
22. A. Zangwill, *Physics at surfaces*. Cambridge: Cambridge University Press, 1988.
23. X. D. Cui, M. Freitag, R. Martel, L. Brus, and P. Avouris, "Controlling energy-level alignments at carbon nanotube/Au contacts," *Nano Letters*, vol. 3, pp. 783–787, 2003.
24. A. Javey, R. Tu, D. B. Farmer, J. Guo, R. G. Gordon, and H. J. Dai, "High performance n-type carbon nanotube field-effect transistors with chemically doped contacts," *Nano Letters*, vol. 5, pp. 345–348, 2005.
25. W. Kim, A. Javey, R. Tu, J. Cao, Q. Wang, and H. J. Dai, "Electrical contacts to carbon nanotubes down to 1 nm in diameter," *Applied Physics Letters*, vol. 87, p. 173101, 2005.
26. Z. H. Chen, J. Appenzeller, J. Knoch, Y. M. Lin, and P. Avouris, "The role of metal-nanotube contact in the performance of carbon nanotube field-effect transistors," *Nano Lett.*, vol. 5, pp. 1497–1502, 2005.
27. F. Leonard and J. Tersoff, "Role of Fermi-level pinning in nanotube Schottky diodes," *Physical Review Letters*, vol. 84, pp. 4693–4696, 2000.
28. F. Leonard and J. Tersoff, "Novel length scales in nanotube devices," *Physical Review Letters*, vol. 83, pp. 5174–5177, 1999.
29. F. Leonard and A. A. Talin, "Size-dependent effects on electrical contacts to nanotubes and nanowires," *Physical Review Letters*, vol. 97, p. 026804, 2006.

30. C. Huckstadt, S. Schmidt, S. Hufner, F. Forster, F. Reinert, and M. Springborg, "Work function studies of rare-gas/noble metal adsorption systems using a Kelvin probe," *Physical Review B*, vol. 73, p. 075409, 2006.
31. M. W. Roberts and C. S. McKee, *Chemistry of the metal-gas interface, monographs on the physics and chemistry of materials*. Oxford University Press, 1978.
32. N. Chakrapani, Y. M. M. Zhang, S. K. Nayak, J. A. Moore, D. L. Carroll, Y. Y. Choi, and P. M. Ajayan, "Chemisorption of acetone on carbon nanotubes," *Journal of Physical Chemistry B*, vol. 107, pp. 9308–9311, 2003.
33. J. J. Zhao, A. Buldum, J. Han, and J. P. Lu, "Gas molecule adsorption in carbon nanotubes and nanotube bundles," *Nanotechnology*, vol. 13, pp. 195–200, 2002.
34. K. Bradley, J. C. P. Gabriel, M. Briman, A. Star, and G. Gruner, "Charge transfer from ammonia physisorbed on nanotubes," *Physical Review Letters*, vol. 91, p. 218301, 2003.
35. J. A. Robinson, E. S. Snow, S. C. Badescu, T. L. Reinecke, and F. K. Perkins, "Role of defects in single-walled carbon nanotube chemical sensors," *Nano Letters*, vol. 6, p. 1747–1751, 2006.
36. C. Y. Lee, S. Baik, J. Q. Zhang, R. I. Masel, and M. S. Strano, "Charge transfer from metallic single-walled carbon nanotube sensor arrays," *Journal of Physical Chemistry B*, vol. 110, pp. 11055–11061, 2006.
37. C. Y. Lee and M. S. Strano, "Understanding the dynamics of signal transduction for adsorption of gases and vapors on carbon nanotube sensors," *Langmuir*, vol. 21, pp. 5192–5196, 2005.
38. Y. C. Tseng, K. Phoa, D. Carlton, and J. Bokor, "Effect of diameter variation in a large set of carbon nanotube transistors," *Nano Letters*, vol. 6, pp. 1364–1368, 2006.
39. J. Y. Park, S. Rosenblatt, Y. Yaish, V. Sazonova, H. Ustunel, S. Braig, T. A. Arias, P. W. Brouwer, and P. L. McEuen, "Electron–phonon scattering in metallic single-walled carbon nanotubes," *Nano Letters*, vol. 4, pp. 517–520, 2004.
40. X. L. Liu, Z. C. Luo, S. Han, T. Tang, D. H. Zhang, and C. W. Zhou, "Band engineering of carbon nanotube field-effect transistors via selected area chemical gating," *Applied Physics Letters*, vol. 86, p. 243501, 2005.
41. H. Chang, J. D. Lee, S. M. Lee, and Y. H. Lee, "Adsorption of NH_3 and NO_2 molecules on carbon nanotubes," *Applied Physics Letters*, vol. 79, pp. 3863–3865, 2001.
42. Y. Takao, K. Miyazaki, Y. Shimizu, and M. Egashira, "High amonia sensitive semiconductor gas sensors with double layer structure and interface electrode," *Journal of Electrochemical Society*, vol. 141, pp. 1028–1034, 1994.
43. Y. Shimizu and M. Egashira, "Basic concepts and challenges of semiconductor gas sensors," *MRS Bulletin*, vol. 24, pp. 18–24, 1999.
44. A. Mandelis and C. Christofides, *Physics, chemistry and technology of solid state gas sensor devices*. New York: Wiley, 1993.
45. Y. J. Lu, C. Partridge, M. Meyyappan, and J. Li, "A carbon nanotube sensor array for sensitive gas discrimination using principal component analysis," *Journal of Electroanalytical Chemistry*, vol. 593, pp. 105–110, 2006.
46. C. Staii and A. T. Johnson, "DNA-decorated carbon nanotubes for chemical sensing," *Nano Letters*, vol. 5, pp. 1774–1778, 2005.
47. E. S. Snow, J. P. Novak, P. M. Campbell, and D. Park, "Random networks of carbon nanotubes as an electronic material," *Applied Physics Letters*, vol. 82, pp. 2145–2147, 2003.
48. Y. J. Lu, J. Li, J. Han, H. T. Ng, C. Binder, C. Partridge, and M. Meyyappan, "Room temperature methane detection using palladium loaded single-walled carbon nanotube sensors," *Chemical Physics Letters*, vol. 391, pp. 344–348, 2004.
49. J. P. Novak, E. S. Snow, E. J. Houser, D. Park, J. L. Stepnowski, and R. A. McGill, "Nerve agent detection using networks of single-walled carbon nanotubes," *Applied Physics Letters*, vol. 83, pp. 4026–4028, 2003.
50. E. S. Snow and F. K. Perkins, "Capacitance and conductance of single-walled carbon nanotubes in the presence of chemical vapors," *Nano Letters*, vol. 5, pp. 2414–2417, 2005.
51. E. S. Snow, F. K. Perkins, and J. A. Robinson, "Chemical vapor detection using single-walled carbon nanotubes," *Chemical Society Reviews*, vol. 35, pp. 790–798, 2006.

Chapter 9
Single–Walled Carbon Nanotubes for High Performance Thin Film Electronics

Qing Cao, Coskun Kocabas, Matthew A. Meitl, Seong Jun Kang, Jang Ung Park and John A. Rogers

9.1 Introduction and Motivation

Although the great majority of work on single-walled carbon nanotube (SWNT) electronics has focused on devices and test structures that incorporate individual tubes as the active components [1–6], it is likely that realistic technology applications will require systems that involve large numbers of tubes, in the form of random networks or aligned arrays or something in between. These types of SWNT based monolayer or sub-monolayer 'films' avoid many of the challenges of single tube devices because they (i) offer attractive statistics that minimize device-to-device variations even with electronically heterogeneous tubes, (ii) provide large active areas and high current outputs, due to the large number of tubes involved in transport, and (iii) do not require, in many cases, precise spatial positioning of individual tubes [7–9]. Potential applications that could derive from a successful effort in SWNT thin film electronics range from enhanced, heterogeneous versions of existing single crystal inorganic semiconductor technologies, e.g., Si complementary metal-oxide-semiconductor (CMOS), to high performance large area electronics as replacements for amorphous Si (α-Si) based systems, e.g., backplanes for liquid crystal display televisions, to newer, mechanically flexible circuits for emerging devices, such as paperlike displays, conformable antenna structures, and structural health monitors [10, 11]. These last two areas, sometimes referred to as *macroelectronics*, might represent the most realistic short/medium term goals [12, 13]. Currently, α-Si, low temperature polycrystalline silicon, and organic semiconductors represent the most widely explored materials for the semiconductor components of these systems [14–16]. The modest carrier mobilities, however, preclude their use for high frequency circuits, such as those needed in large area communication

J.A. Rogers (✉)
Department of Chemistry, Department of Physics, Department of Materials Science and Engineering, Department of Electrical and Computer Engineering, Department of Mechanical Science and Engineering, Beckman Institute and Frederick Seitz Materials Research Laboratory, University of Illinois at Urbana-Champaign, Urbana, IL 61801, USA

A. Javey, J. Kong (eds.), *Carbon Nanotube Electronics*, Series on Integrated Circuits and Systems, DOI 10.1007/978-0-387-69285-2_9,

devices. In addition, they lack other optical and mechanical attributes that might be attractive for transparent and/or stretchable electronic systems [17, 18].

For these sorts of applications, we believe that films of SWNTs have properties, due to the excellent electrical [19–21], optical [22, 23], and mechanical [24–27] characteristics of the individual tubes that make them attractive relative to other materials choices. In addition to good electrical properties that can allow, for example, high current outputs and high frequency operation even at relatively large channel lengths (L_C), the SWNTs are compatible with large area, low cost device fabrication techniques, such as transfer [28–30] and ink-jet printing [31]. Furthermore, since nanotubes can exhibit either metallic or semiconducting properties [4], they can be used as conducting and/or semiconducting films in these systems [32, 33]. The SWNTs are thermally and chemically robust (especially compared to organic semiconductors, for example) [34, 35], and their lack of dangling bonds makes them compatible with a variety of dielectrics including high-k materials [36–39]. At the same time, the electrical properties of SWNTs are sensitive to their surroundings, due to the large surface area to volume ratio, so that charge transfer doping can be used to tune their polarities [40–44] and various sensors based on SWNT films can be constructed [45–48]. The extraordinary mechanical robustness of SWNTs is ideal for flexible/stretchable electronic systems and other devices, such as strain gauges [11, 49]. Small optical absorption cross-sections [50] make SWNT films with low to moderate densities almost invisible to the eye [32, 51–53], which could be important for certain security applications and backlit displays.

In spite of these attractive features, many additional advances are needed in practical aspects of film formation and device fabrication as well as in fundamental theories of transport and percolation and their dependence on tube configuration [54, 55], especially at high frequencies. This chapter describes some research, with an emphasis on our own efforts, that address these and other challenges as carried out during the relatively short time ($\sim$3 years) since the earliest papers in this area [9, 32, 56, 57]. We begin with methods to form SWNT films and then discuss the physics and materials science associated with these systems. Unusual fabrication techniques and materials, developed specifically for SWNT TFTs, and some examples of high performance devices on plastic substrates are then described. We conclude with some discussion of future directions and remaining challenges in SWNT-based high-performance thin film electronics.

9.2 Film Formation Techniques

Approaches to use SWNTs in thin film electronics (e.g., thin-film transistors, TFTs) begin with the formation of monolayer or sub-monolayer films of SWNTs on desired substrates. The techniques for forming these films should provide high levels of control over the tube density (D, as measured in the number of tubes per unit area for random network films or tubes per length for aligned arrays), the overall spatial layout of the films and, preferably, the tube orientation. Ideally, they should also be compatible with large areas and low cost processing, for some of the applications mentioned in the introduction. Known synthetic routes to SWNTs yield collections of tubes that contain distributions of diameters and mixtures of metallic (m-SWNTs) and semiconducting tubes (s-SWNTs), usually in a ratio of 1:2. This feature, especially in random network geometries, leads to electronic properties of the films that depend strongly on *D*. For example, at suitably moderate/low *D*, the s-SWNT, but not the m-SWNTs, can form percolating networks. SWNT films in this case show semiconducting properties, as measured on length scales large compared to the average tube length. We refer to such films as semiconducting carbon nanotube networks (s-CNNs). At high *D*, there are sufficient numbers of m-SWNTs to create purely metallic pathways through the network. Such films behave like conductors and we refer to them as metallic carbon nanotube networks (m-CNNs). The many tube–tube contacts present in random network films of SWNTs may limit charge transport. For this reason, film formation techniques that provide control over tube orientation to enable large scale aligned arrays of SWNTs are important. These arrays avoid or minimize tube–tube contacts, thereby offering the possibility to provide levels of electronic performance that approach those of individual, pristine tubes. Many features of transport in networks and arrays appear in Sections 9.3 and 9.4.

The present section presents two general classes of methods to fabricate SWNT films, both of which enable some degree of control over spatial layout, *D* and orientation. The first involves synthesis of the SWNTs, typically in bulk quantities, using techniques such as high pressure CO (HiPco) synthesis or laser ablation, followed by formation of solution suspensions and casting onto target substrates or onto stamps for printing. The other approach uses direct synthesis of films of SWNT, with techniques such as chemical vapor deposition (CVD), either directly onto device substrates or on other substrates from which they can be physically transferred, using printing-like processes.

9.2.1 Solution Deposition Methods

Depositing SWNTs from solution provides a strategy for film formation that naturally scales to large areas, and is compatible with patterning techniques such as ink jet and soft lithographic printing. This approach typically uses SWNTs that are synthesized by bulk processes (e.g., HiPco, laser-oven, etc.) and then suspended in organic solvents or stabilized in aqueous surfactant solutions followed by some possible purification processes [58–60]. The SWNTs can be deposited from these suspensions onto desired substrates (e.g., device substrates or stamps from which they can be printed to other substrates) by solution casting [61]. A controlled flocculation approach enables single step deposition with controlled densities and orientations on a wide range of substrates, including those with little specific affinity for the SWNTs. The sections below describe this technique and its application to film formation by spin casting, laminar flow patterning, and stamp based printing.

9.2.1.1 Solution Casting via Controlled Flocculation

A successful strategy for casting films of SWNTs should employ a reliable means to form stable solutions of SWNTs and a robust mechanism to remove them from solution, such as by evaporation or by specific interactions between nanotubes, ligands, or surfaces. One approach accomplishes these goals by using known surfactants (e.g., sodium dodecylsulfate, SDS, or sodium dodecylbenzensulfonate, SDBS) to suspend SWNTs in aqueous solutions followed by adding other liquids (e.g., methanol or some other solvent miscible with water) to drive them out of solution during the casting step [61, 62]. Methanol added to a surfactant-stabilized SWNT solution interacts with the surfactant and disrupts its capacity to stabilize the SWNTs, allowing them to flocculate. Van der Waals forces cause neighboring SWNTs to aggregate and to adhere to adjacent surfaces. Confining the fluids close to the surface of a target substrate as they mix produces uniform films of SWNTs. We refer to this approach as controlled flocculation (cF) [62]. The confinement may be accomplished in several different ways. In one example, streams of methanol and SWNT solution simultaneously impinge onto the center of a rapidly spinning substrate that pulls them into a thin liquid film and receives the destabilized SWNTs [61]. The rapid mixing and vertical confinement of the two liquids favor the formation of uniform coatings in the form of individual or minimally bundled SWNTs (Fig. 9.1(a)). In another implementation, the deposition is confined in the plane of the substrate by laminar flows in microfluidic channels [62]. The fluids flow side-by-side in a microchannel and mix by diffusion only in a narrow region near the interface between the two liquids. SWNTs deposit in this region onto the substrate, forming a patterned film (Fig. 9.1(b)) [62]. In both the spinning and laminar flow methods, SWNTs do not deposit without the addition of methanol.

Shear forces associated with fluid flows can align the SWNTs, especially for low D films, in which most of the tubes align to within about 10°, as illustrated in the atomic force microscope (AFM) images of Fig. 9.1(c) [62]. Formation of films with Ds that range from a small fraction of a monolayer to thick, multilayer coatings can be achieved by simply increasing the duration of the procedure or the relative amounts of SWNT suspension and methanol. These techniques can deposit SWNT films onto a wide range of substrates with different surface chemistries, including low energy surfaces, like those of polydimethylsiloxane (PDMS, see Section 9.2.1.2) [61, 62]. The surface chemical properties, however, do influence the quality of the deposited films. For example, the D and uniformity of films deposited onto amine-terminated surfaces are slightly higher than those deposited onto bare silica surfaces [61]. Equally important is the smoothness of the target surface [61, 62]. Smoother surfaces receive the SWNTs more readily, due to stronger surface interactions, resulting in denser films. The effects of surface roughness on the deposition of SWNTs can be quite dramatic [62].

9.2.1.2 Printing Solution-Cast SWNT from a Stamp

The ability to deposit SWNT films onto low energy surfaces makes it possible to print those films simply by contact onto higher energy surfaces [61, 63]. This type of transfer printing approach is attractive because it is a completely additive, dry process in which the target substrate is not exposed to any liquid during the deposition.

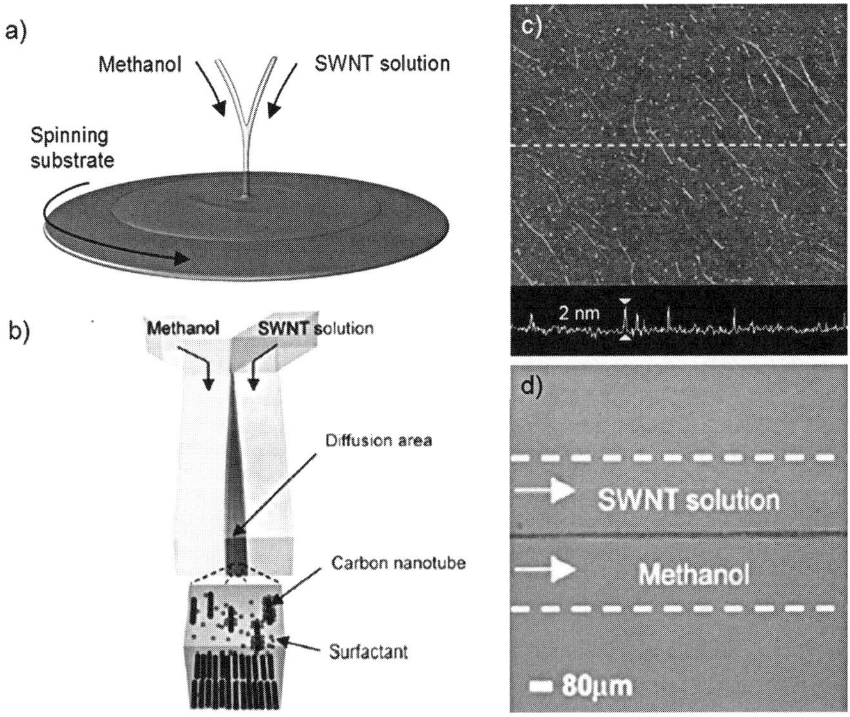

Fig. 9.1 Deposition of films of SWNTs by cF. (**a**) Schematic illustration of the deposition of uniform films by mixing methanol and a solution of SWNT on a rapidly spinning substrate. (**b**) Schematic illustration of the deposition of films in line geometries by mixing methanol and a solution of SWNTs in the inter-diffusion region of a laminar flow microfluidic cell. (**c**) AFM image of a film of aligned SWNTs deposited by cF on a spinning wafer. The SWNTs show radial alignment. The inset (bottom) shows a line trace revealing the heights of individual SWNT and small bundles. Reproduced with permission from [61]. Copyright 2004 American Chemical Society. (**d**) Optical micrograph of a SWNT film in the geometry of a line (*dark grey* in the center of the image) deposited underneath the inter-diffusion region of a microfluidic cell, as illustrated in (**b**). Reproduced with permission from [62]. Copyright 2006 Wiley-VCH

Soft PDMS elastomers are useful materials for stamps, due to their low energy surfaces and their ability to conform to smooth planar or non-planar substrates [64]. Figure 9.2(a) outlines the procedure, in which the methods described in Section 9.2.1.1, form a coating of SWNTs on a PDMS stamp. Contacting a stamp inked in this fashion to a target substrate results, upon removal of the stamp, in the transfer of the tubes. This method defines patterns of SWNT films in geometries defined by the relief features on the stamp. As in the solution methods described in the previous section, the SWNT films can be printed onto a wide variety of substrates. The quality of the printed films depends on the chemistry and surface roughness of the target surface [61]. Figure 9.2(b) shows orthogonal lines of low-D SWNT films formed in two sequential printing steps. The scanning electron microscope (SEM) image of

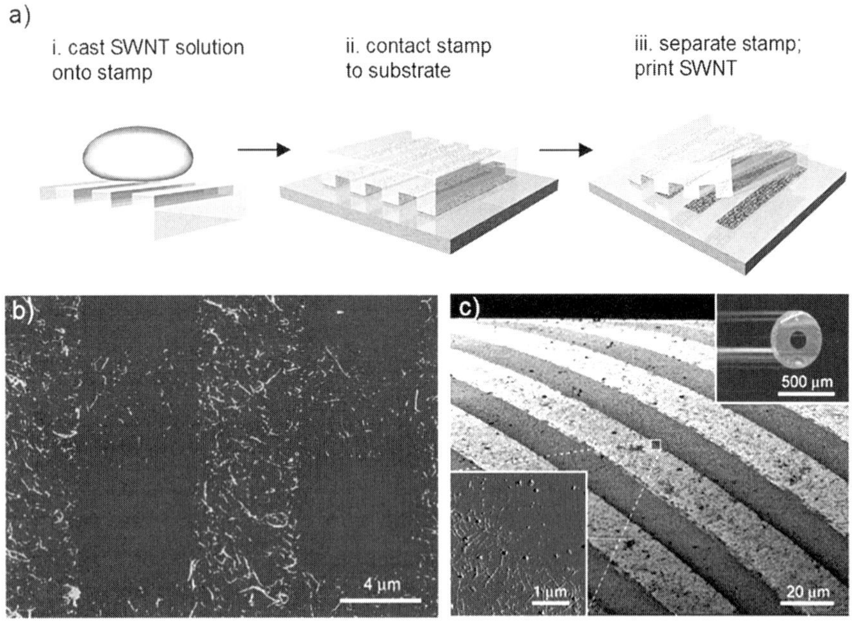

Fig. 9.2 Transfer printing SWNTs using PDMS stamps. (**a**) Schematic illustration of the inking and printing process. (**b**) AFM image of crossed line patterns of SWNT films formed in two printing steps. (**c**) SEM image of lines of SWNT films printed onto a glass capillary tube (outer diameter = 0.5 mm). The insets show a photograph and AFM image. Reproduced with permission from [61]. Copyright 2004 American Chemical Society

Fig. 9.2(c) shows lines of SWNTs printed by rolling a glass capillary tube over a SWNT-coated PDMS stamp [61].

9.2.2 Chemical Vapor Deposition Growth

Nanotube films can also be formed directly by CVD growth [65, 66]. The SWNTs that result from this process exhibit, compared to those derived from the types of solution techniques described in the previous section, a high level of structural perfection, long average tube lengths, high purity and relative absence of tube bundles. The CVD method provides opportunities to control D, morphology, alignment and position at the growth step [67]. The result, as described in detail below, is that it is possible by CVD to achieve levels of alignment and linearity that approach perfection, to an extent that is unlikely to be possible using deposited tubes. Although not as convenient for large area substrates as solution approaches, CVD methods are intrinsically scalable for realistic applications, as evidenced by their widespread use for other materials in various areas of electronics [68, 69]. The following sections describe the growth of films that consist of random networks of SWNT on amorphous substrates, with an emphasis on control of *D*. The latest results on guided growth of aligned arrays of SWNTs on single crystal substrates are also presented.

9.2.2.1 Unguided Growth on Amorphous Substrates

The chemical and morphological properties of the catalysts and the composition of the feeding gas control *D* in CVD films. For example, with ethanol as the carbon feedstock, *D* can be significantly increased, compared to the case of methane, possibly due to the ability of oxygen to tune the ratio between the carbon and hydrogen radicals in the growth environment (Fig. 9.3(a) and (b)) [70]. The nature of the catalyst is also important. For example, catalysts that use Fe/Co/Mo on silica supports [71] yield densities higher than those obtained from discrete iron nanoparticle catalysts [72], due to increased surface area, pore volume, and catalytic activity (Fig. 9.3(b) and (c)) [73]. The concentration of the catalyst in either case provides an additional route to control *D* (Fig. 9.3 (c)–(f)) [9, 29]. By suitable selection of parameters, it is possible to grow high quality films of SWNTs with uniform and defined densities, from less than one tube/μm^2 to hundreds of tubes/μm^2, over large areas. Figure 9.3 shows images of representative results. These capabilities provide the basis for the development of SWNT-based high-performance thin-film electronics.

9.2.2.2 Guided Growth on Certain Crystal Substrates

We recently developed a convenient process for generating large-scale, horizontally aligned arrays of SWNTs [74]. The approach uses guided growth of SWNTs

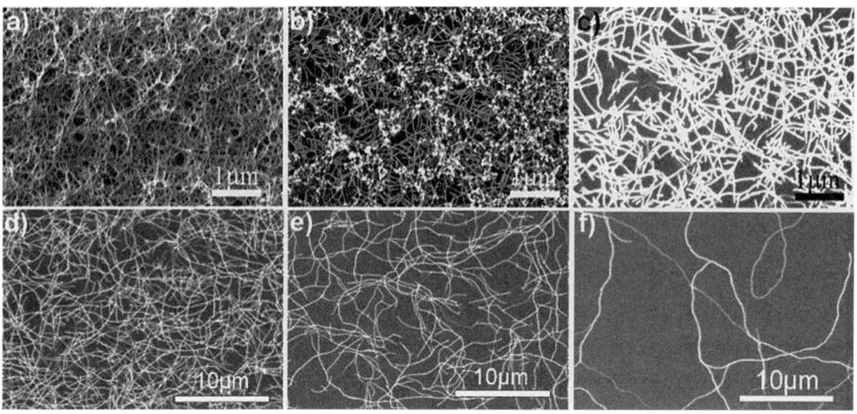

Fig. 9.3 SEM images of SWNT films grown by CVD with ethanol (**a**) and methane (**b**), using Fe/Co/Mo catalysts on silica supports. Images of films formed with methanol using ferritin catalysts deposited by spin coating from solutions with concentrations of 3.8 mg/ml (**c**), 0.38 mg/ml (**d**), 38 µg/ml (**e**), and 7.6 µg/ml (**f**)

by CVD on single-crystal quartz substrates. Studies of the process reveal the dependence of D and tube alignment on the growth conditions and the morphology and crystal cut of the quartz. The layouts of the tubes can be controlled to yield nearly any arrangement, from perfectly aligned arrays of perfectly linear tubes to nearly random networks of meandering tubes. Optimized procedures can yield well aligned arrays over large areas and with D up to several, and perhaps more, SWNTs/μm [74]. The highest levels of alignment and high D are achieved with catalysts patterned into narrow stripes, such that the tubes grow primarily in regions of the substrate that are uncontaminated by unreacted catalyst particles [75]. Random networks of SWNTs appear in the regions of high coverage of catalyst. By combining these features, CVD growth can generate, in a single step, random network films aligned and electrically interfaced to aligned arrays. This capability can be useful for certain devices, such as those that use the networks and arrays for conducting and semiconducting elements, respectively. Figure 9.4 shows some representative results. Note that few SWNTs emerge from edges of the pads that lie along the preferred growth (Fig. 9.4(d)), consistent with a single, strongly preferred growth direction.

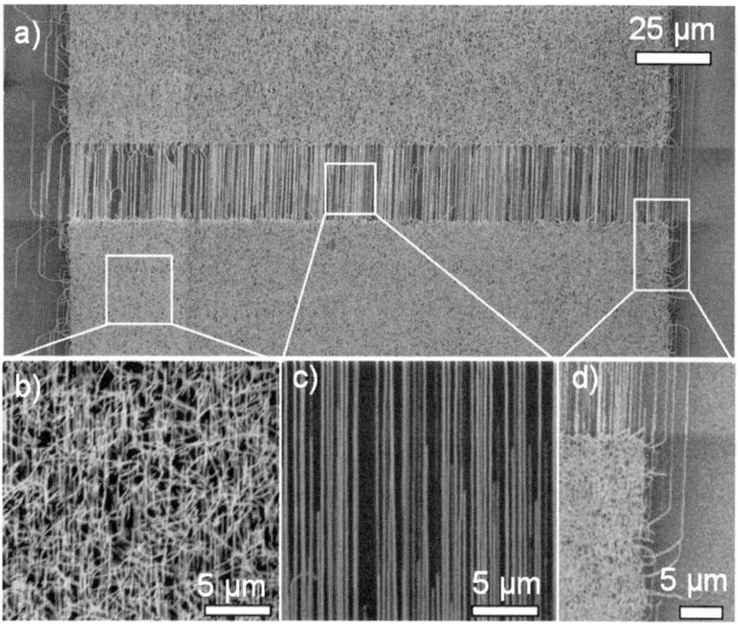

Fig. 9.4 SEM images of random network and aligned array SWNT films grown in a single step by CVD on quartz. The ferritin catalyst exists only in the rectangular regions at the top and bottom of the image in (**a**). High magnification SEM images of network (**b**), and array (**c**) regions of this structure of SWNTs. SEM image of a corner of the network (**d**) which illustrates strongly preferential growth in the vertical direction. Reproduced with permission from [75]. Copyright 2006 American Chemical Society

Perfectly aligned arrays of perfectly linear SWNTs can be generated using optimized versions of this patterned catalyst technique [75]. Figure 9.5(a) and (b) show SEM images of such aligned SWNT films, grown from catalyst patterned into narrow stripes orientated perpendicular to the preferred growth direction. The images show excellent alignment and linearity in tubes with lengths in the range of 100 μm and in uniform densities over large areas (up to 2.5 × 8 cm, limited by the CVD chamber.) Most of the tubes (75%) span the gap between adjacent catalyst stripes. AFM images (Fig. 9.5(c)) give detailed information concerning the linearity and distribution of tube diameters. The data indicate diameters between 0.5 nm and 4 nm, centered at 1 nm. For a given tube, the diameter varies by less than 0.1 nm along its length, limited by the resolution of the AFM. Also within the uncertainty of the AFM, the tubes are perfectly linear. Figure 9.5(d) shows the deviation of the center of a representative tube from a perfect linear shape, evaluated at several locations

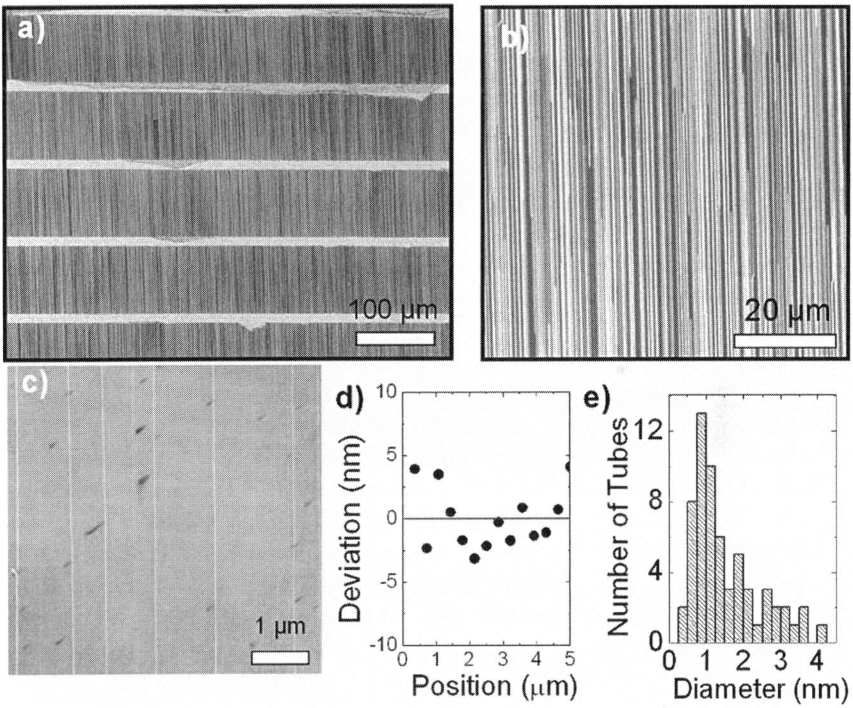

Fig. 9.5 Images and characteristics of SWNT films that consist of perfectly aligned, perfectly linear tubes. (**a**, **b**) Low- and high-resolution SEM images of aligned arrays of SWNTs grown by CVD with methanol and Fe catalyst patterned into 10 μm wide stripes (*bright horizontal lines* in (**a**)) on quartz. (**c**) AFM image of aligned SWNTs in the array. The tube lengths are ∼100 μm, limited by the spacing between the catalyst stripes. (**d**) Deviation of the center of a single tube from a perfect linear shape, as a function of position along its length. To within the uncertainty of the AFM, the shape is perfectly linear. (**e**) Histogram of diameters of SWNTs in the arrays. Reproduced with permission from [89]. Copyright 2007 Nature Publishing Group

along the tube. The maximum deviation is less than 10 nm, comparable to the resolution of the AFM. The tubes in the array are parallel to one another to better than 0.1 degrees.

9.3 Physical Properties and Device Physics

The unusual and promising properties of films of networks and arrays of SWNTs motivate efforts to understand the nature of charge transport through them. This section describes the physics of percolation transport in high *D* conductive SWNT films as well as in low *D* SWNT networks or partially aligned arrays where substantial, non-percolative, transport occurs through individual tubes. Some features associated with the electrostatic coupling of such films to gate electrodes in transistor devices are then summarized, along with approaches to use chemical functionalization/polymer coatings to control the operation of such devices. The section concludes with a summary of some of the unique optical and mechanical properties of SWNT films and devices formed from them.

9.3.1 Conducting Films of SWNTs

The high intrinsic conductivities and aspect ratios of m-SWNTs lead to low percolation thresholds in networks, thereby making such films, with sufficiently high Ds, attractive as conducting layers. Such m-CNN can achieve sheet resistances, R_S, and optical transmittances comparable to those of films of indium-doped tin oxide (ITO), but with superior mechanical properties and the ability to be integrated onto a wide range of substrates [32, 51, 76–79]. Methods described in the preceding sections, as well as those such as vacuum-filtration [32, 63], can be used to form m-CNNs. The long, pristine tubes grown by CVD yield conducting films with excellent properties [77, 80]. Transfer methods, described subsequently, allow these films to be integrated with substrates other than the growth substrate. For example, high *D* SWNT films grown by CVD, can be transferred to transparent plastic substrates to yield conductive transparent nanotube films for transparent TFTs (TTFTs) that use SWNT films or organic thin films for the semiconductor (Fig. 9.6) [77, 80].

The dependence of R_s on *D* can be described by standard percolation theory [78],

$$R_s = k(D - N_c)^{\alpha}$$

where k is a fitting constant; N_c is the percolation threshold and α is a parameter determined by the spatial arrangement of SWNTs in the film. Data suggest that the adhesive force between SWNTs leads to low percolation thresholds by increasing the contact lengths between tubes [81].

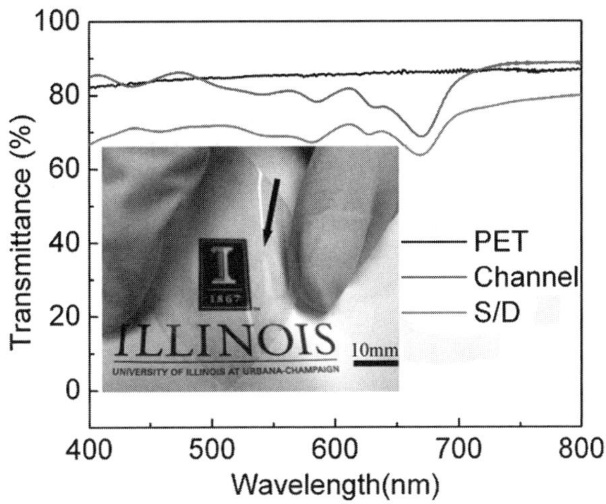

Fig. 9.6 Transmission properties of organic TTFTs that use m-CNN electrodes on plastic. The data show spectra for the (PET) substrate, and a TTFT as measured through the m-CNN source/drain (S/D) electrodes and through the transistor channel. Inset: optical image of an array of organic TTFTs, positioned above printed text on paper to illustrate the degree of optical transparency. The *arrow* indicates the S/D structures, which appear as *faint grey squares*. Reproduced with permission from [80]. Copyright 2006 American Institute of Physics

9.3.2 Semiconducting Films

Fundamental, predictive knowledge of the physics of transport through moderate/low D SWNT films is important to interpret and optimize electrical performance of these thin films when used as the semiconducting component of SWNT TFTs [54, 55, 82, 83]. Transport in these cases involves the combined effects of (i) transport from source to drain electrode through individual tubes that directly connect these electrodes and (ii) percolation transport through multiple tubes via tube/tube junctions that are present when some degree of misaligned tubes existing in the film. For partially aligned tube arrays, the influence of even a small number of misaligned tubes can be significant, since such tubes can electrically contact many aligned tubes. We recently investigated these effects through systematic sets of experiments and theoretical computations [54]. In these studies, we built devices with a range of Ds and degrees of alignment, using CVD growth on quartz substrates. The average tube lengths, $< L_S >$, in these films (Fig. 9.7) were 40, 22, and 5 μm. The degree of alignment, as defined in terms of an anisotropy parameter, R, where $R = L_1/L_\cdot = \sum_{i=1}^{N} \left|L_{S,i}\cos\theta_i\right| \Big/ \sum_{i=1}^{N} \left|L_{S,i}\sin\theta_i\right|$ were 21.4, 6.0, and 2.9, respectively. Top gate transistors in which the films serve as the semiconductor layer provide a means to evaluate gate modulated transport through them. We evaluated the scaling behavior of 'on' and 'off' currents, I_{on} and I_{off}, as a function of channel length L_C. The theoretical results use a percolation model with sticks having

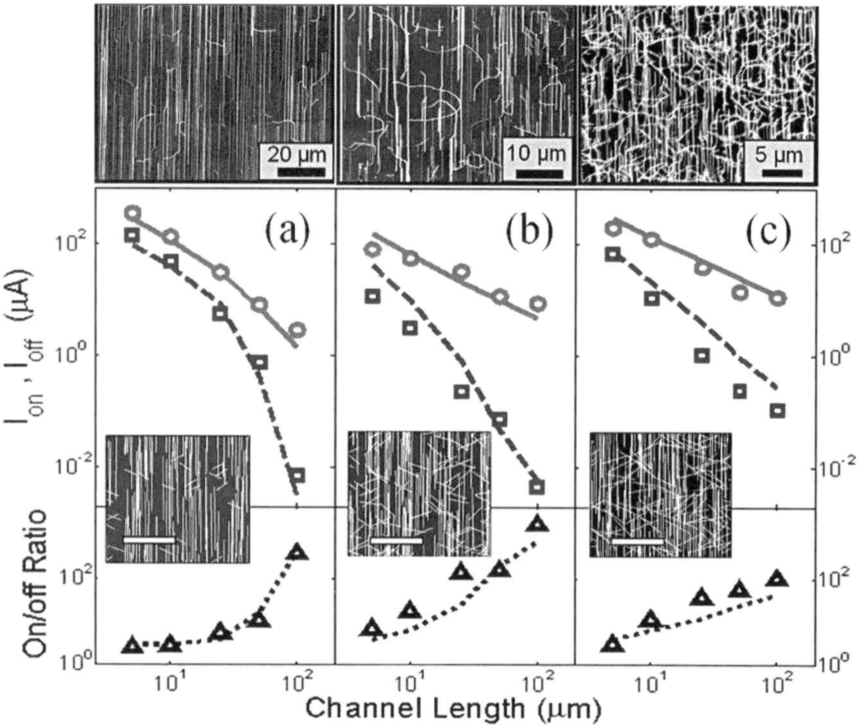

Fig. 9.7 Systematic experimental and theoretical studies of charge transport in SWNT TFTs that use SWNT films with different degrees of alignment and densities. (**a-c**) SEMs (*top frames*) and measured (*symbols*) and computed (*lines*) properties of SWNT TFTs. The films range from well aligned, low coverage (**a**) to partially aligned, high coverage (**c**) cases. The plots show I_{on}, I_{off}, and on/off ratio for (**a**) aligned, (**b**) partially aligned and (**c**) dense partially aligned networks. The insets present images of the simulated networks, where the scale bar has a length of 10 μm. The calculations explain, at a quantitative level, the scaling properties in these devices. Reproduced with permission from [54]. Copyright 2007 American Chemical Society

lengths (L_S) and orientations (R) characterized by a probability density function, all chosen to be consistent with images of the SWNT films. Drift-diffusion theory [82, 84, 85] describes transport within individual sticks, according to $J = q\mu n\, d\varphi/ds$. When combined with current continuity equation, $dJ/ds = 0$, this equation gives the non-dimensional potential φ_i along tube i as $d^2\varphi_i/ds^2 - c_{ij}\,(\varphi_i - \varphi_j) = 0$. Here, s is the length along the tube and $c_{ij} = G_0/G_1$ is the dimensionless charge-transfer coefficient between tubes i and j, and G_0 and $G_1(V_G)$ are mutual and self conductances of the tubes [82]. The network is assumed to contain metallic and semiconducting tubes at a ratio of 1:2. I_{on} and I_{off} are computed by assigning $G_1(\text{metal}) \sim G_1(\text{semi})$, and $G_1(\text{semi}) = 10^{-4} \times G_l(\text{metal})$, respectively. Figure 9.7 shows I_{on}, I_{off}, and the on/off ratio (I_{on} / I_{off}) for aligned (Fig. 9.7(a)), partially aligned (Fig. 9.7(b)) and dense partially aligned (Fig. 9.7(c)) networks. The measured data (symbols) agree with simulation results (lines), including trends such as (i) the on/off ratio increases with L_C (Δ symbols in Fig. 9.7) for all values of R, (ii) the rate of increase in the on/off ratio is steeper for partially aligned networks than for dense partially aligned networks (Fig. 9.7(b) vs. Fig. 9.7(c)) and (iii) I_{on} and I_{off} remain finite for $L_C > L_S$ even for the most well aligned system studied here. These results indicate that coverage and alignment are correlated in their effects on device performance; they must therefore be simultaneously optimized. Also, the trade off between I_{on} (which increases with coverage) and on/off ratio (which increases with decreasing coverage) must be considered for specific applications.

9.3.3 Capacitance Coupling in SWNT TFTs

Although not explicitly considered in the calculations above, the electrostatic coupling of the gate electrode of the transistor to the SWNT film is critically important in device operation. This coupling can be much different, depending on D and separation between the planar gate electrode and the film, from that of a traditional TFT. In particular, the fringing fields and electrostatic screening between neighboring SWNTs can lead to gate capacitances that deviate significantly from the capacitance of a parallel-plate capacitor [86, 87]. A simple model system, consisting of a parallel array of equally spaced SWNTs, can provide a semi-quantitative understanding of the gate capacitance coupling in SWNT TFTs that use films with some degree of misalignment and nonuniform spacings (Fig. 9.8(a)). An analytical expression of gate capacitance, based on single subband quantum limit which assumes that the charge distributes symmetrically around the nanotube, can be obtained, for the case that nanotubes that are fully embedded in a material with the same dielectric constant as the gate dielectric,

$$C_i = \left(\frac{2}{\varepsilon} \log \frac{\Lambda_0}{R_T} \frac{\sinh \pi 2d/\Lambda_0}{\pi} + C_Q^{-1} \right)^{-1} \Lambda_0^{-1}$$

where Λ_0 is the distance between the tubes; d is the dielectric thickness; R_T is the tube radius and $C_Q{}^{-1}$ is quantum capacitance [88]. To explore the classical limit, which allows non-uniform charge distributions on the nanotubes, finite element modeling (FEM) simulation was performed for the same system. The FEM simulations show clearly the screening of electrical field by neighboring tubes (Fig. 9.8(b)). The results obtained in the classical and quantum limits were compared through calculations of the coupling efficiency (Ξ), defined as the ratio between the effective capacitance of the SWNT-array TFT and the plate capacitance $4\pi d/\varepsilon$. In most regimes, these two calculations agree reasonably well (Fig. 9.8(c)) [88]. The validity of these models has been confirmed, qualitatively, through experiments on SWNT TFTs with a range of dielectric thicknesses [88].

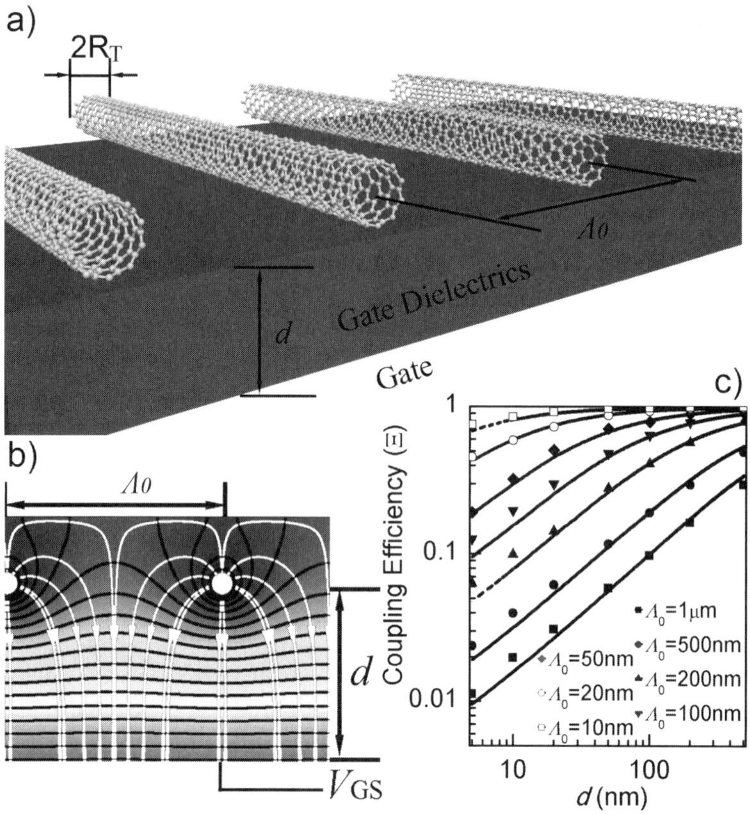

Fig. 9.8 Electrostatic coupling of films of SWNTs to gate electrodes in transistor structures. (**a**) Schematic illustration of the model system used for the calculations. (**b**) FEM computed electrostatics for this system. The *white* and *black lines* correspond to field and equipotential lines, respectively. (**c**) Coupling efficiency (Ξ) versus gate dielectric thicknesses (d) for various inter-tube distances (Λ_0) ranging from 10 nm to 1 μm, computed with FEM (*symbols*) and an analytical expression (*lines*). Reproduced with permission from [88]. Copyright 2007 American Institute of Physics

9.3.4 Control of Electronic Properties

As fabricated, SWNT TFTs with the most well established metals for S/D electrodes exhibit unipolar *p*-channel behavior. For CMOS circuits, it is necessary to achieve *n*-channel operation. In addition, due to the presence of m-SWNT, the on/off ratios are low for systems other than those that involve moderate/low *D* random or partially aligned networks with long channel lengths. This section describes some chemical approaches that address these two issues. In particular, it presents chemistries for selective chemical functionalization of metallic tubes that enable large increases in the on/off ratios of SWNT TFTs. It also describes methods for using polymer coatings to switch the operation of the devices from unipolar p-channel to unipolar n-channel or ambipolar operation.

9.3.4.1 Selective Removal, Functionalization of Metallic Tubes

Separating m-SWNTs from s-SWNTs represents a challenge for nearly all applications of SWNTs. Direct electrical breakdown of m-SWNT is effective in increasing device on/off ratios of SWNT TFTs but this method might not be suitable for complex circuits in which independent electrical access to every TFTs is not possible [9, 61, 89]. Approaches based on electrochemical functionalization have this same limitation [90, 91]. One solution relies on the preferential reaction of a diazonium compound with m-SWNTs, which is thought to be due to stabilization of the charge transfer complex by the finite density of states near Fermi level of the m-SWNTs [92, 93]. Because the functionalization renders the metallic tubes insulating, without altering the properties of the s-SWNTs, this approach can yield SWNT TFTs with high on/off ratios.

This chemistry and its effects on individual tubes in transistor devices and collections of tubes in SWNT TFTs were examined systematically using Raman and electrical measurements [94]. Raman spectra clearly show higher reaction rates in m-SWNTs than in s-SWNTs where the chiralities of nanotubes were assigned based on the peak position of radial breathing mode (RBM) in Raman spectra [95, 96]. At moderate concentrations of diazonium salt, e.g., 5 μM for the conditions studied, the intensity of the disorder mode in m-SWNTs at $\sim$1300 cm^{-1} increases, which suggests an increase in sp^3 carbon in the nanotubes [92, 97]. At the same time, the tangential mode at $\sim$1590 cm^{-1} decreases and at $\sim$169 cm^{-1} disappears, both of which suggest an increase in the level of structural defects [92, 97]. No significant changes appear with s-SWNTs at this concentration. At higher concentrations, e.g., 10 μM for the conditions studied, Raman indicates that similar reactions begin to take place with the s-SWNTs. This observation is consistent with electrical measurements on functioning devices. At moderate concentrations, I_{on} and I_{off} decrease by similar amounts, consistent with selective elimination of conduction pathways through the m-SWNTs. The result is a sharp increase in the on/off ratio without significantly reducing the device mobility (Fig. 9.9(c) and (d)). These observations are promising, but the range of concentrations that lead to reactions with m-SWNTs but not with s-SWNTs is small, especially for devices that use SWNTs with a wide distribution of diameters and chiralities. This delicate balance reduces the practical value of this method. Other similar chemistries might be developed to eliminate this limitation.

9.3.4.2 Chemical Modification of Transport

Transport in SWNTs is known to be sensitive to their surrounding environment due to the high surface to volume ratios [98, 99]. SWNT TFTs that use as-grown or as-deposited nanotube networks/arrays typically exhibit unipolar p-channel behavior when built with high work function metals for S/D contacts due to the presence of

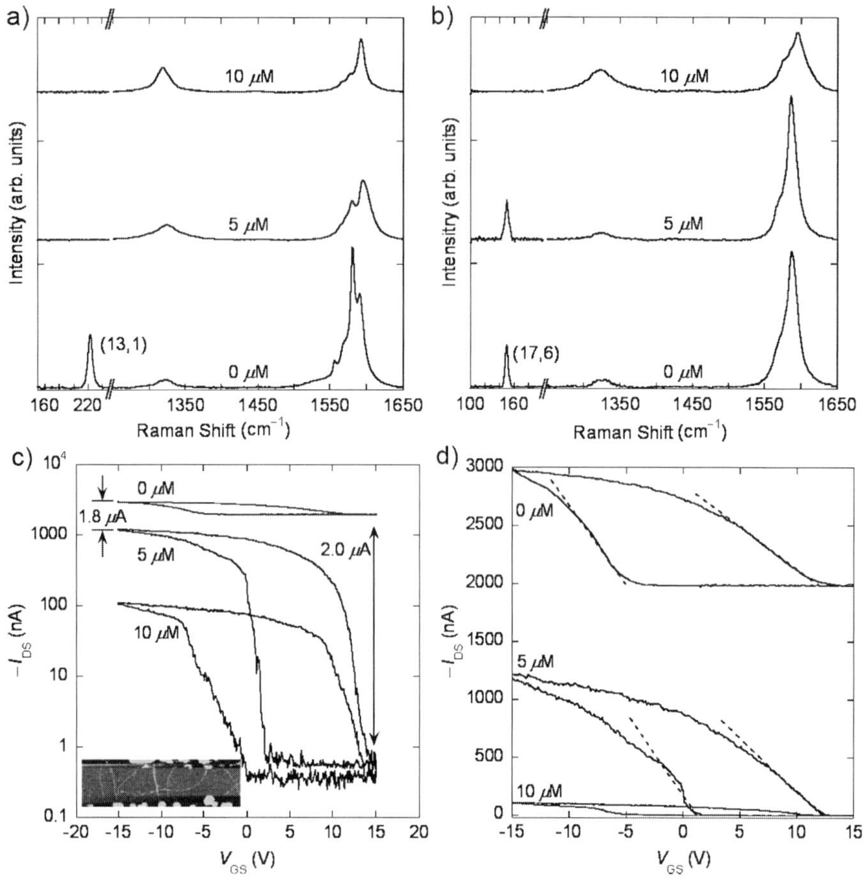

Fig. 9.9 Raman and electrical data from SWNTs and SWNT TFTs before and after functionalization with diazonium salts. Raman spectra of individual m-SWNT (**a**) and s-SWNT (**b**) before and after functionalization at the indicated concentrations. The intensities are normalized to the substrate Si Raman peak at 940 cm^{-1}. The results show chemical modification to the m-SWNT at concentrations lower than those needed to induce similar changes in the s-SWNT. Transfer characteristics of a SWNT TFT before and after functionalization ($V_{DS} = -0.1$ V) plotted in logarithmic scale (**c**) and linear scale (**d**). The inset in (**c**) is an AFM image of the channel region showing that most tubes directly span the S/D electrodes (1 μm separation). The dashed lines illustrate the slope used to extract the transconductance of forward and reverse scan directions. Reproduced with permission from [94]. Copyright 2005 American Chemical Society

Schottky barriers (SBs) at the contact. Such devices can be converted to n-channel or ambipolar modes when annealed and operated in vacuum, but this method for controlling transport cannot be easily used in realistic circuit applications [100]. Similar changes can be achieved by doping with alkali metals, although this approach is also practically unattractive [101, 102]. On the other hand, charge transfer doping with amine containing molecules/polymers provides a convenient means to achieve the same outcome, as initially demonstrated in single tube devices [40, 41, 44]. This strategy also works for SWNT TFTs with conventional gate dielectrics [9, 29] as well as those that use polymer electrolytes [42]. In particular, uniformly coating the channel region with low molecular weight polyethyleneimine (PEI) leads to unipolar n-channel behavior in as-fabricated p-channel devices (Fig. 9.10(a) and (b)). In a similar way, ambipolar operation can be achieved with coatings of polyethylene oxide (PEO) (Fig. 9.10(c)) [29]. These behaviors are thought to arise from changes in the electrical properties of nanotubes themselves, due to the polymer coatings [29, 43]. The effective device mobilities of n-channel devices that result from this process are generally somewhat inferior to those of their p-channel counterparts (Fig. 9.10(d)), possibly because of partial coating/interaction of the PEI with the tubes or residual electron withdrawing species adsorbed onto the devices prior to coating [29]. This ability to control the device polarity by simple application of polymer coatings represents an advantage of SWNT TFTs compared to organic TFTs, where completely different chemistries for the semiconducting materials are typically needed for p-channel and n-channel devices. The disadvantages of this coating approach are that (1) it is not readily compatible with top gate device geometries and (2) the polymers often show effects of degradation near the contacts at high V_{DS}, e.g., $V_{DS}>2V$.

9.3.5 Mechanical and Optical Properties

Although the band gaps of SWNTs are relatively small, SWNT films of the type described in the preceding sections are transparent to visible light because the SWNTs have (i) low, and polarization dependent, optical absorption cross sections [50] due to their small size and high aspect ratio, (ii) low plasma frequency due to low carrier density [32, 85] and (iii) high intrinsic mobilities and conductivities such that even relatively low coverage films provide good electrical properties. For example, m-CNN s that contain tubes grown by CVD to relatively high D show R_S as small as 265 Ω/sq and transmittances larger than 75% in the visible region (Fig. 9.11(a)) [77]. Compared to traditional transparent conductive/semiconducting oxides such as ITO, such SWNT films offer excellent mechanical properties due in part to the intrinsic mechanical properties of the SWNTs, i.e., high elastic moduli (1.36–1.76TP nm/tube diameter) [103] and fracture stresses (100–150GPa) [25]. These features make SWNT films attractive for applications that require high degrees of mechanical bending, such as flexible or conformable electronic systems. To assess the bendability of CVD SWNT networks, changes in currents through TFTs that use s-CNN as the semiconductor and through m-CNN were evaluated as a function of bending of their plastic substrates [104]. For bending radii that create surface strains of up to 1% in compression or tension, little change in the behavior of the TFTs was observed (Fig. 9.11(b)). For radii smaller than these values, the devices failed, due to cracking in the gate electrodes and/or failure in the gate dielectrics. For the resistors, bending radii as small as ~50 μm, corresponding to surface strains as high as ~25%, could be tolerated. These values lie beyond the

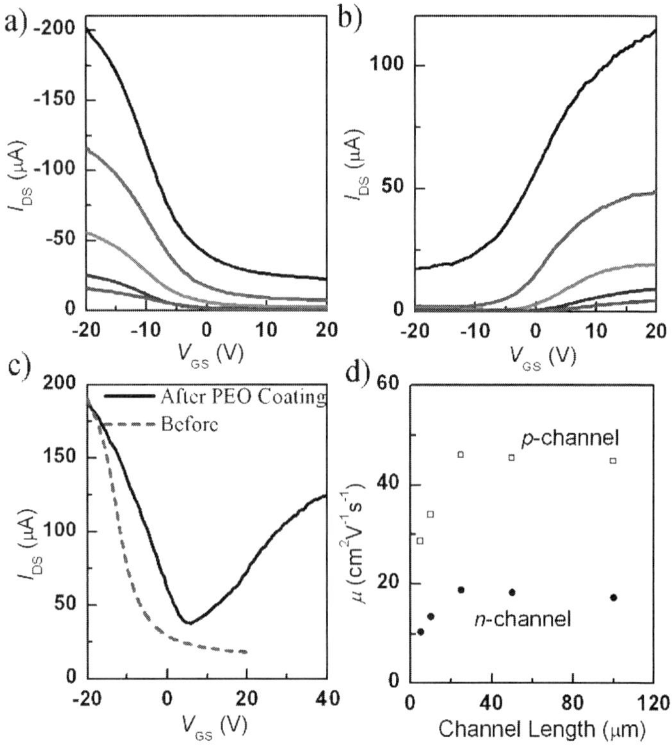

Fig. 9.10 Electrical properties of unipolar p-channel, unipolar n-channel and ambipolar SWNT TFTs achieved with polymer coatings. Transfer curves of a series of SWNT TFTs without coatings (**a**) and SWNT TFTs with uniform coatings of PEI (**b**). The channel lengths were 5 μm, 10 μm, 25 μm, 50 μm, and 100 μm, respectively, from the top to the bottom. (**c**) Transfer curves of a SWNT TFT before (p-channel, *dashed line*) and after (ambipolar, *solid line*) coating with PEO. The channel length was 5 μm. (**d**) Device mobilities (μ) before (p-channel, *open squares*) and after (n-channel, *solid circles*) coating with PEI, for various channel lengths. In all cases V_{DS} was -0.5 V, and the channel widths were 250 μm. Reproduced with permission from [29]. Copyright 2005 American Institute of Physics

limits of plastic deformation in the 25 μm thick poly(ethyleneteraphalate) (PET) substrate, as shown in the inset in Fig. 9.11(b) inset. The current through the m-CNN varied by only a few tens of percent at these high strains [104].

9.4 Devices and Circuits

Despite great progress in fabricating electronic devices and circuits, including transistors [105], solar cells [106], logic gates [107] and ring oscillators [108, 109], that use individual SWNTs, the lack of diameter and position control in the synthesis represents an obstacle to the use of such devices in realistic systems. Films

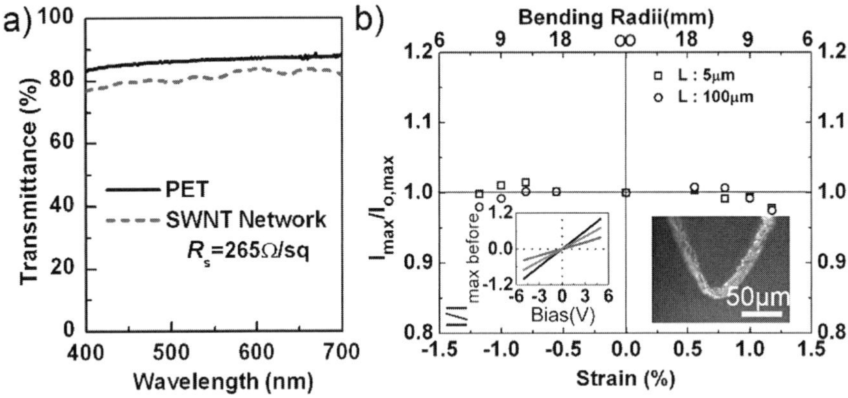

Fig. 9.11 Optical and electrical properties of conductive films of SWNT and of TFTs that use SWNT films as the semiconductor. (**a**) Optical transmittance spectra for a conducting SWNT film on a transparent PET substrate and for the PET. (**b**) Change in normalized current output of a SWNT TFT for various degrees of bending; channel lengths were 5 μm (*squares*) and 100 μm (*circles*). The thickness of the gate dielectric layer (epoxy) was 1.6 μm. Left inset: Current–voltage response of a m-CNN resistor before folding (*top*), in valley folding (compression; *middle*), and in mountain folding (tension; *bottom*). The resistors used SWNT networks transferred onto 25 μm thick PET substrates with electrodes of Ti/Au formed by evaporation through a shadow mask. The length and width of the resistor were 500 μm and 1 mm, respectively. The electrodes are well separated from the fold. Right inset: optical microscope image after sharp folding. Reproduced with permission from [77]. Copyright 2005 American Institute of Physics

of SWNTs, as described in the previous sections, circumvent the assembly problem and provide reproducible averaged electrical properties due to the favorable statistics associated with the large number of active SWNTs in each transistor device. The following sections review some strategies for integrating SWNT films into devices and circuits, and present results on the levels of performance that can be achieved on both rigid and flexible substrates.

9.4.1 Materials and Processing

Although traditional microfabrication techniques and electronic materials are often used to form SWNT TFTs, certain unusual fabrication processes, such as transfer printing techniques, and unconventional materials, such as multilayer nanodielectrics, are needed to meet the special requirements of these devices. This section presents these aspects, as well as the nature of electrical contacts in SWNT TFTs.

9.4.1.1 Transfer Techniques

Most CVD procedures for SWNTs require high temperatures, generally above 800°C, thereby preventing the direct growth of nanotubes on plastic and other potentially interesting materials for substrates. Although solution suspended tubes can be deposited at room temperature [57, 110, 111], their electrical properties are typically observed, by us, to be much worse than those of CVD tubes, for applications in SWNT TFTs due, at least in part, to their relatively short tube lengths, the structural imperfections that can result from the suspension process, and residual surfactant. Transfer printing techniques separate the high temperature CVD synthesis from target substrates, such as plastics, which cannot withstand such temperatures. One transfer technique that can be applied to CVD grown SWNT films on SiO_2/Si wafers uses PDMS stamps to remove the films after releasing them by undercut HF etching of the oxide [104]. This method is simple and has very high efficiency, as shown in Fig. 9.12. The holes (dark areas) visible in the transferred networks represent features associated with the procedures used here, but not with the process itself. The values of D evaluated away from these regions are almost the same as those on the growth substrate. A related method, with uses stamps made of materials other than PDMS, avoids the need to etch the buried oxide layer [112, 113]. Figure 9.12(c) and (d) show single and multiple transfer results obtained with this approach, as applied with aligned arrays of tubes grown on quartz [113]. These transfer techniques not only provide high quality starting materials for the development of high performance electronic devices and circuits on desired substrates, but they also enable further control of D and tube layouts through the application of multiple transfer processes (Fig. 9.12(d)).

9.4.1.2 Dielectrics

High performance SWNT TFTs demand gate dielectrics with large capacitance, low leakage current, good mechanical flexibility (for applications in flexible electronics) and low deposition temperatures. Those stringent requirements cannot be satisfied with the thick oxides (e.g., 100 nm or 500 nm SiO_2) or with the types of polymer dielectrics that are often used for studies of scaling and other device properties [9, 29, 74, 87, 104]. Two classes of dielectric materials that meet many of these requirements have been applied to SWNT TFTs. One consists of three-dimensional (3D) crosslinked organic multilayers ($\sim$16 nm) grown by a self-assembly process (Fig. 9.13(a)) [114]. These layers can have large capacitances ($\sim$170 nF/cm^2), excellent insulating properties (leakage current densities less than 10^{-9}A/cm^2) and smooth surface morphologies with improved yield and structural stability compared to standard self-assembled monolayer dielectrics [114, 115]. Another approach utilizes a bilayer composed of atomic-layer-deposited inorganic oxide (2–5 nm) (Fig. 9.13(b)) with a spin-cast crosslinked epoxy ($\sim$10 nm) (Fig. 9.13(c)) on top. The epoxy serves as an adhesive layer to assist high efficiency transfer printing [116] These bilayer nanodielectrics are compatible with flexible plastic substrates (e.g., PET) and have comparable high capacitance (up to $\sim$330 nF/cm^2) compared to the organic multilayer assemblies [115]. Their other electrical properties, including leakage current density, interface charge density, interface state density, and dissipation factor, are comparable to the organic multilayers and many conventional high k dielectrics [115, 116].

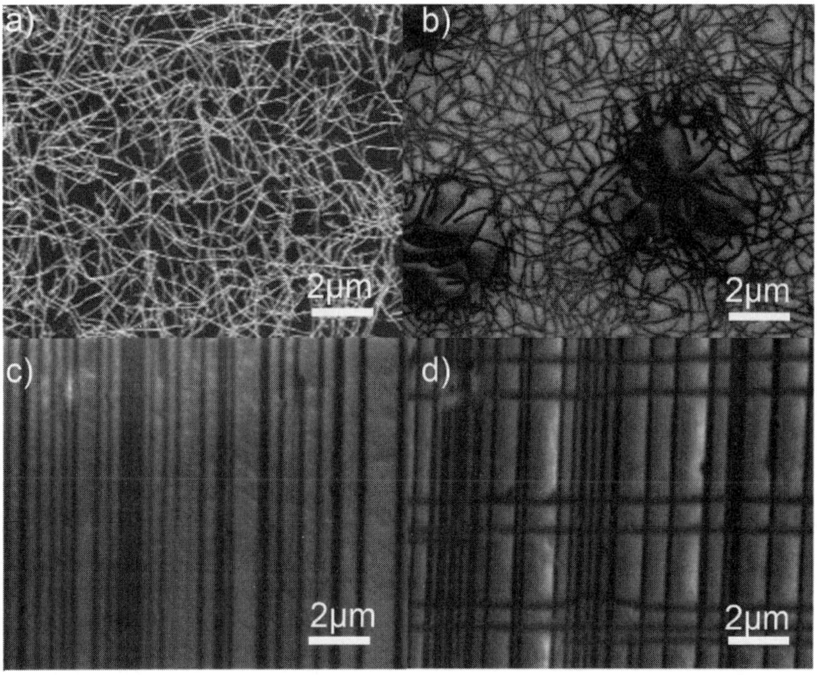

Fig. 9.12 SEM images illustrating transfer processes for films of SWNT. SEM images of a s-CNN before (**a**) and after (**b**) transfer from a SiO_2/Si growth wafer to a plastic substrate. SEM images of aligned tubes transferred from a single crystal quartz growth substrate to a plastic substrate (**c**) and a crossbar array of tubes formed by two consecutive transfer processes (**d**)

9.4.1.3 Contacts

The effects of contacts in short channel, single tube devices and the SBs that can exist have been studied extensively, both theoretically [117, 118] and experimentally [119]. Simple channel length scaling studies suggest that well-formed contacts of Au and Pd have negligible effects on the device performance of most SWNT TFTs for channel lengths that have, thus far, been most fully explored, i.e., $L_C > 1\ \mu m$ [9, 29, 42, 114, 116]. This result is partly caused by the relatively large channel resistances in this range of channel lengths, particularly with SWNT films that consist of random networks of tubes. For high mobility devices built with aligned arrays of nanotubes, the effects of contacts can be prominent. It is important to note that the fabrication conditions and the work functions and chemistries of the metals for the contacts have important effects on the nature of the contacts and on the polarity of devices. With decreasing work function, ambipolar and n-channel behavior can be observed in SWNT TFTs, similar to observations in devices based on individual tubes [120].

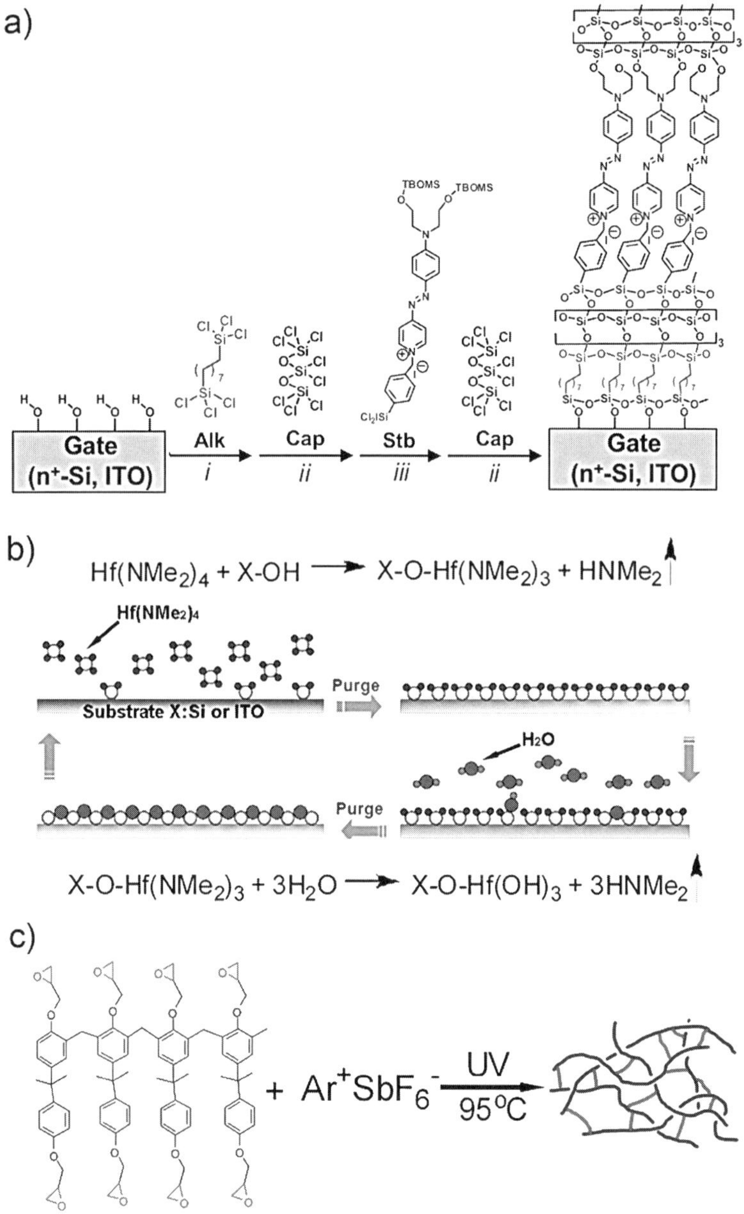

Fig. 9.13 Examples of high capacitance gate dielectrics that have been used successfully in SWNT TFTs. (**a**) Synthetic procedures for forming self-assembled multilayer nanodielectrics. Reproduced with permission from [115]. Copyright 2005 American Academy of Science. (**b**) Atomic layer deposition reaction sequence for HfO_2. (**c**) Polymerization reaction of the epoxy component of a bilayer nanodielectric. Reproduced with permission from [116]. Copyright 2006 Wiley-VCH

9.4.2 Transistors Based on SWNT Networks

SWNT TFTs that use s-CNNs have been built with a range of materials and device designs. These devices can be constructed in large scale arrays, in which electrical isolation is accomplished by patterning and etching the SWNT films through optical lithography and oxygen plasma etching to prevent cross-talk between neighboring devices [9]. From an electrical standpoint, the best performance has been achieved in devices that use the classes of nanodielectrics described previously with tubes grown by CVD [114, 116]. From a mechanical and optical standpoint, extremely flexible TTFTs can be obtained by using network SWNT films for all current carrying layers [77]. For the former, measurements on pristine p-channel devices and on n-channel devices made by PEI coating show very small hysteresis, enabled by the low operating gate voltage range, and large subthreshold swings (S), resulting from the high capacitance gate dielectrics (Fig. 9.14(a) and (b)). I_{on} is proportional with the reciprocal of channel length, consistent with negligible effects of contacts (Fig. 9.14(a) inset).

Scaling studies of such devices show that their effective device mobilities, as extracted using the Shockley model with gate capacitances computed according to procedures described previously [88], are independent of channel length (Fig. 9.15(a)). This result, which is consistent with measurements on other devices that use similar types of SWNT films, indicates that gate modulation of the channel dominates the device behavior for the studied range of channel lengths. The effective mobilities of network TFTs are much higher than those of devices built with polymers or organic small molecule based semiconductors [121] but they are inferior to mobilities of devices that use individual tubes or aligned arrays [20, 89]. Consistent with results of Section 9.3.2, the on/off ratios of SWNT network TFTs increase with channel length (Fig. 9.15(a)). This scaling property is different than that of traditional TFTs, and provides an easy means to achieve high on/off ratio devices. A disadvantage, of course, is that high on/off ratios are difficult to achieve at short channel lengths unless procedures to eliminate m-SWNTs are used. The subthreshold slope, S, is a function of the gate capacitance and the capacitance due to interface traps, and is heavily influenced by the on/off ratio because the off state current is independent of V_{GS} (Fig. 9.15(b)) [83, 116]. At an on/off ratio of 10^3, S as small as 230 mV/dec, has been achieved with bilayer nanodielectrics [116].

The full combination of extraordinary electrical, optical and mechanical properties of SWNT networks has been demonstrated in bendable TTFTs that use random network films of SWNTs for all of the current carrying layers [77]. Figure 9.16(a) shows the schematic of device layout for this type of "all-tube" TTFT. Such devices can be formed through sequential transfer printing of CVD nanotube networks with different densities onto a plastic substrate. High D films form the S/D and gate electrodes, while moderate D films form the semiconductor. Devices with this design have optical transparency as large as 75% even in the most opaque S/D electrode region, as shown in Fig. 9.16(b). This degree of transparency is comparable to TTFTs that use inorganic oxides [122, 123]. When combined with mechanically robust elastomeric dielectrics, the devices can withstand tensile strains up to

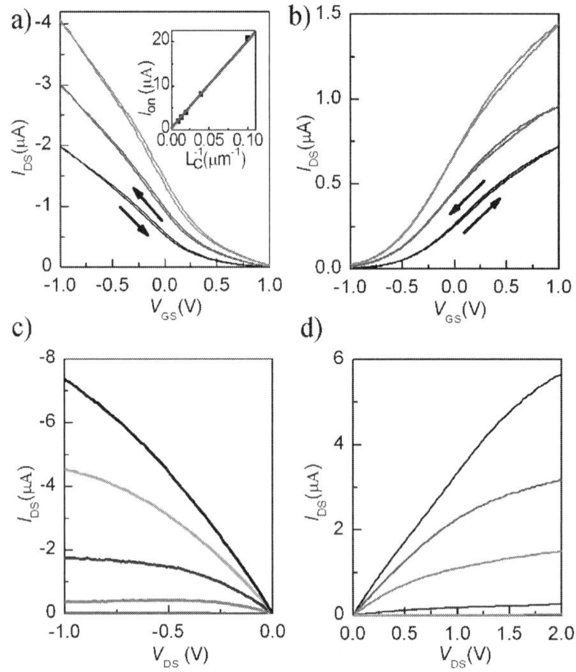

Fig. 9.14 Electrical properties of SWNT TFTs that use random networks of tubes and high capacitance dielectrics of HfO_2/epoxy. Transfer curves of SWNT TFTs on plastic substrates, with channel lengths, from top to bottom, of 50 μm, 75 μm, 100 μm, before (**a**) and after (**b**) uniformly coating the channel regions with PEI. The channel widths are 250 μm. The drain/source voltage (V_{DS}) is −0.2 V. Inset: I_{on} versus the reciprocal of channel length ($1/L_C$). Current–voltage characteristics of a device with channel length of 100 μm and channel width of 250 μm before (**c**) and after (**d**) uniformly coating the channel regions with PEI. The gate voltage varies between –1 V and 1 V in steps of 0.5 V. Reproduced with permission from [116]. Copyright 2006 Wiley-VCH

3.5% (Fig. 9.16(c)) without degradation. Beyond this limit, the dielectrics fail but the SWNT films are still conductive.

9.4.3 Transistors Based on SWNT Arrays

Arrays are of interest because they avoid the tube/tube contacts that can limit transport in networks [124, 125]. In principle, the arrays should enable certain device level characteristics that approach the intrinsic properties of the individual tubes. The device layouts and fabrication processes for building TFTs from aligned arrays of SWNT are almost identical to those based on random networks [89]. Figure 9.17shows a schematic illustration of an aligned tube TFT made by transferring an array of SWNTs to an epoxy coated (150 nm) substrate of SiO_2 (100 nm)/Si where the epoxy/SiO_2 bilayer serves as the gate dielectric, the Si provides a back gate

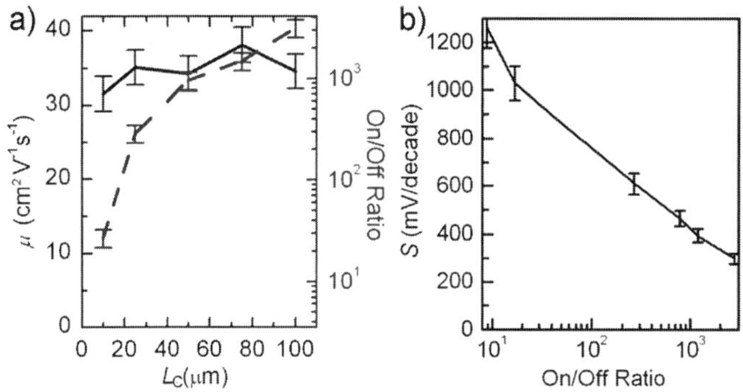

Fig. 9.15 Mobilities and subthreshold properties of SWNT TFTs with HfO_2/epoxy gate dielectrics. (**a**) Effective mobilities (μ, *solid line*) calculated considering the effect of fringing field and on/off ratios (dash line) as a function of channel length (L_C) for a typical set of devices. (**b**) Subthreshold swing (S) versus on/off ratio. Reproduced with permission from [116]. Copyright 2006 Wiley-VCH

and Au (100 nm) is used for the S/D electrodes. Similar devices can also be fabricated in a similar way on flexible plastic substrates and with various other dielectrics, including high-k (Al_2O_3, HfO_2, TiO_2) materials. Figure 9.17(b) shows an SEM image of the channel region. High on/off ratios ($>10^4$) can be obtained by selective electrical breakdown of the m-SWNT in a manner similar to that described previously for the networks, but in a much cleaner and reproducible manner, due to the simple layout of tubes in the arrays [9]. Figure 9.17(c) shows transfer curves

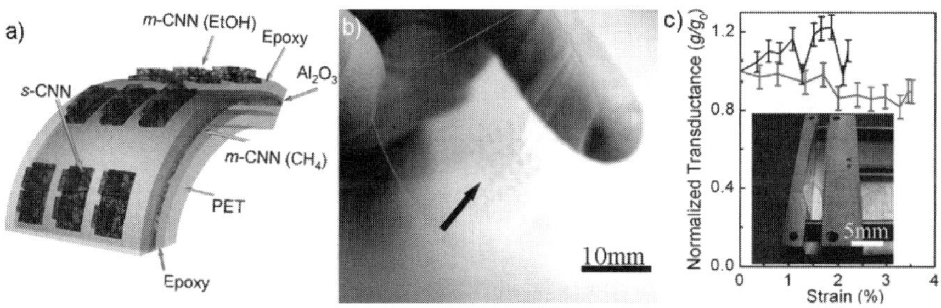

Fig. 9.16 Transparent, mechanically flexible TFTs that use SWNT films for all current carrying layers. (**a**) Schematic illustration of a device. (EtOH) indicates a CVD procedure that uses Fe/Co/Mo tri-metallic catalyst loaded onto a high surface-area silica support and ethanol; (CH_4) indicates a similar growth procedure, but with methane. The s-CNN layer was synthesized by using CVD with ferritin catalyst and methane. (**b**) Array of 'all-tube' TTFTs on a plastic substrate. The arrow indicates the S/D structures, which are faintly visible as arrays of *grey squares* in the center of this image. (**c**) Change of normalized transconductance (g/g_0) for transistors with epoxy (*black line*) and PDMS (*red line*) dielectric layers, as a function of bending induced tensile strains at the surface. Inset: Image showing the extremely high levels of bending that can be achieved with all tube TTFTs that use PDMS gate dielectrics. Reproduced with permission from [77]. Copyright 2006 Wiley-VCH

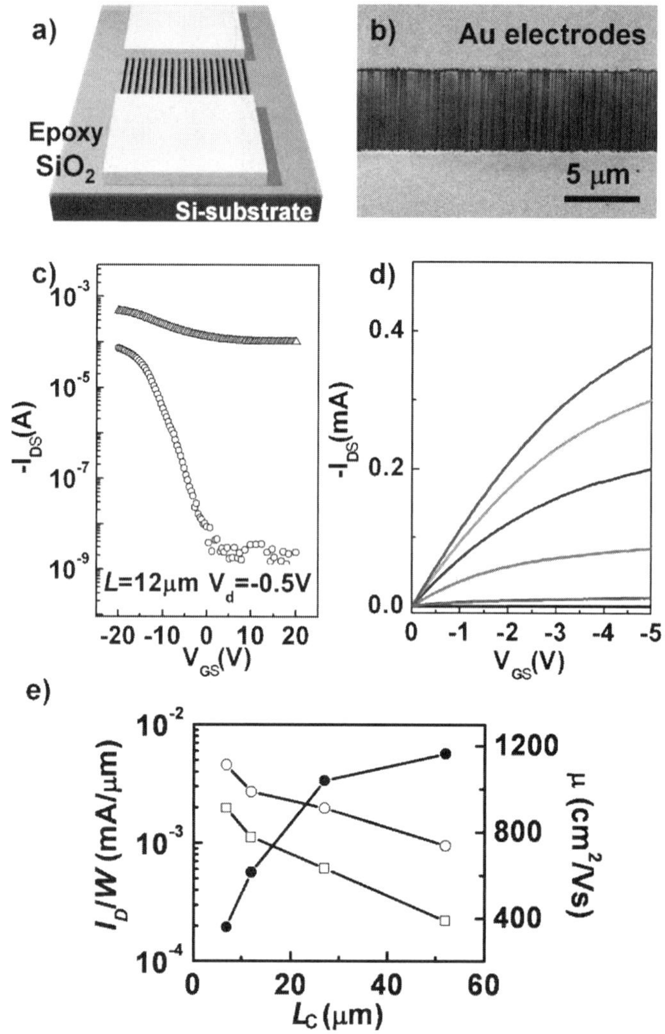

Fig. 9.17 Schematics, images, and data from TFTs that use SWNT films consisting of aligned arrays of tubes. (**a**) Schematic illustration of such a device that uses a dielectric of epoxy/SiO_2 and a doped Si substrate as a back gate. (**b**) SEM image of the channel region. (**c**) Transfer curves before (*triangles*) and after (*circles*) performing an electrical breakdown process that destroys metallic tubes. (**d**) Current–voltage characteristics of a device after electrical breakdown. The gate source voltage (V_{GS}) changes, from *bottom* to *top*, from 5 V to −5 V. (**e**) Width normalized on (*open circles*) and off (*squares*) currents and field effect mobilities (*solid circles*) as a function of L_C for devices fabricated on the quartz growth substrates. Reproduced with permission from [89]. Copyright 2007 Nature Publishing Group

before and after electrical breakdown process. The breakdown process improves the on/off ratio by more than 10,000. Mobilities are in the range of a few hundred cm^2/Vs and, in some cases, higher. Figure 9.17(d) presents the full I–V response. Devices that use source and drain electrodes of Pd, offer the best device performance, with mobilities greater that 1000 cm^2/Vs [89]. Figure 9.17(e) shows I_{on}, I_{off} and device mobility as a function of L_C for such devices.

9.4.4 Inverters and Logic Gates

SWNT TFTs that consist of films of random networks or aligned arrays of tubes, with p- and n-channel operation provide building blocks for circuits of various types. A CMOS type inverter, which represents an important element in digital circuits, can be constructed by integrating a p-channel and an n-channel SWNT TFT in the manner illustrated in Fig. 9.18(a). SWNT TFTs with bilayer nanodielectrics enable gains in such circuits as high as ~8 (Fig. 9.18(b)) [116], which is comparable to SWNT network CMOS inverters fabricated with organic multilayer nanodielectrics [114] and to single tube inverters based on local bottom gated devices [108].

Inverters based on aligned arrays of SWNTs have also been fabricated [89]. Figure 9.19(a) and (b) show the electrical response of (p-channel metal-oxide-semiconductor) PMOS and CMOS devices. The PMOS inverter used two p-channel SWNT TFTs with one as a resistance load (V_{GS} constant). The possibility to fabri-

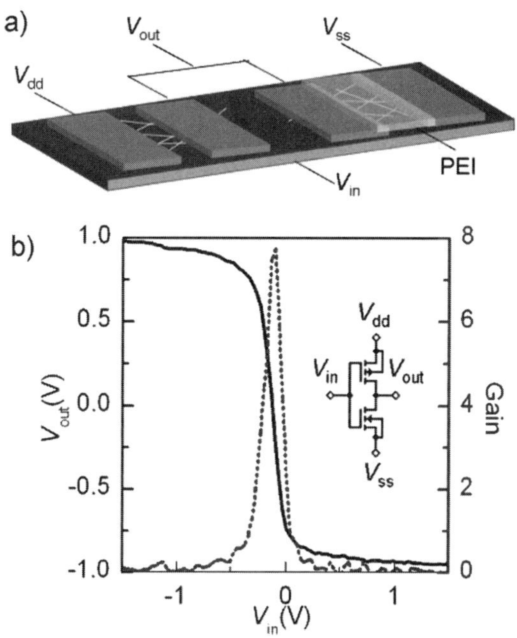

Fig. 9.18 CMOS inverter formed with a pair of SWNT TFTs that use films of random networks of tubes. (**a**) Schematic illustration of a device. The n-channel transistor used a coating of the polymer PEI. (**b**) V_{out} vs. V_{in} for an SWNT network CMOS-type inverter formed with p-channel and n-channel SWNT TFTs based on HfO_2/epoxy dielectrics. The inset provides a circuit diagram. Reproduced with permission from [116]. Copyright 2006 Wiley-VCH

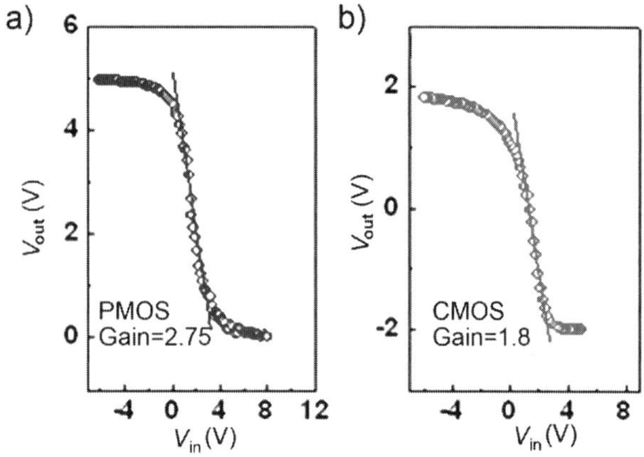

Fig. 9.19 Logic gates formed with transistors that use aligned arrays of SWNTs for the semiconductor. (**a**) PMOS inverter and (**b**) CMOS inverter output curves. The solid lines indicate the slope used to extract the gain. Reproduced with permission from [89]. Copyright 2007 Nature Publishing Group

cate more complex electronic devices, such as logic gates and ring oscillators, with the SWNT arrays as a thin film semiconductor is under investigation.

9.5 Outlook and Conclusions

Research on SWNT thin film electronics over the last ~3 years has yielded significant progress in the fabrication and fundamental understanding of transistor devices and logic gates that use SWNT films in the form of networks and aligned arrays. In particular, device mobilities have increased by two hundred fold, from ~5 $cm^2 V^{-1} s^{-1}$ to >1000 $cm^2 V^{-1} s^{-1}$, driven mainly by the development of techniques to grow dense, perfect arrays. At the same time, effective ways to yield high on/off ratios, in some cases as high as 10^5, including electrical breakdown and selective chemical functionalization, have been developed. The operating gate voltages have decreased from ~20 V to ~1 V and, in related work, the hysteresis has been reduced from levels so large that the transistors could be used effectively as memory devices to values that are nearly negligible. Not only p-channel but also and n-channel and ambipolar devices have been achieved by use of simple polymer coating strategies. The scaling properties of devices based on both aligned arrays and random networks have been defined and modeled, at a quantitative level. Unusual transparent and/or stretchable TFTs based on SWNT films have been also demonstrated in some prototype devices, and PMOS and CMOS logic gates have been achieved. Recently, procedures have been developed for integrating SWNT TFTs into 3D formats and with other inorganic semiconductor devices, such as Si metal-oxide-semiconductor field-effect transistors (MOSFETs), thereby creating new application possibilities [126]. Complex circuits appear possible.

Nevertheless, there still remain significant challenges. First, techniques for separating m-SWNT and s-SWNT in a cost-effective and high throughput manner must be developed. Recent work on solution and chromatographic approaches and on plasma enhanced CVD that can lead to SWNTs with enriched quantities of s-SWNT appear promising [127–132]. Second, improved growth methods are needed to achieve, in aligned arrays, high values of *D*. The latest methods, which involve patterning the catalyst, can achieve perfect alignment and perfectly linear tubes with $D \sim 5$ tubes/μm. New combinations of catalysts and feed gases have a strong potential to lead to further improvements, and multiple transfer printing steps can also be used. Third, and perhaps most important, techniques are needed to dope the contacts. Emerging results from work in single tube devices suggest that chemical strategies to this problem might be effective [133, 134]. Fourth, computational algorithms and compact circuit models will need to be developed for devices and systems for operating frequencies that range from DC to many hundreds of GHz [135].

In spite of these challenges, it worth noting that almost thirty years passed between the demonstration of the first Si transistor to the first microprocessor. Twenty years elapsed between the first organic transistor and realistic demonstrator display systems based on active matrix circuits using these transistors [10]. As benchmarked against these two other material technologies, the progress in SWNT film based electronics is encouraging, particularly relative to the organics. For this reason, we feel that additional basic and applied work in this area is well justified. In our view, existing and emerging data suggest that the selected applications, cost structures, addressable markets and related issues will ultimately determine the success of this approach to electronics, rather than any intrinsic feature of the physics or the materials.

Acknowledgements We thank T. Banks, K. Colravy and D. Sievers for help with the processing. This work was supported by DARPA-funded AFRL-managed Macroelectronics Program Contract FA8650-04-C-7101, the NSF through grant NIRT-0403489, the Frederick Seitz Materials Research Lab and the Center for Microanalysis of Materials in University of Illinois which is funded by U.S. Department of Energy through grant DEFG02-91-ER45439, the Center for Nanoscale Chemical Electrical Mechanical Manufacturing Systems in University of Illinois which is funded by the NSF through grant DMI-0328162, and a graduate fellowship from the Fannie and John Hertz Foundation (M.A.M.).

References

1. P. Avouris, "Molecular Electronics with carbon nanotubes," *Acc. Chem. Res.*, vol. 35, pp. 1026–1034, 2002.
2. P. Avouris, J. Appenzeller, R. Martel, and S. J. Wind, "Carbon nanotube electronics," *Proc. IEEE*, vol. 91, pp. 1772–1784, 2003.
3. P. Avouris, R. Martel, V. Derycke, and J. Appenzeller, "Carbon nanotube transistors and logic circuits," *Physica B*, vol. 323, pp. 6–14, 2002.
4. M. Ouyang, J.-L. Huang, and C. M. Lieber, "Fundamental electronic properties and applications of single-walled carbon nanotubes," *Acc. Chem. Res.*, vol. 35, pp. 1018–1025, 2002.
5. V. N. Popov, "Carbon nanotubes: properties and application," *Mater. Sci. Eng., R*, vol. 43, pp. 61–102, 2004.
6. A. P. Graham, G. S. Duesberg, W. Hoenlein, F. Kreupl, M. Liebau, R. Martin, B. Rajasekharan, W. Pamler, R. Seidel, W. Steinhoegl, and E. Unger, "How do carbon nanotubes fit into the semiconductor roadmap?," *Appl. Phys. A – Mater. Sci. Process.*, vol. 80, pp. 1141–1151, 2005.

7. E. S. Snow, J. P. Novak, M. D. Lay, E. H. Houser, F. K. Perkins, and P. M. Campbell, "Carbon nanotube networks: Nanomaterial for macroelectronic applications," *J. Vac. Sci. Technol. B*, vol. 22, pp. 1990–1994, 2004.
8. R. Seidel, A. P. Graham, E. Unger, G. S. Duesberg, M. Liebau, W. Steinhoegl, F. Kreupl, and W. Hoenlein, "High-current nanotube transistors," *Nano. Lett.*, vol. 4, pp. 831–834, 2004.
9. Y. Zhou, A. Gaur, S.-H. Hur, C. Kocabas, M. A. Meitl, M. Shim, and J. A. Rogers, "p-channel, n-channel thin film transistors and p-n diodes based on single wall carbon nanotube networks," *Nano. Lett.*, vol. 4, pp. 2031–2035, 2004.
10. J. A. Rogers, Z. Bao, K. Baldwin, A. Dodabalapur, B. Crone, V. R. Raju, V. Kuck, H. Katz, K. Amundson, J. Ewing, and P. Drzaic, "Paper-like electronic displays: Large-area rubber-stamped plastic sheets of electronics and microencapsulated electrophoretic inks," *Proc. Natl. Acad. Sci. U.S.A.*, vol. 98, pp. 4835–4840, 2001.
11. I. P. Kang, M. J. Schulz, J. H. Kim, V. Shanov, and D. L. Shi, "A carbon nanotube strain sensor for structural health monitoring," *Smart Mater. Struct.*, vol. 15, pp. 737–748, 2006.
12. R. H. Reuss, B. R. Chalamala, A. Moussessian, M. G. Kane, A. Kumar, D. C. Zhang, J. A. Rogers, M. Hatalis, D. Temple, G. Moddel, B. J. Eliasson, M. J. Estes, J. Kunze, E. S. Handy, E. S. Harmon, D. B. Salzman, J. M. Woodall, M. A. Alam, J. Y. Murthy, S. C. Jacobsen, M. Olivier, D. Markus, P. M. Campbell, and E. Snow, "Macroelectronics: Perspectives on technology and applications," *Proc. IEEE*, vol. 93, pp. 1239–1256, 2005.
13. R. H. Reuss, D. G. Hopper, and J. G. Park, "Macroelectronics," *MRS Bull.*, vol. 31, pp. 447–450, 2006.
14. S. Lee, B. Koo, J. G. Park, H. Moon, J. Hahn, and J. M. Kim, "Development of high-performance organic thin-film transistors for large-area displays," *MRS Bull.*, vol. 31, pp. 455–459, 2006.
15. P. van der Wilt, M. G. Kane, A. B. Limanov, A. H. Firester, L. Goodman, J. Lee, J. Abelson, A. M. Chitu, and J. S. Im, "Low-temperature polycrystalline silicon thin-film transistors and circuits on flexible substrates," *MRS Bull.*, vol. 31, pp. 461–465, 2006.
16. S. R. Forrest, "The path to ubiquitous and low-cost organic electronic appliances on plastic," *Nature*, vol. 428, pp. 911–918, 2004.
17. G. Thomas, "Invisible circuits," *Nature*, vol. 389, pp. 907–908, 1997.
18. R. F. Service, "Materials science – Inorganic electronics begin to flex their muscle," *Science*, vol. 312, pp. 1593–1594, 2006.
19. T. Durkop, S. A. Getty, E. Cobas, and M. S. Fuhrer, "Extraordinary mobility in semiconducting carbon nanotubes," *Nano Lett.*, vol. 4, pp. 35–39, 2004.
20. X. J. Zhou, J. Y. Park, S. M. Huang, J. Liu, and P. L. McEuen, "Band structure, phonon scattering, and the performance limit of single-walled carbon nanotube transistors," *Phys. Rev. Lett.*, vol. 95, p. 146805, 2005.
21. A. Javey, J. Guo, M. Paulsson, Q. Wang, D. Mann, M. Lundstrom, and H. J. Dai, "High-field quasiballistic transport in short carbon nanotubes," *Phys. Rev. Lett.*, vol. 92, 2004.
22. J. Chen, V. Perebeinos, M. Freitag, J. Tsang, Q. Fu, J. Liu, and P. Avouris, "Bright infrared emission from electrically induced excitons in carbon nanotubes," *Science*, vol. 310, pp. 1171–1174, 2005.
23. J. A. Misewich, R. Martel, P. Avouris, J. C. Tsang, S. Heinze, and J. Tersoff, "Electrically induced optical emission from a carbon nanotube FET," *Science*, vol. 300, pp. 783–786, 2003.
24. R. H. Baughman, C. X. Cui, A. A. Zakhidov, Z. Iqbal, J. N. Barisci, G. M. Spinks, G. G. Wallace, A. Mazzoldi, D. De Rossi, A. G. Rinzler, O. Jaschinski, S. Roth, and M. Kertesz, "Carbon nanotube actuators," *Science*, vol. 284, pp. 1340–1344, 1999.
25. M. R. Falvo, G. J. Clary, R. M. Taylor, V. Chi, F. P. Brooks, S. Washburn, and R. Superfine, "Bending and buckling of carbon nanotubes under large strain," *Nature*, vol. 389, pp. 582–584, 1997.
26. D. Bozovic, M. Bockrath, J. H. Hafner, C. M. Lieber, H. Park, and M. Tinkham, "Plastic deformations in mechanically strained single-walled carbon nanotubes," *Phys. Rev. B*, vol. 67, p. 033407, 2003.

27. S. B. Cronin, A. K. Swan, M. S. Unlu, B. B. Goldberg, M. S. Dresselhaus, and M. Tinkham, "Measuring the uniaxial strain of individual single-wall carbon nanotubes: Resonance Raman spectra of atomic-force-microscope modified single-wall nanotubes," *Phys. Rev. Lett.*, vol. 93, p. 167401, 2004.
28. Y. L. Loo, R. L. Willett, K. W. Baldwin, and J. A. Rogers, "Interfacial chemistries for nanoscale transfer printing," *J. Am. Chem. Soc.*, vol. 124, pp. 7654–7655, 2002.
29. S.-H. Hur, C. Kocabas, A. Gaur, M. Shim, O. O. Park, and J. A. Rogers, "Printed thin film transistors and complementary logic gates that use polymer coated single-walled carbon nanotube networks," *J. Apply. Phys.*, vol. 98, p. 114302, 2005.
30. B. D. Gates, Q. Xu, J. C. Love, D. B. Wolfe, and G. M. Whitesides, "Unconventional nanofabrication," *Annu. Rev. Mater. Res.*, vol. 34, pp. 339–372, 2004.
31. P. Calvert, "Inkjet printing for materials and devices," *Chem. Mater.*, vol. 13, pp. 3299–3305, 2001.
32. Z. Wu, Z. Chen, X. Du, J. M. Logan, J. Sippel, M. Nikolou, K. Kamaras, J. R. Reynolds, D. B. Tanner, A. F. Hebard, and A. G. Rinzler, "Transparent, conductive carbon nanotube films," *Science*, vol. 305, pp. 1273–1276, 2004.
33. M. Zhang, S. L. Fang, A. A. Zakhidov, S. B. Lee, A. E. Aliev, C. D. Williams, K. R. Atkinson, and R. H. Baughman, "Strong, transparent, multifunctional, carbon nanotube sheets," *Science*, vol. 309, pp. 1215–1219, 2005.
34. U. J. Kim, H. R. Gutierrez, J. P. Kim, and P. C. Eklund, "Effect of the tube diameter distribution on the high-temperature structural modification of bundled single-walled carbon nanotubes," *J. Phys. Chem. B*, vol. 109, pp. 23358–23365, 2005.
35. A. Hirsch, "Functionalization of single-walled carbon nanotubes," *Angew. Chem. Int. Ed.*, vol. 41, pp. 1853–1859, 2002.
36. A. Javey, J. Guo, D. B. Farmer, Q. Wang, D. W. Wang, R. G. Gordon, M. Lundstrom, and H. J. Dai, "Carbon nanotube field-effect transistors with integrated ohmic contacts and high-*k* gate dielectrics," *Nano Lett.*, vol. 4, pp. 447–450, 2004.
37. A. Javey, H. Kim, M. Brink, Q. Wang, A. Ural, J. Guo, P. McIntyre, P. McEuen, M. Lundstrom, and H. J. Dai, "High-kappa dielectrics for advanced carbon-nanotube transistors and logic gates," *Nat. Mater.*, vol. 1, pp. 241–246, 2002.
38. M. H. Yang, K. B. K. Teo, L. Gangloff, W. I. Milne, D. G. Hasko, Y. Robert, and P. Legagneux, "Advantages of top-gate, high-k dielectric carbon nanotube field-effect transistors," *Appl. Phys. Lett.*, vol. 88, pp. 113507, 2006.
39. B. M. Kim, T. Brintlinger, E. Cobas, M. S. Fuhrer, H. M. Zheng, Z. Yu, R. Droopad, J. Ramdani, and K. Eisenbeiser, "High-performance carbon nanotube transistors on SrTiO3/Si substrates," *Appl. Phys. Lett.*, vol. 84, pp. 1946–1948, 2004.
40. M. Shim, A. Javey, N. W. S. Kam, and H. J. Dai, "Polymer functionalization for air-stable n-type carbon nanotube field-effect transistors," *J. Am. Chem. Soc.*, vol. 123, pp. 11512–11513, 2001.
41. G. P. Siddons, D. Merchin, J. H. Back, J. K. Jeong, and M. Shim, "Highly efficient Gating and doping of carbon nanotubes with polymer electrolytes," *Nano Lett.*, vol. 4, pp. 927–931, 2004.
42. T. Ozel, A. Gaur, J. A. Rogers, and M. Shim, "Polymer electrolyte gating of carbon nanotube network transistors," *Nano Lett.*, vol. 5, pp. 905–911, 2005.
43. M. Shim, T. Ozel, A. Gaur, and C. J. Wang, "Insights on charge transfer doping and intrinsic phonon line shape of carbon nanotubes by simple polymer adsorption," *J. Am. Chem. Soc.*, vol. 128, pp. 7522–7530, 2006.
44. C. Klinke, J. Chen, A. Afzali, and P. Avouris, "Charge transfer induced polarity switching in carbon nanotube transistors," *Nano Lett.*, vol. 5, pp. 555–558, 2005.
45. E. S. Snow, F. K. Perkins, E. J. Houser, S. C. Badescu, and T. L. Reinecke, "Chemical detection with a single-walled carbon nanotube capacitor," *Science*, vol. 307, pp. 1942–1945, 2005.
46. E. S. Snow and F. K. Perkins, "Capacitance and conductance of single-walled carbon nanotubes in the presence of chemical vapors," *Nano Lett.*, vol. 5, pp. 2414–2417, 2005.

47. H. R. Byon and H. C. Choi, "Network single-walled carbon nanotube-field effect transistors (SWNT-FETs) with increased Schottky contact area for highly sensitive biosensor applications," *J. Am. Chem. Soc.*, vol. 128, pp. 2188–2189, 2006.
48. A. Star, E. Tu, J. Niemann, J. C. P. Gabriel, C. S. Joiner, and C. Valcke, "Label-free detection of DNA hybridization using carbon nanotube network field-effect transistors," *Proc. Natl. Acad. Sci. U S A*, vol. 103, pp. 921–926, 2006.
49. I. P. Kang, Y. Y. Heung, J. H. Kim, J. W. Lee, R. Gollapudi, S. Subramaniam, S. Narasimhadevara, D. Hurd, G. R. Kirikera, V. Shanov, M. J. Schulz, D. L. Shi, J. Boerio, S. Mall, and M. Ruggles-Wren, "Introduction to carbon nanotube and nanofiber smart materials," *Compos. Pt. B – Eng.*, vol. 37, pp. 382–394, 2006.
50. M. F. Islam, D. E. Milkie, C. L. Kane, A. G. Yodh, and J. M. Kikkawa, "Direct measurement of the polarized optical absorption cross section of single-wall carbon nanotubes," *Phys. Rev. Lett.*, vol. 93, p. 037404, 2004.
51. M. Kaempgen, G. S. Duesberg, and S. Roth, "Transparent carbon nanotube coatings," *Appl. Surf. Sci.*, vol. 252, pp. 425–429, 2005.
52. A. D. Pasquier, H. E. Unalan, A. Kanwal, S. Miller, and M. Chhowalla, "Conducting and transparent single-wall carbon nanotube electrodes for polymer-fullerene solar cells," *Appl. Phys. Lett.*, vol. 87, p. 203511, 2005.
53. K. Lee, Z. Wu, Z. Chen, F. Ren, S. J. Pearton, and A. G. Rinzler, "Single Wall Carbon Nanotubes for p-Type Ohmic Contacts to GaN Light-Emitting Diodes," *Nano Lett.*, vol. 4, pp. 911–914, 2004.
54. C. Kocabas, N. Pimparkar, O. Yesilyurt, M. A. Alam, and J. A. Rogers, "Experimental and theoretical studies of transport through large scale, partially aligned arrays of single walled carbon nanotubes in thin film type transistors," *Nano Lett.,* vol. 7, pp. 1195–1202, 2007.
55. M. A. Alam, N. Pimparkar, S. Kumar, and J. Murthy, "Theory of nanocomposite network transistors for macroelectronics applications," *MRS Bull.*, vol. 31, pp. 466–470, 2006.
56. E. S. Snow, J. P. Novak, P. M. Campbell, and D. Park, "Random networks of carbon nanotubes as an electronic material," *Appl. Phys. Lett.*, vol. 82, pp. 2145–2147, 2003.
57. K. Bradley, J. C. P. Gabriel, and G. Grüner, "Flexible nanotube electronics," *Nano Lett.*, vol. 3, pp. 1353–1355, 2003.
58. R. Bandyopadhyaya, E. Nativ-Roth, O. Regev, and R. Yerushalmi-Rozen, "Stabilization of individual carbon nanotubes in aqueous solutions," *Nano Lett.*, vol. 2, pp. 25–28, 2002.
59. R. C. Haddon, J. Sippel, A. G. Rinzler, and F. Papadimitrakopoulos, "Purification and separation of carbon nanotubes," *Mrs Bull.*, vol. 29, pp. 252–259, 2004.
60. D. E. Johnston, M. F. Islam, A. G. Yodh, and A. Johnson, "Electronic devices based on purified carbon nanotubes grown by high-pressure decomposition of carbon monoxide," *Nature Mater.*, vol. 4, pp. 589–592, 2005.
61. M. A. Meitl, Y. X. Zhou, A. Gaur, S. Jeon, M. L. Usrey, M. S. Strano, and J. A. Rogers, "Solution casting and transfer printing single-walled carbon nanotube films," *Nano Lett.*, vol. 4, pp. 1643–1647, 2004.
62. J. U. Park, M. A. Meitl, S. H. Hur, M. L. Usrey, M. S. Strano, P. J. A. Kenis, and J. A. Rogers, "In situ deposition and patterning of single-walled carbon nanotubes by Laminar flow and controlled flocculation in microfluidic channels," *Angew. Chem.-Int. Edit.*, vol. 45, pp. 581–585, 2006.
63. Y. X. Zhou, L. B. Hu, and G. Grüner, "A method of printing carbon nanotube thin films," *Appl. Phys. Lett.*, vol. 88, p. 123109, 2006.
64. Y. N. Xia, J. A. Rogers, K. E. Paul, and G. M. Whitesides, "Unconventional methods for fabricating and patterning nanostructures," *Chem. Rev.*, vol. 99, pp. 1823–1848, 1999.
65. D. S. Bethune, C. H. Kiang, M. S. Devries, G. Gorman, R. Savoy, J. Vazquez, and R. Beyers, "Cobalt-catalyzed growth of carbon nanotubes with single-atomic-layerwalls," *Nature*, vol. 363, pp. 605–607, 1993.

66. H. J. Dai, "Nanotube growth and characterization," in *Carbon Nanotubes: Synthesis, Structure, Properties, and Applications*, M. S. Dresselhaus, G. Dresselhaus, and P. Avouris, Eds. Berlin Heidelberg: Springer-Verlag, 2001, pp. 29–53.
67. H. J. Dai, J. Kong, C. W. Zhou, N. Franklin, T. Tombler, A. Cassell, S. S. Fan, and M. Chapline, "Controlled chemical routes to nanotube architectures, physics, and devices," *J. Phys. Chem. B*, vol. 103, pp. 11246–11255, 1999.
68. A. M. Cassell, J. A. Raymakers, J. Kong, and H. J. Dai, "Large scale CVD synthesis of single-walled carbon nanotubes," *J. Phys. Chem. B*, vol. 103, pp. 6484–6492, 1999.
69. H. M. Cheng, F. Li, G. Su, H. Y. Pan, L. L. He, X. Sun, and M. S. Dresselhaus, "Large-scale and low-cost synthesis of single-walled carbon nanotubes by the catalytic pyrolysis of hydrocarbons," *Appl. Phys. Lett.*, vol. 72, pp. 3282–3284, 1998.
70. G. Zhang, D. Mann, L. Zhang, A. Javey, Y. Li, E. Yenilmez, Q. Wang, J. P. McVittie, Y. Nishi, J. Gibbons, and H. Dai, "Ultra-high-yield growth of vertical single-walled carbon nanotubes: Hidden roles of hydrogen and oxygen " *Proc. Natl. Acad. Sci. U. S. A.*, vol. 102, pp. 16141–16145, 2005.
71. Y. Murakami, S. Chiashi, Y. Miyauchi, M. H. Hu, M. Ogura, T. Okubo, and S. Maruyama, "Growth of vertically aligned single-walled carbon nanotube films on quartz substrates and their optical anisotropy," *Chem. Phys. Lett.*, vol. 385, pp. 298–303, 2004.
72. Y. Li, W. Kim, Y. Zhang, M. Rolandi, D. Wang, and H. Dai, "Growth of single-walled carbon nanotubes from discrete catalytic nanoparticles of various sizes," *J. Phys. Chem. B*, vol. 105, pp. 11424–11431, 2001.
73. M. Su, B. Zheng, and J. Liu, "A scalable CVD method for the synthesis of single-walled carbon nanotubes with high catalyst productivity," *Chem. Phys. Lett.*, vol. 322, pp. 321–326, 2000.
74. C. Kocabas, S. H. Hur, A. Gaur, M. A. Meitl, M. Shim, and J. A. Rogers, "Guided growth of large-scale, horizontally aligned arrays of single-walled carbon nanotubes and their use in thin-film transistors," *Small*, vol. 1, pp. 1110–1116, 2005.
75. C. Kocabas, M. Shim, and J. A. Rogers, "Spatially selective guided growth of high-coverage arrays and random networks of single-walled carbon nanotubes and their integration into electronic devices," *J. Am. Chem. Soc.*, vol. 128, pp. 4540–4541, 2006.
76. N. Saran, K. Parikh, D. S. Suh, E. Munoz, H. Kolla, and S. K. Manohar, "Fabrication and characterization of thin films of single-walled carbon nanotube bundles on flexible plastic substrates," *J. Am. Chem. Soc.*, vol. 126, pp. 4462–4463, 2004.
77. Q. Cao, S.-H. Hur, Z.-T. Zhu, Y. Sun, C. Wang, M. Meitl, M. Shim, and J. A. Rogers, "Highly bendable, transparent thin film transistors that use carbon nanotube based conductors and semiconductors with elastomeric dielectrics," *Adv. Mater.*, vol. 18, pp. 304–309, 2006.
78. L. Hu, D. S. Hecht, and G. Grüner, "Percolation in transparent and conducting carbon nanotube networks," *Nano Lett.*, vol. 4, pp. 2513–2517, 2004.
79. M. W. Rowell, M. A. Topinka, M. D. McGehee, H. J. Prall, G. Dennler, N. S. Sariciftci, L. B. Hu, and G. Gruner, "Organic solar cells with carbon nanotube network electrodes," *Appl. Phys. Lett.*, vol. 88, 2006.
80. Q. Cao, Z. T. Zhu, M. G. Lemaitre, M. G. Xia, M. Shim, and J. A. Rogers, "Transparent flexible organic thin-film transistors that use printed single-walled carbon nanotube electrodes," *Appl. Phys. Lett.*, vol. 88, p. 113511, 2006.
81. B. Vigolo, C. Coulon, M. Maugey, C. Zakri, and P. Poulin, "An experimental approach to the percolation of sticky nanotubes," *Science*, vol. 309, pp. 920–923, 2005.
82. S. Kumar, J. Y. Murthy, and M. A. Alam, "Percolating conduction in finite nanotube networks," *Phys. Rev. Lett.*, vol. 95, p. 066802, 2005.
83. S. Kumar, N. Pimparkar, J. Y. Murthy, and M. A. Alam, "Theory of transfer characteristics of nanotube network transistors," *Appl. Phys. Lett.*, vol. 88, p. 123505, 2006.
84. S. D. Li, Z. Yu, C. Rutherglen, and P. J. Burke, "Electrical properties of 0.4 cm long single-walled carbon nanotubes," *Nano Lett.*, vol. 4, pp. 2003–2007, 2004.

85. V. Perebeinos, J. Tersoff, and P. Avouris, "Mobility in semiconducting carbon nanotubes at finite carrier density," *Nano Lett.*, vol. 6, pp. 205–208, 2006.
86. J. Guo, S. Goasguen, M. Lundstrom, and S. Datta, "Metal-insulator-semiconductor electrostatics of carbon nanotubes," *Appl. Phys. Lett.*, vol. 81, pp. 1486–1488, 2002.
87. E. S. Snow, P. M. Campbell, M. G. Ancona, and J. P. Novak, "High-mobility carbon-nanotube thin-film transistors on a polymeric substrate," *Appl. Phys. Lett.*, vol. 86, p. 033105, 2005.
88. Q. Cao, M. G. Xia, C. Kocabas, M. Shim, S. V. Rotkin, and J. A. Rogers, "Gate Capacitance Coupling of SWNT Thin-film Transistor," *Appl. Phys. Lett.*, vol. 90, p. 023516, 2007.
89. S. J. Kang, C. Kocabas, T. Ozel, M. Shim, S. V. Rotkin, and J. A. Rogers, "High performance electronics based on dense, perfectly aligned arrays of single walled carbon nanotubes," *Nat. Nanotech.* vol 2, pp. 230–236, 2007.
90. K. Balasubramanian, R. Sordan, M. Burghard, and K. Kern, "A selective electrochemical approach to carbon nanotube field-effect transistors," *Nano. Lett.*, vol. 4, pp. 827–830, 2004.
91. K. Balasubramanian, M. Friedrich, C. Jiang, Y. Fan, A. Mews, M. Burghard, and K. Kern, "Electrical transport and confocal raman studies of electrochemically modified individual carbon nanotubes," *Adv. Mater.*, vol. 15, pp. 1515–1518, 2003.
92. M. S. Strano, C. A. Dyke, M. L. Usrey, P. W. Barone, M. J. Allen, H. Shan, C. Kittrell, R. H. Hauge, J. M. Tour, and R. E. Smalley, "Electronic structure control of single-walled carbon nanotube functionalization," *Science*, vol. 301, pp. 1519–1522, 2003.
93. C. A. Dyke, M. P. Stewart, F. Maya, and J. M. Tour, "Diazonium-based functionalization of carbon nanotubes: XPS and GC-MS analysis and mechanistic implications," *Synlett*, pp. 155–160, 2004.
94. C. Wang, Q. Cao, T. Ozel, A. Gaur, J. A. Rogers, and M. Shim, "Electronically selective chemical functionalization of carbon nanotubes: correlation between raman spectral and electrical responses," *J. Am. Chem. Soc.*, vol. 127, pp. 11460–11468, 2005.
95. A. Kukovecz, C. Kramberger, V. Georgakilas, M. Prato, and H. Kuzmany, "A detailed Raman study on thin single-wall carbon nanotubes prepared by the HiPCO process," *Eur. Phys. J. B*, vol. 28, pp. 223–230, 2002.
96. C. Fantini, A. Jorio, M. Souza, M. S. Strano, M. S. Dresselhaus, and M. A. Pimenta, "Optical transition energies for carbon nanotubes from resonant Raman spectroscopy: environment and temperature effects," *Phys. Rev. Lett.*, vol. 93, p. 147406, 2004.
97. M. S. Strano, "Probing chiral selective reactions using a revised Kataura plot for the interpretation of single-walled carbon nanotube spectroscopy," *J. Am. Chem. Soc.*, vol. 125, pp. 16148–16153, 2003.
98. J. Kong, N. R. Franklin, C. W. Zhou, M. G. Chapline, S. Peng, K. J. Cho, and H. J. Dai, "Nanotube molecular wires as chemical sensors," *Science*, vol. 287, pp. 622–625, 2000.
99. P. G. Collins, K. Bradley, M. Ishigami, and A. Zettl, "Extreme oxygen sensitivity of electronic properties of carbon nanotubes," *Science*, vol. 287, pp. 1801–1804, 2000.
100. V. Derycke, R. Martel, J. Appenzeller, and P. Avouris, "Controlling doping and carrier injection in carbon nanotube transistors," *Appl. Phys. Lett.*, vol. 80, pp. 2773–2775, 2002.
101. J. Kong, C. W. Zhou, E. Yenilmez, and H. J. Dai, "Alkaline metal-doped n-type semiconducting nanotubes as quantum dots," *Appl. Phys. Lett.*, vol. 77, pp. 3977–3979, 2000.
102. A. Javey, R. Tu, D. B. Farmer, J. Guo, R. G. Gordon, and H. J. Dai, "High performance n-type carbon nanotube field-effect transistors with chemically doped contacts," *Nano Lett.*, vol. 5, pp. 345–348, 2005.
103. B. Yakobson and P. Avouris, "Mechanical properties of carbon nanotubes," *Topics Appl. Phys.*, vol. 80, pp. 287–327, 2001.
104. S.-H. Hur, O. O. Park, and J. A. Rogers, "Extreme bendability in thin film transistors that use carbon nanotubes transferred from high temperature growth substrates," *Appl. Phys. Lett.*, vol. 86, p. 243502, 2005.
105. S. J. Tans, A. R. M. Verschueren, and C. Dekker, "Room-temperature transistor based on a single carbon nanotube," *Nature*, vol. 393, pp. 49–52, 1998.

106. J. U. Lee, "Photovoltaic effect in ideal carbon nanotube diodes," *Appl. Phys. Lett.*, vol. 87, p. 073101, 2005.
107. A. Bachtold, P. Hadley, T. Nakanishi, and C. Dekker, "Logic circuits with carbon nanotube transistors," *Science*, vol. 294, pp. 1317–1320, 2001.
108. A. Javey, Q. Wang, A. Ural, Y. M. Li, and H. J. Dai, "Carbon nanotube transistor arrays for multistage complementary logic and ring oscillators," *Nano Lett.*, vol. 2, pp. 929–932, 2002.
109. Z. H. Chen, J. Appenzeller, Y. M. Lin, J. Sippel-Oakley, A. G. Rinzler, J. Y. Tang, S. J. Wind, P. M. Solomon, and P. Avouris, "An integrated logic circuit assembled on a single carbon nanotube," *Science*, vol. 311, pp. 1735–1735, 2006.
110. N. P. Armitage, J. C. P. Gabriel, and G. Gruner, "Quasi-Langmuir-Blodgett thin film deposition of carbon nanotubes," *J. App. Phys*, vol. 95, pp. 3228–3230, 2004.
111. M. D. Lay, J. P. Novak, and E. S. Snow, "Simple route to large-scale ordered arrays of liquid-deposited carbon nanotubes," *Nano Lett.*, vol. 4, pp. 603–606, 2004.
112. M. A. Meitl, Z. T. Zhu, V. Kumar, K. J. Lee, X. Feng, Y. Y. Huang, I. Adesida, R. G. Nuzzo, and J. A. Rogers, "Transfer printing by kinetic control of adhesion to an elastomeric stamp," *Nat. Mater.*, vol. 5, pp. 33–38, 2006.
113. S. J. Kang, C. Kocabas, H.-S. Kim, Q. Cao, M. A. Meitl, D.-Y. Khang and J. A. Rogers, "Printed multilayer superstructures of aligned single-walled carbon nanotubes for electronic applications," *Nano Lett.*, vol.7, pp. 3343–3348, 2007.
114. S. H. Hur, M. H. Yoon, A. Gaur, M. Shim, A. Facchetti, T. J. Marks, and J. A. Rogers, "Organic nanodielectrics for low voltage carbon nanotube thin film transistors and complementary logic gates," *J. Am. Chem. Soc.*, vol. 127, pp. 13808–13809, 2005.
115. M. H. Yoon, A. Facchetti, and T. J. Marks, "Sigma-pi molecular dielectric multilayers for low-voltage organic thin-film transistors," *Proc. Natl. Acad. Sci. U. S. A.*, vol. 102, pp. 4678–4682, 2005.
116. Q. Cao, M.-G. Xia, M. Shim, and J. A. Rogers, "Bilayer organic/inorganic gate dielectrics for high performance, low-voltage single walled carbon nanotube thin-film transistors, complementary logic gates and p-n diodes on plastic substrates," *Adv. Func. Mater.*, vol. 16, pp. 2355–2362, 2006.
117. S. Heinze, J. Tersoff, R. Martel, V. Derycke, J. Appenzeller, and P. Avouris, "Carbon nanotubes as Schottky barrier transistors," *Phys. Rev. Lett.*, vol. 89, p. 106801, 2002.
118. J. Appenzeller, Y. M. Lin, J. Knoch, Z. H. Chen, and P. Avouris, "Comparing carbon nanotube transistors – The ideal choice: A novel tunneling device design," *IEEE Trans. Electron Devices*, vol. 52, pp. 2568–2576, 2005.
119. Z. H. Chen, J. Appenzeller, J. Knoch, Y. M. Lin, and P. Avouris, "The role of metal-nanotube contact in the performance of carbon nanotube field-effect transistors," *Nano Lett.*, vol. 5, pp. 1497–1502, 2005.
120. Y. Nosho, Y. Ohno, S. Kishimoto, and T. Mizutani, "n-Type carbon nanotube field-effect transistors fabricated by using Ca contact electrodes," *Appl. Phys. Lett.*, vol. 86, 2005.
121. V. C. Sundar, J. Zaumseil, V. Podzorov, E. Menard, R. L. Willett, T. Someya, M. E. Gershenson, and J. A. Rogers, "Elastomeric transistor stamps: reversible probing of charge transport in organic crystals," *Science*, vol. 303, pp. 1644–1646, 2004.
122. K. Nomura, H. Ohta, K. Ueda, T. Kamiya, M. Hirano, and H. Hosono, "Thin-film transistor fabricated in single-crystalline transparent oxide semiconductor," *Science*, vol. 300, pp. 1269–1272, 2003.
123. E. Fortunato, P. Barquinha, A. Pimentel, A. Goncalves, A. Marques, L. Pereira, and R. Martins, "Fully transparent ZnO thin-film transistor produced at room temperature," *Adv. Mater.*, vol. 17, pp. 590–594, 2005.
124. M. S. Fuhrer, J. Nygard, L. Shih, M. Forero, Y. G. Yoon, M. S. C. Mazzoni, H. J. Choi, J. Ihm, S. G. Louie, A. Zettl, and P. L. McEuen, "Crossed nanotube junctions," *Science*, vol. 288, pp. 494–497, 2000.
125. A. A. Odintsov, "Schottky barriers in carbon nanotube heterojunctions," *Phys. Rev. Lett.*, vol. 85, pp. 150–153, 2000.

126. J.-H. Ahn, H.-S. Kim, K. J. Lee, S. Jeon, S. J. Kang, Y. Sun, R. G. Nuzzo, and J. A. Rogers, "Heterogeneously integrated, three dimensional electronics by use of printed semiconductor nanomaterials, " *Science,* vol. 314, pp. 1754–1757, 2006.
127. D. A. Heller, R. M. Mayrhofer, S. Baik, Y. V. Grinkova, M. L. Usrey, and M. S. Strano, "Concomitant length and diameter separation of single-walled carbon nanotubes," *J. Am. Chem. Soc.*, vol. 126, pp. 14567–14573, 2004.
128. D. Chattopadhyay, I. Galeska, and F. Papadimitrakopoulos, "A route for bulk separation of semiconducting from metallic single-wall carbon nanotubes," *J. Am. Chem. Soc.*, vol. 125, pp. 3370–3375, 2003.
129. Z. Chen, X. Du, M.-H. Du, C. D. Rancken, H.-P. Cheng, and A. G. Rinzler, "Bulk separative enrichment in metallic or semiconducting single-walled carbon nanotubes," *Nano Lett.*, vol. 3, pp. 1245–1249, 2003.
130. R. Krupke, F. Hennrich, H. v. Lohneysen, and M. M. Kappes, "Separation of metallic from semiconducting single-walled carbon nanotubes," *Science*, vol. 301, pp. 344–347, 2003.
131. S. R. Lustig, A. Jagota, C. Khripin, and M. Zheng, "Theory of structure-based carbon nanotube separations by ion-exchange chromatography of DNA/CNT hybrids," *J. Phys. Chem. B*, vol. 109, pp. 2559–2566, 2005.
132. Y. Li, D. Mann, M. Rolandi, W. Kim, A. Ural, S. Hung, A. Javey, J. Cao, D. Wang, E. Yenilmez, Q. Wang, J. F. Gibbons, Y. Nishi, and H. Dai, "Preferential growth of semiconducting single-walled carbon nanotubes by a plasma enhanced CVD method," *Nano. Lett.*, vol. 4, pp. 317–321, 2004.
133. S. Auvray, J. Borghetti, M. F. Goffman, A. Filoramo, V. Derycke, J. P. Bourgoin, and O. Jost, "Carbon nanotube transistor optimization by chemical control of the nanotube-metal interface," *Appl. Phys. Lett.*, vol. 84, pp. 5106–5108, 2004.
134. J. Chen, C. Klinke, A. Afzali, and P. Avouris, "Self-aligned carbon nanotube transistors with charge transfer doping," *Appl. Phys. Lett.*, vol. 86, p. 123108, 2005.
135. P. J. Burke, "AC performance of nanoelectronics: towards a ballistic THz nanotube transistor," *Solid-State Electron.*, vol. 48, pp. 1981–1986, 2004.

Chapter 10
Circuits, Applications and Outlook

Ali Keshavarzi and Arijit Raychowdhury

10.1 Introduction

In this chapter, we summarize the opportunities and challenges in the integration of carbon nanotubes into circuits and systems for electronic applications, and we present an outlook for the field. First, the promise of nanotube transistors for future digital circuits is discussed in Section 10.2 while presenting a framework for benchmarking their performance limits as compared to the Si technology. Nanotube transistor design considerations for circuit integration are also discussed. In Section 10.3, we cover a range of extended nanotube applications beyond digital circuits and present a discussion of the short-term exploratory applications and products based on nanotube devices. Finally, the materials, processing, and device challenges associated with nanotube electronics are discussed in Section 10.4 followed by the concluding remarks in Section 10.5.

10.2 Nanotubes for Digital Electronics

10.2.1 Scaling of FETs

Aggressive scaling of the CMOS technology continues in nanoscale (Figs. 10.1 and 10.2) in spite of tremendous technology development barriers, design challenges, and prohibitive costs. Currently, the 45 nm technology node is transitioning to high volume manufacturing in companies that rely on high-performance devices. To continue this scaling path, technologists are trying to reduce the effective oxide thickness, improve the channel mobility, and minimize the parasitics. In future, non-planar device structures, such as tri-gate and FinFET thin body transistors, may be incorporated to improve the device electrostatics to alleviate short channel effects. However, optimizing the parasitics of such tri-gate non-planar transistors (such as

---

A. Keshavarzi (✉)
Intel Corporation, Hillsboro, OR 97124, USA

A. Javey, J. Kong (eds.), *Carbon Nanotube Electronics*, Series on Integrated Circuits and Systems, DOI 10.1007/978-0-387-69285-2_10,

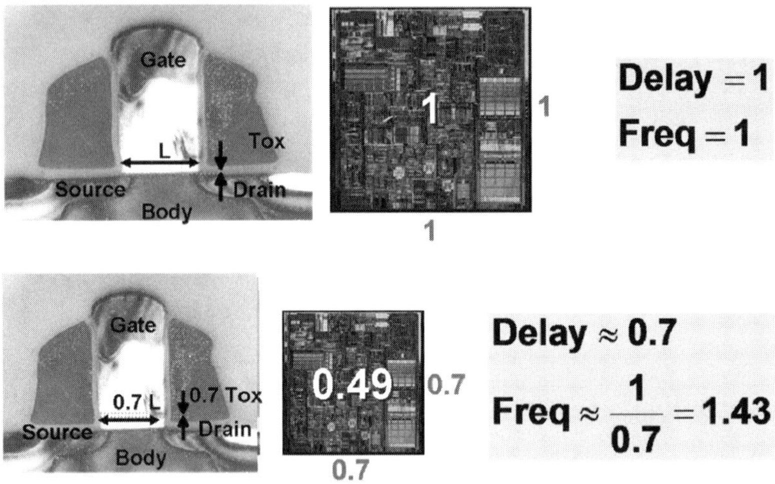

Fig. 10.1 Moore's Law: Scaling of transistor dimensions and switching delay. Each technology generation reduces the chip area by 50% and the switching delay by 30%

Fig. 10.2 Technology node and physical gate length scaling over the last 15 years

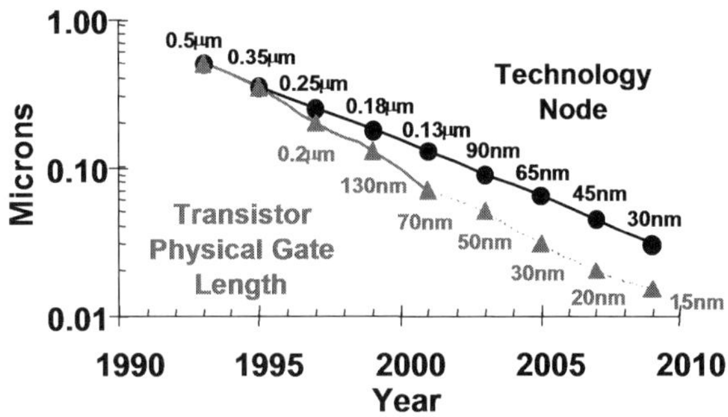

series resistance) will require a significant engineering development effort to convert their improved short channel effect capability into delivering higher drive current. Along with materials, technological, and device challenges, the design of ICs in these scaled technologies also faces growing limitations [1–3]. For instance, it is increasingly difficult to sustain supply and threshold voltage scaling and still provide the required performance enhancement, low energy consumption, and reliability. Furthermore, as the top-down manufacturing and lithography is pushed to their limits, we face the problem of increased device variations and leakage currents (Fig. 10.3) that will impact circuit and system designs negatively. These challenges are to such an extent that it may be virtually impossible to design around them with high yield.

In spite of all the problems, scaling of the silicon technology is expected to continue through research and innovation. Researchers are not only investigating

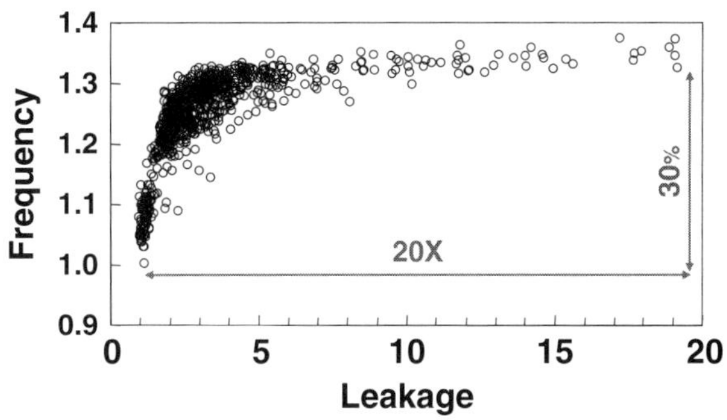

Fig. 10.3 Process induced variation in transistor leakage and switching frequency at the 130 nm technology node

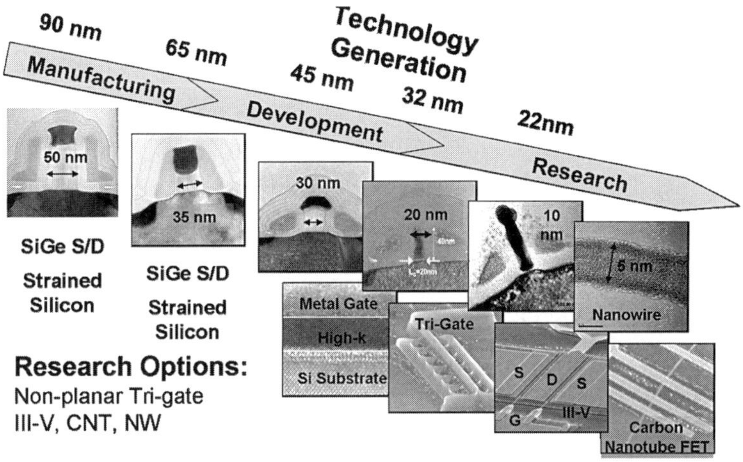

Fig. 10.4 Transistor research for future technology generations (courtesy: Intel Corporation)

non-planar transistor structures, such as FinFETs [4] and tri-gates [5], but they are also looking at means to improve the channel mobility by incorporating strain (and scaling it by increasing the strain) [6] or incorporate compound semiconductors inside the transistor channel [7] as shown in Fig. 10.4. More futuristic research has also started in the earnest to investigate alternative device and circuit architectures in a sub-10 nm transistor era of post-2015 time frame. Several futuristic (and revolutionary) devices have attracted the attention of device/circuit and system engineers worldwide. Carbon nanotube field-effect transistors is one such non-Si based technology which has been the subject of this book. While the evolutionary devices such as non-planar Si devices and strained channel materials promise to mitigate some of the mentioned problems associated with conventional, planar Si MOSFETs, the quest for new materials and structures continues in order to realize faster and fundamentally superior binary switches. Of all the different materials that are being investigated, single-walled carbon nanotubes [8, 9], despite numerousmaterial (and fabrication) research questions and yield concerns, appear to be the most promising due to their high intrinsic carrier mobility, atomically well-defined surfaces, and miniaturized dimensions.

In the light of technology scaling, the impact of Moore's Law remains, univocally, the most significant. The phenomenal success of Moore's Law lies in the high integration density and hence lower cost that technology scaling offers while enabling better performances. Moore's Law, which is based on a 1965 prediction by Intel's co-founder, Gordon E. Moore, states that "innovations in technology would allow a doubling of the number of transistors in a given space every year" (which was subsequently, changed to every two years). This prediction has held true for over a quarter of century and the result of this predicted growth has led to the microelectronics revolution. A good figure of merit capturing the essence of the Moore's Law for evaluating these promised novel materials, devices, and ideas is the area normalized frequency of a circuit. This metric, called Frequency Over Area (FOA), is calculated by dividing the frequency of operation of a circuit by its physical area as shown below:

$$\text{FOA (Frequency Over Area)} = \frac{\text{Freq}}{\text{Area}} \tag{10.1}$$

Historically, frequency has increased by ~1.4X and area has shrunk by ~2X for every process technology generation, leading to an enhancement of ~2.8X in the FOA (Fig. 10.5). Data in Fig. 10.5 has been obtained by considering a fan-out of 4 (FO4) ring oscillator (RO) with the interconnect parasitics of the particular technology node. In the future, FOA enhancement of at least 2X is desired for each new technology generation to justify its development. This can be attained by scaling the area by half even if the frequency of operation is not enhanced. In Section 10.2.2, we will revisit this metric and discuss how the FOA of SWNT devices compares with Si MOSFETs.

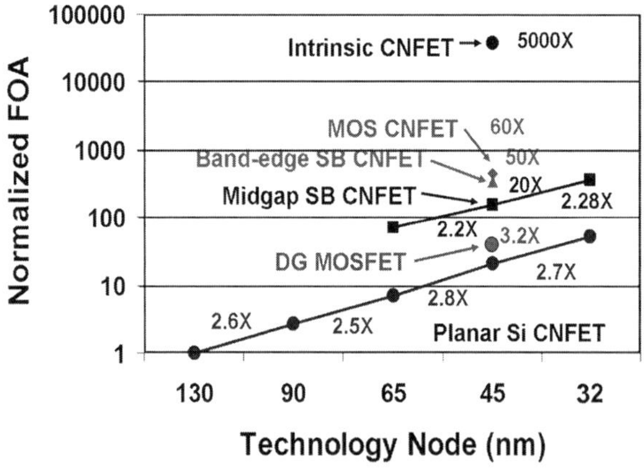

Fig. 10.5 Benchmarking the scalability of SWNT-FETs against planar and double-gate (DG) Si technologies. The circuit under investigation is a 5-stage FO4 ring oscillator. The predicted DG MOSFET data was derived from the experimental results [9] and the foot-print was estimated using [10]. (Reproduced with permission from Ref. [12] © 2006 IEEE)

10.2.2 The Potential of Nanotube Transistors

In this section, we use the FOA metric as an effective mean to compare Si MOSFETs (both planar and non-planar) with SWNT-based devices (both SB-FET and MOSFET configurations). For the comparison purposes, SWNT arrays with tube–tube separation of 1.6 nm (pitch of ~3 nm) were assumed and the corresponding parasitic capacitance and source/drain resistance were estimated as per the discussion in [11]. The parasitic capacitances include the overlap and fringe capacitances. The interconnect capacitance were derived from the corresponding bulk technology node. A clear FOA advantage is observed for SWNT-FETs as compared to Si MOSFETs as depicted in Fig. 10.5. For instance, a 45 nm node, mid-gap nanotube SB-FET enables a ~20X enhancement of FOA as compared to a planar Si MOSFET with similar length scales. A larger enhancement of ~ 60X is observed for the more optimal SWNT-MOSFET, which is expected owing to its higher ON current capability as compared to a mid-gap SB device (see Chapter 3). This dramatic increase in FOA is possible due to the unique electrical properties of SWNTs. This, of course, requires dense arrays of SWNTs with scaled gate dielectric thicknesses (2 nm of HfO_2 used in the simulations) and ohmic source/drain contacts. Furthermore, we have studied the impact of technology scaling (i.e., feature size scaling – 65, 45 and 32 nm nodes) on the FOA metric for mid-gap SB-SWNT-FETs (Fig. 10.5) [12]. The interconnect parasitics are extracted based on the particular technology node. It can be seen that as a result of scaling (both channel length as well as oxide thickness), the FOA improves by 2.2–2.3X for mid-gap SB-SWNT-FETs for every generation node. To put this in perspective, this is less than the historic FOA enhancement of ~ 2.8X, but better than the FOA corresponding to only area scaling (~ 2X) as discussed earlier in this Chapter. Since ballistic transport has been assumed, length scaling does not increase the drain current significantly (only a second order short channel effect is observed). However, the decrease in effective capacitance (both device as well as interconnects) due to the scaling, increases the frequency of operation and hence FOA. Consequently, SB-SWNT-FETs' scalability trend is similar to Si MOSFETs although the benefits of scaling (as manifested by our proposed FOA metric) may be lower.

Our proposed FOA metric for evaluating SWNT-FETs is in addition to other well established device metrics [13] that are used for benchmarking technologies. These other metrics include CV/I delay as a function of transistor channel length for studying intrinsic speed of a proposed technology, energy-delay product as a function of transistor channel length for optimizing switching energy and power consumption, investigating transistor subthreshold slope behavior as a function of transistor channel length that is critical for establishing scalability, and quantifying performance/speed and leakage trade-off by looking at CV/I versus I_{ON}/I_{OFF} [13]. For logic applications, low CV/I (i.e., small delay) is required with a high I_{ON}/I_{OFF} (i.e., small leakage).

Having discussed the different metrics for benchmarking the various technologies, we should point out that the scalability of SWNT-FETs needs to be studied experimentally in greater details. Overall, a careful study of the scaling impacts of the dimensions and V_{DD} for sub-50 nm SWNT-FET-based designs is needed. Furthermore, engineering and optimization of device parasitics, source/drain contacts, and the overlap capacitances need to be carried out for optimal circuit performance.

10.2.3 SWNT-FET Design Considerations for Digital Circuits

In this section, we discuss the design considerations of nanotube Schottky barrier FETs (SB-FETs or SB-MOSFETs) for digital circuits [14–19]. In Chapter 3 and also in references [20, 11], it was discussed that the SB-FETs have ambipolar current–voltage characteristics with the diameter playing a major role in both I_{ON} and I_{OFF} [21–25, 20, 11, 26–28]. The diameter (i.e., bandgap) has a direct impact on the Schottky barrier height at the metal contacts. The I_{ON} and the I_{OFF} are intrinsically tied to the Schottky barrier height and hence to the diameter. Figure 10.6 illustrates the I_{DS}–V_{GS} characteristics of several nanotube SB-FETs with varying diameters. It can be noted that a smaller diameter (larger bandgap) produces a higher Schottky barrier, therefore, resulting in an exponentially lower I_{ON} and also lower I_{OFF}. On the other hand, large diameter SWNTs ($\sim$>2 nm) have significantly higher I_{ON} at the cost of high I_{OFF} and poor I_{ON}/I_{OFF} ratio. The impact of V_{DD} on the I_{OFF} is significant for nanotube SB-FETs. In nanoscale Si MOSFETs, the applied voltage affects the I_{OFF} through drain induced barrier lowering (DIBL) and hence has a second order effect. In clear contrast to planar MOSFETs, in SB-SWNT-FETs, the OFF current is exponentially proportional to the V_{DD} through 1D electrostatics of the Schottky barriers. This is particularly problematic for ultrathin gate

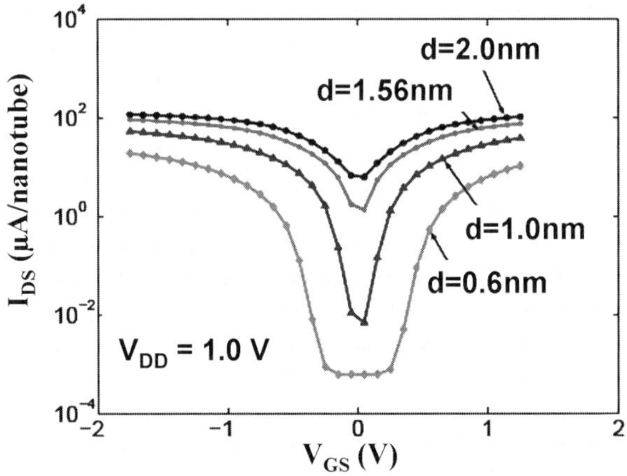

Fig. 10.6 The role of nanotube diameter, d, on the I_{DS}–V_{GS} characteristics of SB-SWNT-FETs. The structure under consideration is top-gated with a 2 nm HfO_2 gate dielectric and mid-gap Schottky source/drain contacts. The SWNT is assumed to be ballistic. The simulation results demonstrate that from an I_{ON} and I_{OFF} perspective, d = 1–1.5 nm is best suited for digital operation. (Reproduced with permission from Ref. [20] © 2006 IEEE)

dielectrics as the thickness of the SBs are directly proportional to the thickness of the gate dielectric (assuming an overlap of the gate with the SB contacts). Therefore, careful consideration is needed when choosing the optimal diameter for specific applications.

To further investigate the SB-SWNT-FET design considerations, DC analysis was performed for the noise margin and voltage swing of an inverter as a function of tube diameter and circuit power supply voltage (Fig. 10.7). Since, the voltage swing depends on the ratio of I_{ON} and I_{OFF}, we observe that the voltage swing degrades at high supply voltages (V_{DD}>700#x00A0;mV) and also with large SWNT diameter. An ideal inverter should have a noise margin of $0.5V_{DD}$. However, large CNT diameters (>1.5 nm) and high supply voltages (~1 V) result in poor noise margins (<$0.5V_{DD}$, see Fig. 10.7). Therefore, when designing the SB-SWNT-FETs, careful attention must be paid in choosing the appropriate nanotube diameter and V_{DD} to enable desirable I_{ON} as well as voltage swing and noise margin. As discussed in Chapter 3, fabricating nanotube transistors with metal ohmic contacts is challenging

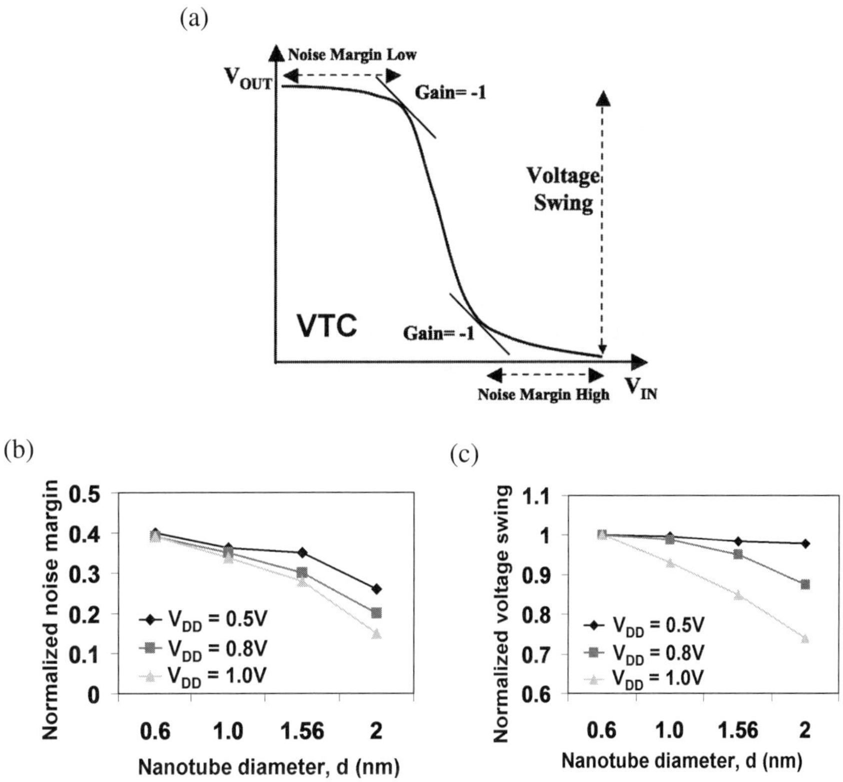

Fig. 10.7 (**a**) The voltage transfer characteristic (VTC) of an inverter. (**b**) Voltage swing and (**c**) Noise margin (both normalized to V_{DD}) as a function of SWNT diameter and supply voltage. (Reproduced with permission from Ref. [20] © 2006 IEEE)

for tube diameters <1.5 nm. This is an area that needs further research investigation to enable SB-FETs with high I_{ON} and yet low leakage currents. SWNT-MOSFETs (with heavily doped contacts [29] rather than metal contacts, refer to Chapter 3) on the contrary, do not show significant ambipolar conduction, and do not exhibit a first-order V_{DD} dependence of I_{OFF}, resulting in lower leakage currents (lower I_{OFF}). Thus, SWNT-MOSTETs may provide a more optimal device structure as compared to SB-SWNT-FETs for digital applications.

Transient AC analysis is required in order to evaluate the power-performance trade-off of SWNT-FET-based digital logic [11]. Power-performance trade-off shown in Fig. 10.8 is an important metric for high performance circuits. We perform our analysis on a ring oscillator (RO) made of nanotube SB-FETs, where each inverter chain has a fan-out of four (FO4). Figure 10.8 illustrates the power and throughput (=1/delay) of a single stage for different SWNT diameters (0.5–2 nm). It can be seen that the optimal SWNT diameter range for attaining the highest throughput is 1–1.5 nm. The very small diameter (<1 nm) nanotubes show lower throughput due to the low current drive arising from non-ohmic contacts. On the other hand, while large diameter (~>2 nm) nanotubes deliver high I_{ON}, they also exhibit low throughput due to poor I_{ON}/I_{OFF} ratio. Consequently, during switching, because of contention between the PMOS and the NMOS, the delay increases and the short circuit power is considerably higher too. So, even with higher switching currents, the large diameter SB-FETs have no advantage in terms of delay and are, practically, undesirable in complementary digital circuits. From these simulations, a diameter of ~1 nm operating at ~0.5–0.6 V appears best suited for digital circuit design.

Digital applications demand driving large capacitive loads (interconnect and more) that require transistors capable of providing high current densities. This digital circuit requirement necessitates fabricating transistor structures where the transistor channel incorporates an array of multiple parallel SWNTs as shown in Fig. 10.9. The arrayed-multiple-channel transistor architecture is required for

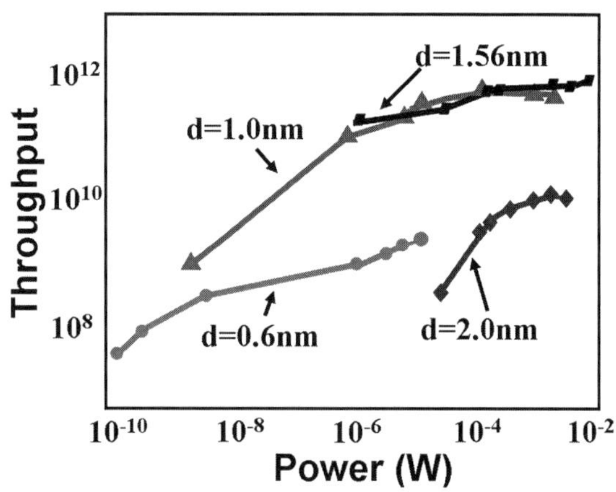

Fig. 10.8 Performance versus power as a function of SWNT diameter (d). An increasing diameter leads to higher I_{ON} and higher throughput. However, for very large diameters (d~>2 nm), the throughput goes down due to increased leakage and more contention between the PMOS and the NMOS. The circuit under test is a five-stage FO4 ring oscillator. (Reproduced with permission from Ref. [11] © 2006 IEEE)

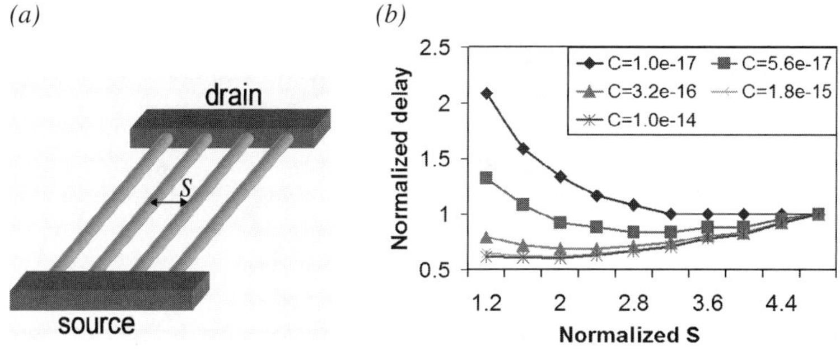

Fig. 10.9 (**a**) A well-ordered SWNT array transistor for digital applications. (**b**) The role of inter-tube spacing, S (normalized to the nanotube diameter) on the switching delay for different parasitic load capacitance values, *C*. (Reproduced with permission from Ref. [11] © 2006 IEEE)

driving large capacitive loads in digital circuit applications. Figure 10.9 illustrates how the delay of an inverter driving an identical inverter depends on the inter-nanotube spacing, S, for different values of the parasitic capacitance. It can be noted that the optimal spacing, S (corresponding to the minimum delay) depends on the parasitic capacitance (which is expressed per 100 nm of transistor width). Intuitively, for very low parasitics, the minimum delay is achieved by placing the nanotubes far apart (to increase $I_{\text{PER TUBE}}$). As the parasitic capacitance begins to play a role, the optimal S becomes smaller and tighter pitch helps in providing more total charging current (we should pack more SWNTs in 100 nm even though we sacrifice current per nanotube, $I_{\text{PER TUBE}}$) [11]. Therefore, when extrinsic load capacitance is much larger than intrinsic load capacitance (this is the case when driving interconnect capacitances), then the spacing has two opposite effects. On one hand, decreasing S leads to an increase of the number of nanotubes in 100 nm of width and hence more total drive current to charge up the load capacitance, but on the other hand the current per nanotube decreases. Thus the optimal spacing depends on value of the load capacitance.

Another critical question regarding the nanotube array transistor is the tube packing density needed to out-perform the Si technology. Figure 10.10 and Table 10.1 attempt to address this question. Figure 10.10 focuses on array density by studying I_{ON} per footprint as a function of spacing between the SWNTs. Two scenarios are considered: (1) a simulated, scaled SWNT-FET with a nominal ON current per tube of 50 μA (assumed ballistic), and (2) an experimentally achieved value of 20 μA [17, 30]. I_{ON} for the nanotube array transistor increases linearly as the SWNTs are brought closer together (Fig. 10.10). Notably, at high nanotube densities ($<$10 nm spacing), a non-linearity is observed in I_{ON} versus spacing which is due to the increased crossovers between the neighboring nanotubes (hence we loose some of the I_{ON} in an ultra-dense array). Assuming current per footprint of 1 mA/μm for state-of-the-art Si technology, our data suggests that SWNTs should be packed denser than 50 nm apart to be better than Si (20 nm apart for best experimentally reported SWNT-FETs). This nanotube density is experimentally feasible

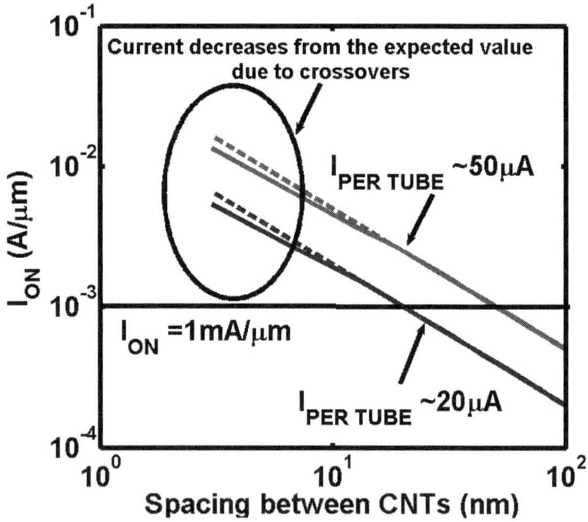

Fig. 10.10 The role of nanotube packing density on the resultant current density, I_{ON} of a parallel array nanotube FET. $I_{PER\ TUBE}$, the ON current per tube, is assumed to be either 50 μA (corresponding to the simulated value for a ballistic nanotube with $d = 1.4$ nm and $t_{ox} = 1.5$ nm HfO_2) or 20 μA (corresponding to the best experimental data [17]). (Reproduced with permission from Ref. [37] © 2008 IEEE)

Table 10.1 Current per unit footprint in SWNT-FETs for varying array packing densities

	I_{ON}per unit footprint		
Spacing (nm)	$I_{PER\ TUBE}$~50 μA	$I_{PER\ TUBE}$~20 μA	Comments
100	0.523	0.2	Spacing feasible today
54	1	–	Current density matching the Si technology
20	–	1	
3	17	7	High performance array, outperforming the Si technology

as preliminary works have demonstrated nanotube pitch (~ separation) of <50 nm achieved by aligned growth of nanotubes on quartz and sapphire substrates [31, 32] and the Langmuir–Blodgett method [33].

Although it is highly desirable to fabricate perfectly ordered and dense arrays of SWNTs with high uniformity, in practice, processing challenges may lead to a number of non-idealities, including diameter, chirality, and orientation variation [34]. To examine the impact of these non-idealities, simulations were conducted for each scenario. Our analysis suggests that the nanotube array FET cannot tolerate any metallic nanotube component, where the presence of only 0.05% metallic tubes results in $I_{ON}/I_{OFF} < 1000$ (Fig. 10.11), which is not as high as that of today's CMOS technology ($I_{ON}/I_{OFF} > 10{,}000$). High I_{ON}/I_{OFF} is required for good signal to noise ratio in circuit operation. Research should continue toward better purity semiconducting tubes. Recently, density gradient ultracentrifugation separation of nanotubes have been shown to yield ~99% purity in chirality [35, 36]. Even for this

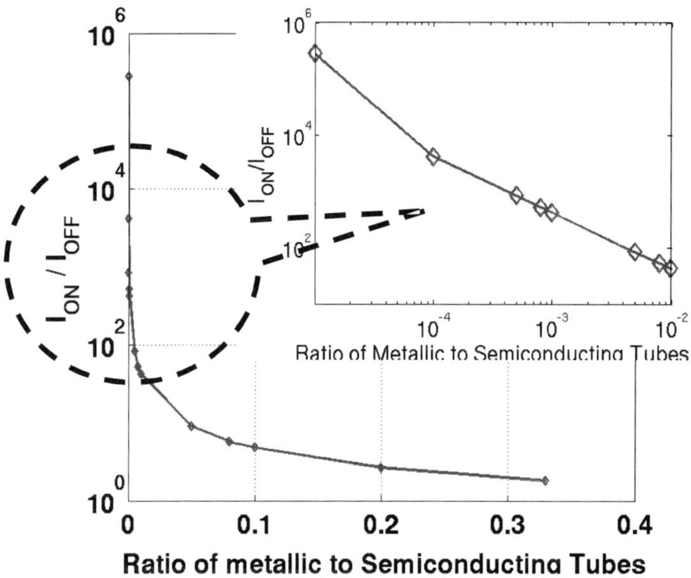

Fig. 10.11 Impact of metallic tubes in I_{ON}/I_{OFF} of a nanotube array FET. (Reproduced with permission from Ref. [37] © 2008 IEEE)

purity level, Fig. 10.11 shows that the I_{ON}/I_{OFF} ratio will still be compromised to a value well below 100. On the other hand, our preliminary analysis suggests that the nanotube array FET exhibit a higher tolerance in diameter, orientation, and angle variation for attaining high-performance switching. A detailed analysis of this topic can be found in reference [37].

10.3 Other Applications and Exploratory Products

It is too early to speculate on the potential CNT (both SWNT and MWNT) device and circuit applications for large-scale manufacturing; but as this book is being published there are several companies and research centers that are actively exploring products utilizing CNT devices. Beside digital electronics, CNTs may play an important role in other applications such as sensors, thin film transistors (TFTs), RF circuits, heat sinks, electrostatic discharge and electromagnetic interference protections, and electrodes for battery, supercapacitors, and fuel cells. In this section, we would like to discuss three other emerging applications. Recently, CNTs have been used as transparent metal contacts replacing Indium Tin Oxide (ITO) that is widely used in touch screens, LCD displays, solar cells and solid-state OLED lighting systems. Unidym is trying to commercialize low cost CNT-based transparent electrodes. A problem with ITO-based transparent electrode is that it is mechanically inflexible and is brittle. Furthermore, they are not cost effective. Consequently, they are not suited for low-cost flexible screens.

CNTs are also being used for field emission displays (FEDs). FED is a new type of flat-panel display in which electron emitters, arranged in a grid, function individually to generate colored light. FED is considered as a breakthrough technology because it can accommodate a thin panel like liquid crystal display (LCD) and offers a wider field-of-view and higher image quality. Field-emission displays use much less power than plasma displays because they are intrinsically more efficient. CNT-FEDs are targeted for clearer, sharper, brighter, and higher resolution displays of the future, potentially replacing or competing with LCDs and Plasma displays. Samsung is developing a FED with a 30-inch diagonal screen based on CNTs [38]. The nanotube-based cathodes are made of MWNTs and SWNTs mixed into a photosensitive resin. The resin is then screen-printed onto the cathode backplane and photo exposed to define the cathode regions. Samsung's device differs from most field-emission displays using so-called lateral field emitters. Fundamentally, the small diameter of CNTs in addition to their chemical stability and mechanical strength have opened up great potential for application of electron field emitters and several researchers have reported on the field emitters based on carbon nanotubes.

Fujitsu has reported that they are pursuing tools to fabricate CNTs for via connections in VLSI interconnect systems [39]. They are growing dense vertical metallic tubes inside the interconnect via to be integrated with the metal interconnects used in ICs. Beside better conductivity, an important advantage of using metallic CNTs for interconnects is their electromigration immunity. The strong C–C sp^2 bonding in nanotubes prevents electromigration that is often observed in ultra-narrow interconnects.

Before concluding this Chapter, we would like to capture a summary of a number of companies that are exploring the commercialization of CNT products in Table 10.2.

Table 10.2 Exploratory and other electronic products based on CNTs

Companies and research centers	Product space
Unidym (and CNI)	Transparent electrode, TFT, and fuel cells
Eikos	CNT formulations for coatings, displays
Samsung	CNT-FED (field emission display)
NEC	CNT-based displays, AFM probes, interconnects, sensors, transistors, and others
NEDO	CNT-based capacitors
Nanomix	CNT Sensors
Monano	Sensors and CVD-grown CNT solutions
Nantero	NRAM and dense nonvolatile memory
RFNano	CNT RF devices
Xintex, Inc	CNT-based X-ray sources
Nano Integris	Chirality purified CNTs for electronics

10.4 Challenges

In order for the nanotubes to come to the forefront of mainstream commercial use, significant research and development has to continue. There are numerous barriers in the field, and Table 10.3 enumerates only some of the major challenges that need to be addressed. Mainly, new strategies need to be developed to (1) remove metallic tubes and purify the semiconducting mixture of CNTs and (2) assemble nanotubes with controlled pitch and placement. Besides these materials challenges, developing new processing and device fabrication technologies will be the key for enabling self-aligned transistors with low parasitics and high performances.

Depending on the maturity of the SWNT fabrication process in the future, SWNTs may be used in ICs either by themselves (homogeneous) or in a hybrid (heterogeneous) configuration. However, regardless of the approach, it is envisioned that Si will remain as the main substrate for processing, mechanical support, and heat transfer with the nanotubes serving as a thin layer of high mobility material for active and/or passive elements.

Table 10.3 Major challenges for device/circuit fabrication with SWNTs

Challenges and future tasks
1. Produce/synthesize highly pure nanotubes with deterministic diameter and chirality
2. Fabricate dense and regular arrays of nanotubes to achieve high current drive capability with high packing density
3. Chemically doped SWNTs with air-stable and robust molecular species to make unipolar devices for traditional circuits
4. Fabricate complementary devices
5. Develop self-aligned device strategies
6. Reduce parasitics and optimize the device structure
7. Develop ohmic, nano-scale contacts to SWNTs
8. Heat removal and managing the high power density [40]

10.5 Conclusions

Nanotechnology research can impact the semiconductor industry either by directly integrating high mobility nano devices and circuits, or by impacting the scaling of silicon CMOS technology by for instance, self-assembly and lower manufacturing cost. We have shown that SWNT-FETs provide a number of performance advantages over Si MOSFETs. For instance, in an optimal geometry, from simulation, we find that SWNT-FETs may provide a 60X improvement in FOA over 65 nm Si technology node. Most of the simulation results presented in this Chapter pertain to idealistic scenarios where precise control on the process of manufacturing down to the nm-scale is assumed. Without sounding either overtly pessimistic or optimistic about the future of SWNT-FET technology, the authors would like to remind the readers that at the time this book was put together, the state-of-the-art manufacturing process of SWNT-FETs was far from what the simulations promise.

Significant technological advances, including removal of metallic tubes, reliable growth of well ordered arrays, maintaining control on the diameter and chirality are only a few of the breakthroughs necessary to make this technology mature enough for the semiconductor industry to adopt. Unlike the process of top-down fabrication where the lessons learnt over the last four decades enable precise fabrication of nanoscale transistors with well-defined dimensions and performances (of course, with an increasing degree of variation), the bottom-up processes always bear the intrinsic uncertainties, unknown parameters and insurmountable testing challenges. Hence, as long as the micro-architecture remains unaltered, carbon nanotubes like many other molecular technologies need to overcome the uncertainties and challenges of predictable growth, for them to become a viable post-Si technology. The field of carbon nanotube electronics, however, without a doubt, has evolved quite rapidly over the past decade and remains highly exciting with potential to dramatically revolutionize a number of different technological applications owing to their unique structure and superb properties. As various engineering and integration challenges are being actively tackled by researchers around the world, one may expect a significant progress in their eventual commercialization in a wide range of technologies and applications. It would be interesting after reading this book to follow this dynamic field of research in order to observe how it will evolve in the next few years.

References

1. K. Roy, S. Mukhopadhyay, H. Mahmoodi-Meimandi, "Leakage current mechanisms and leakage reduction techniques in deep-submicron CMOS circuits," IEEE Proceedings, Feb. 2003, pp. 305–327.
2. S. Borkar, "Technology trends and design challenges for microprocessor design," Proceedings of the 24th European Solid-State Circuits Conference, 1998. (ESSCIRC '98) 22–24 Sept. 1998, pp. 7–8.
3. A. Keshavarzi, J. W. Tschanz, S. Narendra, V. De, W. R. Daasch, K Roy, M. Sachdev, C. F. Hawkins, "Leakage and process variation effects in current testing on future CMOS circuits," IEEE Design and Test of Computers, 19(5), Sept.–Oct. 2002, pp. 36–43.
4. D. Hisamoto, W.-C. Lee, J. Kedzierski, H. Takeuchi, K. Asano, C. Kuo, E. Anderson, T.-J. King, J. Bokor, Chenming Hu, "FinFET-a self-aligned double-gate MOSFET scalable to 20 nm," IEEE Transactions on Electron Devices, 47(12), Dec. 2000, pp. 2320–2325.
5. B. Doyle, B. Boyanov, S. Datta, M. Doczy, S. Hareland, B. Jin, J. Kavalieros, T. Linton, R. Rios, R. Chau, "Tri-Gate fully-depleted CMOS transistors: fabrication, design and layout," Digest of Technical Papers VLSI Technology Symposium, June 2003, pp. 133–134.
6. S. Thompson, M. Armstrong, C. Auth, M. Alavi, M. Buehler, R. Chau, S. Cea, T. Ghani, G. Glass, T. Hoffman, C.-H. Jan, C. Kenyon, J. Klaus, K. Kuhn, Zhiyong Ma, B. McIntyre, K. Mistry, A. Murthy, B. Obradovic, R. Nagisetty, P. Nguyen, S. Sivakumar, R. Shaheed, L. Shifren, B. Tufts, S. Tyagi, M. Bohr, Y. El-Mansy, "A 90-nm logic technology featuring strained-silicon," IEEE Transactions on Electron Devices, 51(11), Nov. 2004, pp. 1790–1797.
7. S. Datta, T. Ashley, R. Chau, K. Hilton, R. Jefferies, T. Martin, T. Phillips, "85 nm gate length enhancement and depletion mode InSb quantum well transistors for ultra high speed and very low power digital logic applications," Technical Digest of International Electron Device Meeting, Dec. 2005, pp. 783–786.

8. P. L. McEuen, M. S. Fuhrer, H. Park, "Single-walled carbon nanotube electronics," IEEE Transactions on Nanotechnology, 1, March 2002, pp. 78–85.
9. Ph. Avouris, "Supertubes [carbon nanotubes]" IEEE Spectrum, 41(8), Aug. 2004, pp. 40–45.
10. Ph. Avouris, J. Appenzeller, V. Derycke, R. Martel, S. Wind, "Carbon nanotube electronics," Digest of International Electron Device Meet, Dec. 2002, pp. 281–284.
11. A. Keshavarzi, A. Raychowdhury, J. Kurtin, K. Roy, V. De, "Carbon nanotube field effect transistors for high performance digital circuits – transient analysis, parasitics, and scalability," IEEE Transactions on Electron Devices, 35, Nov. 2006, pp. 2718–2726.
12. A. Keshavarzi, A. Raychowdhury, J. Kurtin, K. Roy, and V. De, "Scalability of carbon nanotube FET-based circuits," Proceeding of IEEE Asian Solid-State Circuit Conference, Nov. 2006, pp. 415–418.
13. R. Chau, S. Datta, M. Doczy, B. Doyle, B. Jin, J. Kavalieros, A. Majumdar, M. Metz, and M. Radosavljevic, "Benchmarking nanotechnology for high-performance and low-power logic transistor applications," IEEE Transactions on Nanotechnology, 4(2), March 2005, pp. 153–158.
14. V. Derycke, R. Martel, J. Appenzeller, Ph. Avouris, "Carbon nanotube inter- and intramolecular logic gates," Nano Letters, 1(9), pp. 453–456.
15. A. Bachtold, P. Hadley, T. Nakanishi, C. Dekker, "Logic circuits with carbon nanotube transistors," Science, 294, 2001, pp. 1317–1320.
16. M. Freitag, M. Radosavljevic, Y. Zhou, A. T. Johnson, W. F. Smith, "Controlled creation of a carbon nanotube diode by a scanned gate," Applied Physics Letters, 79(20), Nov. 2001, pp. 3326–3328.
17. A. Javey, J. Guo, Q. Wang, M. Lundstrom, H. Dai, "Ballistic carbon nanotube field-effect transistors," Nature, 427, 2003, pp. 654–657.
18. J. Appenzeller, J. Knoch, M. Radosavljevic, Ph. Avouris, "Multimode transport in Schottky-Barrier carbon-nanotube field-effect transistors," Physical Review Letters, 92, June 2004, p. 226802.
19. M. Radosavljevic, S. Heinze, J. Tersoff, Ph. Avouris, "Drain voltage scaling in carbon nanotube transistors," Applied Physics Letters, 83, 2003, p. 2435.
20. A. Raychowdhury, A. Keshavarzi, J. Kurtin, V. De, K. Roy, "Carbon nanotube field effect transistors for high performance digital circuits – DC analysis and modeling toward optimum transistor structure," IEEE Transactions on Electron Device, 35, Nov. 2006, pp. 2711–2717.
21. Y.-M. Lin, J. Appenzeller, J. Knoch, P. Avouris, "High-performance carbon nanotube field-effect transistor with tunable polarities," IEEE Transactions on Nanotechnology, 4(5), Sept. 2005, pp. 481–489.
22. J. Guo, A. Javey, H. Dai, S. Datta, M. Lundstrom, "Predicted performance advantages of carbon nanotube transistors with doped nanotubes as source/drain," cond-mat, 0309039.
23. J. Appenzeller, Y.-M. Lin, J. Knoch, Ph. Avouris, "Band-to-band tunneling in carbon nanotube field-effect transistors," Physical Review Letters, 93(19), Nov. 2004, p. 196805.
24. S. O. Koswatta, D. E. Nikonov, M. S. Lundstrom, "Computational study of carbon nanotube p-i-n tunnel FETs," Technical Digest of International Electron Device Meeting, Dec. 2005, pp. 525–528.
25. A. Raychowdhury, X. Fong, Q. Chen, K. Roy, "Analysis of super cut-off transistors for ultralow power digital logic circuits," Proc. of ISLPED, 1, 2006, pp. 1–8.
26. Simulation were performed by Purdue Emerging Technology Evaluator on NanoHUB – http://nanohub.org
27. A. Raychowdhury, J. Kurtin, K. Roy, V. De, A. Keshavarzi, "Digital circuits with carbon nanotube transistors," 2007 Proceeding of International Conference on Solid State Devices and Materials, Sept. 2007, pp. 1162–1163.
28. M. E. Hwang, A. Raychowdhury, and K. Roy, "Effectiveness of energy recovery techniques in reducing on-chip power density in molecular nano-technologies," Proceedings of the 2004 International Symposium on Circuits and Systems, 3, May 2004, pp. 709–712.
29. J. Chen, C. Clinke, A. Afzali, P. Avouris, "Air-stable chemical doping of carbon nanotube transistors," Proceedings of the Device Research Conference, 2004, pp. 137–138.

30. A. Javey, J. Guo, D. Farmer, Q. Wang, E. Yenilmez, R. Gordon, M. Lundstrom, H. Dai, "Self-aligned ballistic molecular transistors and electrically parallel nanotube arrays," Nanoletters, 4, 2004, pp. 1319–1322.
31. C. Kocabas, S. J. Kang, T. Ozel, M. Shim, and J. A. Rogers, "Improved synthesis of aligned arrays of single-walled carbon nanotubes and their implementation in thin film type transistors," Journal of Physical Chemistry C, 111(48), 2007, 17879–17886.
32. K. Ryu, A. Badmaev, L. Gomez, F. Ishikawa, B. Lei, and C. Zhou. "Syn-thesis of aligned single-walled carbon nanotubes using catalysts de-fined by nanosphere lithography", Journal of American Chemical Society, 129, 2007, 10104–10105.
33. G. Zhang, X. Wang, X. Li, Y. Lu, A. Javey, and H. Dai, "Carbon nanotubes: from growth, placement, and assembly control to 60 mV/decade and Sub-60 mV/decade tunnel transistors", IEEE IEDM Technical Digest, 2006.
34. N. Pimparkar, J. Guo, M. Alam, "Performance assessment of sub-percolating nanobundle network transistors by an analytical model," Digest of IEDM, 1, 2005, pp. 120–125.
35. X. Li, X. Tu, S. Zaric, K. Welsher, W. S. Seo, W. Zhao, and H. Dai, "Selective synthesis combined with chemical separation of single-walled carbon nanotubes for chirality selection," Journal of American Chemical Society, 129(51), 2007, 15770–15771.
36. L. Zhang, S. Zaric, X. Tu, W. Zhao, and H. Dai, "Assessment of chemically separated carbon nanotubes for nanoelectronics," Journal of American Chemical Society, 130(8), 2008, 2686–2691.
37. A. Raychowdhury, J. Kurtin, S. Borkar, V. De, K. Roy, and A. Keshavarzi, "Theory of multi-tube carbon nanotube transistor for high speed variation-tolerant circuits," 2008 Device Research Conference, June 23–25, 2008, Santa Barbara, California.
38. N. S. Lee, D. S. Chung, I. T. Han, J. H. Kang, Y. S. Choi, H. Y. Kim, S. H. Park, Y. W. Jin, W. K. Yi, M. J. Yun, J. E. Jung, C. J. Lee, J. H, You, S. H. Jo, C. G. Lee, and J. M. Kim, "Application of carbon nanotubes to field emission displays," Diamond and Related Materials, 10(2), Feb. 2001, pp. 265–270.
39. M. Nihei, M. Horibe, A. Kawabata, and A. Yuji, "Carbon nanotube vias for future LSI interconnects," Proceedings of the IEEE 2004 International Interconnect Technology Conference, June 2004, pp. 251–253.
40. M. E. Hwang, A. Raychowdhury, and K. Roy, "Energy-recovery techniques to reduce on-chip power density in molecular nanotechnologies," IEEE Transaction on Circuits and Systems I: Fundamental Theory and Applications, 52(8), Aug. 2005, pp. 1580–1589.
41. A. Javey, R. Tu, D. B. Farmer, J. Guo, R. G. Gordon, H. Dai, "High performance n-type carbon nanotube field-effect transistors with chemically doped contacts," Nano Letters, 5(2), 2005, pp. 345–348.
42. J. A. Misewich, R. Martel, Ph. Avouris, J. C. Tsang, S. Heinze, and J. Tersoff, "Electrically induced optical emission from a carbon nanotube FET," Science, 300(5620), May 2003, pp. 783–786.
43. M. Shim and G. P. Siddons, "Photoinduced conductivity changes in carbon nanotube transistors," Applied Physics Letters, 83(17), Oct. 2003, pp. 3564–3566.
44. Y. M. Li et al. "Preferential growth of semiconducting single-walled carbon nanotubes by a plasma enhanced CVD method," Nano Letters, 4(2), Feb. 2004, pp. 317–321.
45. C. Berger, Z. Song, T. Li, X. Li, A. Y. Ogbazghi, R. Feng, Z. Dai, A. N. Marchenkov, E. H. Conrad, P. N. First, and W. A. de Heer, "Ultrathin Epitaxial Graphite: 2D electron gas properties and a route toward graphene-based nanoelectronics," Journal of Physical Chemistry B, 108, 2004, p. 19912.
46. K. Ryu, A. Badmaev, L. Gomez, F. Ishikawa, B. Lei, and C. Zhou, "Symthesis of algined single-walled carbon nanotubes using catalysts defined by nanosphere lithography," Journal of American Chemical Society, 129, 2007, pp. 10104–10105.

Index

A

Ab initio calculations, 5
AC characterization, 94, 96–99
AC gain, 89, 93–94
Acoustic phonons, 31–32, 33–34, 35, 156, 168–169
Aharonov-Bohm effect, 21, 23
Amplifier, 92, 93, 96, 97, 98
Armchair SWNT, 9, 10, 11, 24–25, 27–28, 30
Atomic force microscope (AFM), 24, 28, 47, 94, 214
Atomic layer deposition, 71, 232

B

Ballistic, 15, 16, 20, 29, 67, 76, 77, 78, 87, 88, 96, 107, 108, 111–122, 123, 125, 133–137, 143, 144, 147, 149, 154–156, 170, 251, 252, 255, 256
Ballistic transmission, 16
Band-to-band tunneling, 64, 81, 115, 118–119, 139, 144–145, 147–148, 155
Bandgap modulation, 22, 23, 27–28
Band structures, 1, 6–7, 15, 27–28, 176
Bandwidth, 93–94, 97–98, 99, 186–187, 188
Benchmarking, 247, 250, 251–252
Bloch functions, 2–3
Brillouin zone, 2, 4, 7, 8, 10

C

Capacitance
  based sensing, 204, 206
  coupling, 188, 223
Carrier scattering, 16, 29, 36, 67, 136, 140
Challenges, 247–259
Charge transfer, 73, 191, 195, 197, 199–200, 207, 212, 222, 225, 226
Charging energy, 17, 18–20
Chemical doping, 64, 73–74, 80, 103, 112
Chemical-mechanical-polishing (CMP), 181
Chemical vapor deposition (CVD), 44, 94–95, 213, 216–217
Chirality, 7, 15, 36, 43, 48, 52–53, 57, 155, 157, 163, 171, 176, 256, 258, 259, 260
  distribution, 48, 53–55
Circuit simulation, 133–160
Common gate configuration, 87
Common source configuration, 87
Compact
  model/modeling, 142–143, 154–160, 164, 187
Complementary metal-oxide-semiconductor (CMOS), 48, 57, 63, 67, 80, 114, 127, 133, 142, 143, 145, 158, 159, 160, 211, 224, 237–238, 247, 256, 259
Conductance quantum GQ, 16
Conduction Channel, 176, 177, 178, 180, 187
Contact(s), 64–70, 76–79, 111–115, 136–142, 144, 149–154, 167–169, 183–184, 197–203, 205, 207, 231–232, 252, 259
Controlled flocculation, 213–214
Coulomb
  blockade, 17–18, 20–21
  oscillations, 18, 19
Critical length, 51, 172, 181, 187
Crosstalk
  power, 90
Curvature-induced mixing, 11
Cut-off frequency, 120–123

D

DC characterization, 95–96
Delay, 78, 158, 159, 163, 165, 170, 172, 174, 175, 183–186, 188, 248, 251, 254–255
Delocalized states, 2
Density of states, 4, 9–11, 72–73, 111, 121, 125, 136, 148, 165–166, 173, 225
Design considerations, 160, 247, 252–257

Device
- modeling, 133–160
- simulation, 107–127

Diameter, 52–53, 70, 102, 113, 135, 172, 177, 179, 186, 216, 252, 254, 255
- dependent measurements, 70

Dielectrophoresis, 43, 199
Dynamic delay, 183

E

Electromigration, 163–164, 181–182, 258
Electron beam lithography, 45
Electron-phonon coupling, 33–34, 126
Electrostatic capacitance, 146, 167, 169, 172–173, 180
Energy
- dispersion, 4–8, 15, 25, 26, 35
- level spacing, 17, 19–20

Equivalent circuit, 89, 119–120, 168–171, 180, 187
Exciton, 15, 123, 127
Exploratory products, 257–258

F

Fermi level, 1–2, 4, 7, 10, 11, 13, 15–20, 26, 30, 36, 56, 72–73, 79, 110, 115, 116, 134–135, 138, 146–148, 150, 151, 165–168, 176, 201, 225
- Dirac distribution, 17, 135–136, 148, 176
- golden rule, 29–30, 34
- pinning, 63–65, 67, 193
- velocity, 4, 35, 145, 166

Field-effect
- mobility, 34–35
- transistors, 63–82

Finite element modeling, 223
Flexible circuits, 211
Frequency, 34, 55, 77–78, 87–94, 96–99, 103–104, 119–123, 126, 127, 166, 197, 211–212, 227, 249–251
Frequencyover area (FOA), 250–251, 259

G

Gigascale integration, 163–164, 182
Global Interconnects, 185–188
Graphene, 1–9, 10, 11, 21, 25, 26, 31, 35, 138, 163, 166, 177, 187
Growth mechanism, 45, 52, 57

H

Hall mobility, 34
High-bias conductance, 32–33
High frequency, 77, 78, 87, 88, 91, 93, 211–212
- performance limits, 107, 119–123, 126

High-κ
- gate dielectrics, 63, 70–72, 73, 76, 78, 82, 136
- materials, 77, 82, 107, 117, 122, 212, 235

Homodyne detection, 90–91
Hot filament CVD, 47–48
Hysteresis, 64, 74–76, 233, 238

I

Indium-doped tin oxide, 220
Interconnect, 163–188
Inverter, 74, 134, 141, 142, 143, 157–159, 237–238, 253–255
ION/IOFF, 31, 70, 78–81, 89, 141, 159, 221, 222, 223, 225, 233, 234, 237, 251–257

J

Jitter, 163

K

Kinetic inductance, 164–166, 170–171, 174, 180, 187
Kondo effect, 20, 127

L

Landauer formula, 15, 16, 32
Langmuir
- constant, 196, 201, 203
- isotherm, 196

Large scale integration, 43, 48, 82, 127, 175
Leakage, 249, 251
- current, 64, 70–72, 78–81, 99, 112–115, 118–119, 124, 141, 147, 148, 230, 248, 254

Lift-off, 45, 46, 49
Liquid crystal display, 211, 258
Local interconnects, 180–185, 188
Logic gate, 74, 180–181, 228, 237–238
Low-bias conductance, 32, 34, 168–169, 172, 177
Luttinger liquid, 20–21, 36

M

Macroelectronics, 211, 239
Magnetic inductance, 100, 165, 169–170, 174–175, 180, 187
Matthiessen's rule, 168
Mean free path, 29, 30, 32–34, 35, 76, 101–102, 149, 151, 156, 163, 165, 167–168, 172, 179, 183–185, 187–188
Mesoscopic, 17, 107, 134, 136
Metal contacts, 20, 32, 34, 65–67, 70, 78, 79, 82, 111, 150, 152, 156, 193, 199–200, 203, 205, 207, 252, 254, 257–258

Metallic SWNT, 7–13, 16, 19–20, 22, 23, 27–35, 53–54, 56–57, 147–148, 158–159, 168–169, 172–174, 183–185, 187–188
Metal-oxide-semiconductor field-effect transistor, 238
Metal-semiconductor transition, 22
Microfluidic channels, 214
Microwave, 87, 88, 91–93, 104, 166
Mid-gap, 112, 113, 137, 140–142, 251, 252
Mixed-mode simulations, 141–142
Mixing, 11, 13, 23–24, 90–91, 214, 215
Mobility, 29, 34–36, 63, 71–72, 73–74, 76, 87, 107, 115, 119, 225, 231, 237, 247–250, 259
Mode space basis set, 110–111

N
Nanocrystal, 17, 52
Nanofabrication, 48
Nanoparticle, 17, 44–45, 48–50, 52–53, 54, 205, 217
Nanotube
  array, 255–257
  diameter, 69, 71, 79, 113–114, 179, 252, 253, 255
  spacing, 255
Nearest neighbor interactions, 5–6
Negative resist, 49
Noise Margin, 253
Non-equilibrium Green's function (NEGF), 107–109, 123, 126, 133–134

O
OFF current, 78–80, 113–114, 141, 142, 159, 221–222, 252–253
ON current, 34, 66, 68, 70, 71, 76, 78–79, 81–82, 89, 96, 112–118, 122, 140, 145, 154, 157, 159, 251, 255, 256
On/off ratio, 104, 141, 222–225, 233, 235, 237, 238
Ohmic contacts, 64, 67, 70, 76, 104, 253–254
Optical absorption cross sections, 212, 227
Optical phonons, 31, 76–77, 115, 140, 156
Optoelectronic, 123–127
Outlook, 247–259
Output power, 89, 90

P
Palladium contacts, 64
Parabolic subband, 15
Parasitic capacitance, 72, 89, 97, 99, 102, 104, 120–122, 144–145, 150, 152, 154, 251, 255
Passivation, 74–76, 199–200
Peierls distortion, 9
Pentagon-heptagon pairs, 30
Percolation thresholds, 220
Periodic boundary condition, 6, 9, 21, 26, 110
Phonon
  absorption, 32, 76
  assisted tunneling, 47, 81, 119
  emission, 34, 125, 156
  scattering, 31–34, 56, 66, 76–77, 99, 107, 115–119, 122–123, 125–126, 144, 147, 149, 154–156, 163, 169
Photo-lithography, 45
Photoluminescence (PL), 10, 14, 54
Piecewise model, 169
Plasma-enhanced CVD, 48, 54
P orbitals, 1–3, 109–110
Power, 20, 63, 66, 76, 81, 87–92, 96–97, 113, 114, 124, 147, 160, 163, 169, 183–184, 188, 207, 251, 253–254, 258, 259
Ps hybridization, 1, 97, 110

Q
Quantum
  capacitance, 72–73, 122, 123, 136, 146, 152, 167, 170, 173, 180, 223
  dots (QD), 17
  interference, 20, 21, 126
  transport, 15, 16, 36, 119

R
Radial breathing mode, 54, 225
Radio frequency, 74–76
Raman spectroscopy, 14, 31, 54
Random network film, 212, 213, 218, 233
Rayleigh scattering, 14
Real space basis set, 109–110
Reciprocal lattice, 2
Resistance quantum (RQ), 16
Ring oscillator, 88, 119–120, 141, 228–229, 238, 250, 254

S
Scaling, 63, 78, 113–114, 222, 230, 231, 233, 238, 247–251, 259
Scanning tunneling microscopy/spectroscopy (STM/STS), 10, 11, 28, 30, 31, 47
Scattering, 29–34, 115–119
Schottky
  barrier(s), 64, 68, 94, 96, 111, 134, 136–139, 141, 142, 191, 226, 252–253
    height, 134, 141, 252
    modulation, 193–195
    transistor, 111
  contacts, 252

Self-assembled monolayer, 230
Semiconducting SWNT, 7, 8, 9, 12–14, 20, 22, 24, 27, 28, 34, 35, 54, 56, 63, 144, 147, 148, 151, 158, 159, 191, 201, 206
Semi-global interconnects, 184, 185, 188
Single electron tunneling, 18
Site-binding model, 196, 197
Small-signal model, 87, 100, 119–120, 156
Sticking coefficient, 197, 203
Subband, 6–7, 10, 15–16, 29–30, 73, 116, 117–119, 121, 125–126, 135, 146–148, 151, 168–169, 176, 223
Subthreshold swings, 71, 81, 233

T
Technology node, 70, 78, 79, 158, 170, 182, 247–251, 259
Temperature dependent measurements, 66
Tensile strain, 25–28, 233, 235
Thermal buoyancy, 51
Thermal fluctuation, 17
Thin film electronics, 211–239
Tight binding approximations, 2
Time domain measurement, 93, 97–98
Time-of-flight, 185–186
Tomonaga-Luttinger liquid, 20–21
Torsional strain, 27
Transfer printing, 214, 216, 229–230, 233, 239
Transient response, 143, 148, 197, 200, 201–203
Transmission probability, 16, 65, 73, 138–139, 148, 152
Transparent thin film transistors, 223, 226, 228, 229–230, 237, 257, 258
Transport properties, 1–2, 6, 7, 15–18, 21, 23, 28–31, 34, 36, 63, 64, 66, 76, 115, 119
Two-tone measurement, 91–93

U
Uniaxial strain, 27–28

V
Van der Waals interaction, 28–29
Van Hove singularity, 10
Voltage swing, 253
Voltage transfer characteristics (VTC), 253

W
Work function, 64–69, 102, 136, 157, 192–194, 205, 225, 231
Wrapping index, 7, 15

Z
Zeeman splitting, 20
Zener tunneling, 81
Zigzag SWNT, 6, 7, 12, 23–27, 125
Zone boundary phonons, 31–34, 168

Continued from page ii

Ultra-Low Voltage Nano-Scale Memories
Kiyoo Itoh, Masashi Horiguchi and Hitoshi Tanaka
ISBN 978-0-387-33398-4, 2007

Routing Congestion in VLSI Circuits: Estimation and Optimization
Prashant Saxena, Rupesh S. Shelar, Sachin Sapatnekar
ISBN 978-0-387-30037-5, 2007

Ultra-Low Power Wireless Technologies for Sensor Networks
Brian Otis and Jan Rabaey
ISBN 978-0-387-30930-9, 2007

Sub-Threshold Design for Ultra Low-Power Systems
Alice Wang, Benton H. Calhoun and Anantha Chandrakasan
ISBN 978-0-387-33515-5, 2006

High Performance Energy Efficient Microprocessor Design
Vojin Oklibdzija and Ram Krishnamurthy (Eds.)
ISBN 978-0-387-28594-8, 2006

Abstraction Refinement for Large Scale Model Checking
Chao Wang, Gary D. Hachtel, and Fabio Somenzi
ISBN 978-0-387-28594-2, 2006

A Practical Introduction to PSL
Cindy Eisner and Dana Fisman
ISBN 978-0-387-35313-5, 2006

Thermal and Power Management of Integrated Systems
Arman Vassighi and Manoj Sachdev
ISBN 978-0-387-25762-4, 2006

Leakage in Nanometer CMOS Technologies
Siva G. Narendra and Anantha Chandrakasan
ISBN 978-0-387-25737-2, 2005

Statistical Analysis and Optimization for VLSI: Timing and Power
Ashish Srivastava, Dennis Sylvester, and David Blaauw
ISBN 978-0-387-26049-9, 2005